Studies in Fuzziness and Soft Computing

Volume 329

Series editor

Janusz Kacprzyk, Polish Academy of Sciences, Warsaw, Poland
e-mail: kacprzyk@ibspan.waw.pl

About this Series

The series "Studies in Fuzziness and Soft Computing" contains publications on various topics in the area of soft computing, which include fuzzy sets, rough sets, neural networks, evolutionary computation, probabilistic and evidential reasoning, multi-valued logic, and related fields. The publications within "Studies in Fuzziness and Soft Computing" are primarily monographs and edited volumes. They cover significant recent developments in the field, both of a foundational and applicable character. An important feature of the series is its short publication time and world-wide distribution. This permits a rapid and broad dissemination of research results.

More information about this series at http://www.springer.com/series/2941

Gleb Beliakov · Humberto Bustince Sola
Tomasa Calvo Sánchez

A Practical Guide
to Averaging Functions

 Springer

Gleb Beliakov
School of Information Technology
Deakin University
Burwood, VIC
Australia

Humberto Bustince Sola
Departamento de Automática y
  Computación
Universidad Pública de Navarra
Pamplona
Spain

Tomasa Calvo Sánchez
Departamento de Ciencias de la
  Computación
Universidad de Alcalá
Alcalá de Henares, Madrid
Spain

ISSN 1434-9922   ISSN 1860-0808   (electronic)
Studies in Fuzziness and Soft Computing
ISBN 978-3-319-37207-5   ISBN 978-3-319-24753-3   (eBook)
DOI 10.1007/978-3-319-24753-3

Printed on acid-free paper

Springer International Publishing AG Switzerland is part of Springer Science+Business Media
(www.springer.com)

To
Gelui Patricia, Chaquen and Sofia

Gleb Beliakov

Juan María Urruticoechea

Humberto Bustince Sola

My parents, sister and brother

Tomasa Calvo Sánchez

Preface

Averaging is ubiquitous in many sciences, engineering, and everyday practice. The notions of the arithmetic, geometric, and harmonic means developed by the ancient Greeks are in widespread use today. When thinking of an average, most people would use arithmetic mean, "the average", or perhaps its weighted version in order to associate the inputs with the degrees of importance. While this is certainly the simplest and most intuitive averaging function, its use is often not warranted. For example, when averaging the interest rates, it is the geometric and not the arithmetic mean which is the right method. On the other hand, the arithmetic mean can also be biased for a few extreme inputs, and hence can convey false meaning. This is the reason why real estate markets report the median and not the average prices (which could be biased by one or a few outliers), and why judges' marks in some Olympic sports are trimmed of the smallest and the largest values.

The world of averages (also called means) is very rich in both mathematical and practical senses. There exist averages that allow one to incorporate not only the weights of importance but also various interactions among the inputs; averages that are robust to a few, or even many outlying values; averages that model various types of majority, necessary, desirable, and sufficient inputs; averages that model a representative (in various senses) input, and so on.

The theory of aggregation functions, which includes most averages, became an established area of research in the last 30 years. Theoretical advances are complemented by numerous applications in decision sciences, artificial intelligence, fuzzy systems, and image processing. Several monographs and edited volumes dedicated to this topic provide a comprehensive analysis of both theory and applications, and regular conferences and special sessions on aggregation provide a forum for presenting the latest achievements.

However, it has been 7 years since the publication of the most recent monograph in the field of aggregation, and we think it is time to provide an update on the most recent developments in this area. Our specific focus is on averaging functions. These functions, whose prototypical example is the arithmetic mean, are most often used in decision sciences as they provide compensatory properties: low values of some inputs are compensated by high values of the others, and the output is always bounded by the smallest and the largest input. The result of the averaging is a value representative of the inputs.

The target audience of this book is computer scientists, system architects, knowledge engineers, and programmers, as well as decision scientists and mathematicians, who face a problem of combining various inputs into a single output. Our intent is to provide these people with an easy-to-use guide about possible ways of averaging input values given on a numerical scale, and ways of choosing/ constructing aggregation functions for their specific applications. All relevant mathematical notions are explained in the book (in the introduction or as footnotes).

Chapter 1 gives a broad introduction to the topic of aggregation functions. It covers important general properties and lists the most important prototypical examples: means, ordered weighted averaging (OWA) functions, and Choquet integrals, as well as non-averaging aggregation functions such as triangular norms and conorms, uninorms and nullnorms. It addresses the problem of choosing the right aggregation function, and also introduces a number of basic numerical tools: methods of interpolation and smoothing, linear and nonlinear optimization, which will be used to construct aggregation functions from empirical data. Prototypical applications are presented.

The remaining chapters focus on different types of averaging functions. Chapter 2 presents classical averages, which include the arithmetic mean, power means, quasi-arithmetic means, medians, and several other functions. Some less known averages such as Gini, Bonferroni, Heronian, and logarithmic means are also presented. Chapter 3 discusses the ordered weighted averaging and focuses on the issue of OWA weights. Chapter 4 is dedicated to fuzzy integrals, in particular the Choquet and Sugeno integrals. Each class of these functions has many distinct families which are treated in separate sections. We give formal definitions, discuss important properties of the averaging functions and their interpretation, and also present specific methods for fitting a particular family to the empirically collected data. We also provide various generalizations, advanced constructions, and pointers to specific literature.

Chapters 5–8 focus on more advanced and recently developed methods of averaging. In Chap. 5 we present a view of averaging functions from the perspective of minimizing some sort of a penalty. Such a penalty can be interpreted from the point of view of inputs disagreement. If all inputs coincide, then the average is that common value, whereas when the inputs differ we impose a price on the difference between the output and each input. The value that minimizes the total penalty is the output of the averaging function.

It turns out that all averaging functions allow such an interpretation through a properly chosen penalty, and that the classical averages such as the arithmetic means and the median are the minimizers of the sum of squared and absolute differences between the inputs and the output, respectively. One can define new averages by specifying a particular suitable penalty function.

In Chap. 6 we present several construction methods for averaging functions and discuss some of their useful properties. The construction methods include idempotization, composition, fitting averaging functions to empirical data, and graph-based construction. We discuss overlap and grouping functions, and present in detail generalizations of the Bonferroni mean. We also treat the issues of consistency and stability of the families of averaging functions.

In Chap. 7 we extend the class of averaging aggregation functions, which are by definition monotone increasing, to include averaging functions that are not always monotone. We present a weaker notion of monotonicity, which requires the output not to decrease if all the inputs are increased by the same amount. Weak monotonicity makes sense when averaging data containing significant noise, the outlying values. The outliers should not contribute to the average, or even worse, drag the average in their direction. On the contrary, the average must be robust to often unavoidable gross errors in the data, and the price for such robustness is lack of monotonicity. We study various robust averages (also called robust estimators of location) and establish weak monotonicity of several classes of averaging functions. We also develop some new weakly monotone averaging methods using penalty-based approach and extend the mode operation. We present directionally monotone functions and define pre-aggregation functions.

The final Chap. 8 presents some topics related to averages on lattices, in particular product lattices. We focus on averaging functions for Atanassov intuitionistic fuzzy sets and interval-valued fuzzy sets. We present general construction methods and then treat some special cases, such as the weighted means, OWA, and Bonferroni means. A special attention is paid to the medians on lattices, and several alternative constructions are explored.

Melbourne   Gleb Beliakov
Pamplona   Humberto Bustince Sola
Alcalá de Henares   Tomasa Calvo Sánchez
June 2015

Contents

1 **Review of Aggregation Functions** 1
   1.1   Aggregation Functions 1
   1.2   Applications of Aggregation Functions 5
   1.3   Classification and General Properties 9
        1.3.1   Main Classes......................... 9
        1.3.2   Main Properties 10
        1.3.3   Duality 18
        1.3.4   Comparability 21
        1.3.5   Continuity and Stability 21
   1.4   Main Families and Prototypical Examples 24
        1.4.1   Min and Max 24
        1.4.2   Means........................... 24
        1.4.3   Medians 25
        1.4.4   Ordered Weighted Averaging.................. 26
        1.4.5   Choquet and Sugeno Integrals 26
        1.4.6   Conjunctive and Disjunctive Functions 27
        1.4.7   Mixed Aggregation..................... 28
   1.5   Composition and Transformation of Aggregation Functions.... 29
   1.6   How to Choose an Aggregation Function 32
   1.7   Supplementary Material: Some Methods for Approximation
       and Optimization 35
        1.7.1   Univariate Approximation and Smoothing.......... 35
        1.7.2   Approximation with Constraints................ 37
        1.7.3   Multivariate Approximation.................. 38
        1.7.4   Convex and Non-convex Optimization 41
        1.7.5   Main Tools and Libraries 47
   References 50

2   Classical Averaging Functions.   55
   2.1   Semantics   55
        2.1.1   Measure of Orness   56
   2.2   Classical Means.   58
        2.2.1   Arithmetic Mean   58
   2.3   Weighted Quasi-arithmetic Means   65
        2.3.1   Definitions   65
        2.3.2   Main Properties   66
        2.3.3   Examples   69
        2.3.4   Calculation   73
        2.3.5   Weighting Triangles   73
        2.3.6   Weights Dispersion   76
        2.3.7   How to Choose Weights   77
   2.4   Other Means   82
        2.4.1   Gini Means   82
        2.4.2   Bonferroni Means   83
        2.4.3   Heronian Mean   85
        2.4.4   Generalized Logarithmic Means   86
        2.4.5   Cauchy and Lagrangean Means   88
        2.4.6   Mean of Bajraktarevic   89
        2.4.7   Mixture Functions   89
        2.4.8   Compound Means   90
        2.4.9   Extending Bivariate Means to More Than
                 Two Arguments.   91
   References   97

3   Ordered Weighted Averaging   101
   3.1   Definitions   101
   3.2   Main Properties   102
        3.2.1   Orness Measure   103
        3.2.2   Entropy.   105
   3.3   Other Types of OWA Functions.   106
        3.3.1   Neat OWA   106
        3.3.2   Generalized OWA   107
        3.3.3   Weighted OWA.   109
   3.4   How to Choose Weights in OWA   114
        3.4.1   Methods Based on Data   114
        3.4.2   Methods Based on a Measure of Dispersion   117
        3.4.3   Methods Based on Weight Generating Functions   120
        3.4.4   Fitting Weight Generating Functions.   123
        3.4.5   Choosing Parameters of Generalized OWA   126
   3.5   Induced OWA   130
        3.5.1   Definition   130
        3.5.2   Properties   130

	3.5.3	Induced Generalized OWA	133
	3.5.4	Choices for the Inducing Variable	134
3.6	Medians and Order Statistics		137
	3.6.1	Median	137
	3.6.2	Order Statistics	141
References			141

4 Fuzzy Integrals ... 145
4.1	Choquet Integral		145
	4.1.1	Semantics	145
	4.1.2	Definitions and Properties	147
	4.1.3	Types of Fuzzy Measures	153
	4.1.4	Interaction, Importance and Other Indices	160
	4.1.5	Special Cases of the Choquet Integral	164
	4.1.6	Fitting Fuzzy Measures	166
	4.1.7	Generalized Choquet Integral	173
4.2	Sugeno Integral		174
	4.2.1	Definition and Properties	174
	4.2.2	Special Cases	175
4.3	Induced Fuzzy Integrals		177
References			178

5 Penalty Based Averages ... 183
5.1	Motivation and Definitions		183
5.2	Types of Penalty Functions		186
	5.2.1	Faithful Penalty Functions	186
	5.2.2	Restricted Dissimilarity Functions	188
	5.2.3	Minkowski Gauge Based Penalties	191
5.3	Examples		193
	5.3.1	Quasi-arithmetic Means, OWA and Choquet Integral	193
	5.3.2	Deviation Means	194
	5.3.3	Entropic Means	195
	5.3.4	Bregman Loss Functions	195
5.4	New Penalty Based Aggregation Functions		197
5.5	Relation to the Maximum Likelihood Principle		201
5.6	Representation of Averages		203
References			204

6 More Types of Averaging and Construction Methods ... 207
6.1	Some Construction Methods		207
	6.1.1	Idempotization	207
	6.1.2	Means Defined by Using Graduation Curves	208
	6.1.3	Aggregation Functions with Flying Parameter	210
	6.1.4	Construction of Shift-Invariant Functions	212
	6.1.5	Interpolatory Constructions	212

6.2   Other Types of Aggregation and Properties   216
       6.2.1   Bi-capacities   216
       6.2.2   Linguistic Aggregation Functions   217
       6.2.3   Multistage Aggregation   217
       6.2.4   Migrativity   218
6.3   Overlap and Grouping Functions   218
       6.3.1   Definition of Overlap Functions
                and Basic Properties   219
       6.3.2   Characterization of Overlap Functions   220
       6.3.3   Homogeneous Overlap Functions   221
       6.3.4 k-Lipschitz Overlap Functions   222
       6.3.5 n-dimensional Overlap Functions   224
       6.3.6   Grouping Functions   225
6.4   Generalized Bonferroni Mean   227
       6.4.1   Main Definitions   227
       6.4.2   Properties of the Generalized Bonferroni Mean   229
       6.4.3   Replacing the Outer Mean   230
       6.4.4   Replacing the Inner Mean   231
       6.4.5   Replacing the Product Operation   232
       6.4.6   Extensions to $B_{\mathrm{M}}^{\mathrm{K}}$ 233
       6.4.7   Boundedness of the Generalized Bonferroni Mean   234
       6.4.8 k-intolerance Boundedness   237
       6.4.9   Generated t-norm and Generated Quasi-arithmetic
                Means as Components of B_{M}   238
6.5   Consistency and Stability   239
       6.5.1   Motivation   239
       6.5.2   Strictly Stable Families   240
       6.5.3 R-strict Stability   241
       6.5.4   Learning Consistent Weights   245
       6.5.5   Consistency and Global Monotonicity   246
References   247

7   Non-monotone Averages   251
    7.1   Motivation   251
    7.2   Weakly Monotone Functions   252
           7.2.1   Basic Properties of Weakly Monotone Functions   254
    7.3   Robust Estimators of Location   256
           7.3.1   Mode   257
           7.3.2   Shorth   257
           7.3.3   Least Median of Squares (LMS)   258
           7.3.4   Least Trimmed Squares (LTS)   258
           7.3.5   Least Trimmed Absolute Deviations (LTA)   259
           7.3.6   The Least Winsorized Squares Estimator   260
           7.3.7   OWA Penalty Functions   260

7.4   Lehmer and Gini Means 261
      7.4.1   Lehmer Means........................... 261
      7.4.2   Gini Means 263
7.5   Mixture Functions 265
      7.5.1   Some Special Cases of Weighting Functions 266
      7.5.2   Affine Weighting Functions. 267
      7.5.3   Linear Combinations of Weighting Functions 269
      7.5.4   The Duals of Lehmer Mean and Other Mixture
              Functions 270
7.6   Density Based Means and Medians 271
      7.6.1   Density Based Means 271
      7.6.2   Density Based Medians and ML Estimators. 272
      7.6.3   Modified Weighting Functions. 273
7.7   Mode-Like Averages 275
7.8   Spatial-Tonal Filters 277
7.9   Transforms 278
7.10 Cone Monotone Functions. 281
      7.10.1 Formal Definitions and Properties 281
      7.10.2 Verification of Cone Monotonicity............... 285
      7.10.3 Construction of Cone Monotone Lipschitz
              Functions 286
7.11 Monotonicity with Respect to Coalitions................ 288
      7.11.1 Simple Majority........................... 288
      7.11.2 Majority and Preferential Inputs 290
      7.11.3 Coalitions 291
7.12 Directional Monotonicity. 293
      7.12.1 Properties of r-Monotone functions 294
      7.12.2 The Set of Directions of Increasingness. 295
7.13 Pre-aggregation Functions 296
      7.13.1 Definitions and Properties 296
      7.13.2 Construction of Pre-aggregation Functions
              by Composition 297
      7.13.3 Choquet-Like Construction Method
              of Pre-aggregation Functions 299
      7.13.4 Sugeno-Like Construction Method
              of Pre-aggregation Functions 300
      References 302

8   Averages on Lattices 305
8.1   Aggregation of Intervals and Intuitionistic Fuzzy Values...... 305
      8.1.1   Preliminary Definitions...................... 306
      8.1.2   Aggregation on Product Lattices 306
      8.1.3   Arithmetic Means and OWA for AIFV 309
      8.1.4   Alternative Definitions of Aggregation Functions
              on AIFV................................ 313

      8.1.5   Consistency with Operations on Ordinary
                  Fuzzy Sets 316
      8.1.6   Medians for AIFV 319
      8.1.7   Bonferroni Means 320
  8.2   Medians on Lattices 327
      8.2.1   Medians as Penalty Based Functions. 327
      8.2.2   Median Graphs and Distributive Lattices. 328
      8.2.3   Medians on Infinite Lattices and Fermat Points 330
      8.2.4   Medians Based on Distances Between Intervals 331
      8.2.5   Numerical Comparison 333
  8.3   Penalty Functions on Cartesian Products of Lattices 336
  References 342

Index 347

Notations

$\mathbb{R}$	The set of real numbers
$\mathbb{I}$	An interval $[a, b]$, often the unit interval $[0,1]$
$\mathcal{N}$	The set $\{1, 2, \ldots, n\}$
2^X	The power set (i.e., the set of all subsets of the set X)
$\mathcal{A}^c$	The complement of the set $\mathcal{A}$
$\mathbf{x}$	n-dimensional real vector, usually from $\mathbb{I}^n$
$\langle \mathbf{x}, \mathbf{y} \rangle$	Scalar (or dot) product of vectors $\mathbf{x}$ and $\mathbf{y}$
$\mathbf{x}_\searrow$	Permutation of the vector $\mathbf{x}$ which arranges its components in non-increasing order
$\mathbf{x}_\nearrow$	Permutation of the vector $\mathbf{x}$ which arranges its components in non-decreasing order
f_n	A function of n variables, usually $f_n : \mathbb{I}^n \to \mathbb{I}$
F	An extended function $F : \bigcup_{n \in \{1,2,\ldots\}} \mathbb{I}^n \to \mathbb{I}$
$f \circ g$	The composition of functions f and g
g^{-1}	The inverse of the function g
$g^{(-1)}$	The pseudo-inverse of the function g
N	A strong negation function
v	A fuzzy measure
$\mathbf{w}$	A weighting vector
$\log$	The natural logarithm
P	A penalty function

Averaging Functions

$M(\mathbf{x})$	Arithmetic mean of $\mathbf{x}$
$M_\mathbf{w}(\mathbf{x})$	Weighted arithmetic mean of $\mathbf{x}$ with the weighting vector $\mathbf{w}$
$M_w(\mathbf{x})$	Mixture function of $\mathbf{x}$ with the weighting function w
$M_{\mathbf{w},[r]}(\mathbf{x})$	Weighted power mean of $\mathbf{x}$ with the weighting vector $\mathbf{w}$ and exponent r

$M_{\mathbf{w},g}(\mathbf{x})$	Weighted quasi-arithmetic mean of $\mathbf{x}$ with the weighting vector $\mathbf{w}$ and generator g
$Med(\mathbf{x})$	Median of $\mathbf{x}$
$Med_a(\mathbf{x})$	a-median of $\mathbf{x}$
$Q_{\mathbf{w}}(\mathbf{x})$	Weighted quadratic mean of $\mathbf{x}$ with the weighting vector $\mathbf{w}$
$H_{\mathbf{w}}(\mathbf{x})$	Weighted harmonic mean of $\mathbf{x}$ with the weighting vector $\mathbf{w}$
$HR(\mathbf{x})$	Heronian mean of $\mathbf{x}$
$G(\mathbf{x})$	Geometric mean of $\mathbf{x}$
$G_{\mathbf{w}}(\mathbf{x})$	Weighted geometric mean of $\mathbf{x}$ with the weighting vector $\mathbf{w}$
$B^{p,q}(\mathbf{x})$	Bonferroni mean of $\mathbf{x}$
$B_{\mathbb{M}}(\mathbf{x})$	Generalized Bonferroni mean of $\mathbf{x}$ with $\mathbb{M} = \langle M_1, M_2, C \rangle$
$G_{\mathbf{w}}^{p,q}(\mathbf{x})$	Weighted Gini mean of $\mathbf{x}$
$L_{\mathbf{w}}^{p}(\mathbf{x})$	Weighted Lehmer mean of $\mathbf{x}$
$L^{p}(x,y)$	Generalized logarithmic mean of (x, y)
$C^{g,h}(x,y)$	Cauchy mean of (x, y)
$AGM(x,y)$	Arithmetico-geometric mean of (x, y)
$A \otimes B(x,y)$	Compound mean of (x, y)
$OWA_{\mathbf{w}}(\mathbf{x})$	Ordered weighted average of $\mathbf{x}$ with the weighting vector $\mathbf{w}$
$IOWA_{\mathbf{w}}(\langle \mathbf{x}, \mathbf{z} \rangle)$	Induced ordered weighted average of $\mathbf{x}$ with the weighting vector $\mathbf{w}$ and inducing variable $\mathbf{z}$
$C_v(\mathbf{x})$	Discrete Choquet integral of $\mathbf{x}$ with respect to the fuzzy measure v
$S_v(\mathbf{x})$	Discrete Sugeno integral of $\mathbf{x}$ with respect to the fuzzy measure v

Conjunctive and Disjunctive Functions

T	Triangular norm
S	Triangular conorm
T_P, T_L, T_D	The basic triangular norms (product, Łukasiewicz, and drastic product)
S_P, S_L, S_D	The basic triangular conorms (dual product, Łukasiewicz, and drastic sum)

Acronyms and Abbreviations

LAD	Least absolute deviation
LP	Linear programming
LS	Least squares
LMS	Least median of squares
LTS	Least trimmed squares
OWA	Ordered weighted averaging
QP	Quadratic programming

AIFV	Atanassov intuitionistic fuzzy value
IVFV	Interval valued fuzzy value
t-norm	Triangular norm
t-conorm	Triangular conorm
s.t.	Subject to
w.r.t.	With respect to
WOWA	Weighted ordered weighted averaging

Chapter 1
Review of Aggregation Functions

Abstract This chapter introduces aggregation functions and some of their applications, and then discusses the main properties and prototypical examples of aggregation functions. Some construction methods are discussed and strategies for choosing suitable aggregation functions are outlined. This chapter ends with an overview of some methods of approximation and optimization that will be used for fitting aggregation functions to empirical data.

1.1 Aggregation Functions

Aggregation is the process of combining several values into a single value. Mathematical functions which provide a mechanism for doing so are called *aggregation functions*. Aggregation functions are functions with special properties we shall discuss later on. In this book we mainly consider aggregation functions that take real arguments from a closed interval $\mathbb{I} = [a, b] \subset \mathbb{R}$ such as $[0, 1]$ and produce a real value in $[0, 1]$.[1] This is usually denoted as $f : [0, 1]^n \to [0, 1]$ or $f : \mathbb{I}^n \to \mathbb{I}$ for functions that take arguments with n components.

One of the simplest aggregation function is the arithmetic mean, or average, which produces an output value in some sense representative of its inputs. We shall see, however, that there are many alternatives which are better suited to specific applications than the arithmetic mean. Before we proceed with formal definitions, let us look at some examples.

Example 1.1 (A multicriteria decision making problem) There are two (or more) alternatives, and n criteria to evaluate each alternative (or rather a preference for each alternative). Denote the scores (preferences) by $x_1, x_2, \ldots, x_n$ and $y_1, y_2, \ldots, y_n$ for the alternatives x and y respectively. The goal is to combine these scores using some aggregation function f, and to compare the values $f(x_1, x_2, \ldots, x_n)$ and $f(y_1, y_2, \ldots, y_n)$ to decide on the winning alternative.

---

[1]The interval $[0, 1]$ can be substituted with any interval $[a, b]$ using a simple transformation, see Sect. 1.3.

© Springer International Publishing Switzerland 2016
G. Beliakov et al., *A Practical Guide to Averaging Functions*,
Studies in Fuzziness and Soft Computing 329,
DOI 10.1007/978-3-319-24753-3_1

Example 1.2 (*Connectives in fuzzy logic*) An object x has partial degrees of membership in n fuzzy sets, denoted by $\mu_1, \mu_2, \ldots, \mu_n$. The goal is to obtain the overall membership value in the combined fuzzy set $\mu = f(\mu_1, \mu_2, \ldots, \mu_n)$. The combination can be the set operation of union, intersection, or a more complicated (e.g., composite) operation.

Example 1.3 (*A group decision making problem*) There are two (or more) alternatives, and n decision makers or experts who express their evaluation of each alternative as $x_1, x_2, \ldots, x_n$. The goal is to combine these evaluations using some aggregation function f, to obtain a global score $f(x_1, x_2, \ldots, x_n)$ for each alternative.

Example 1.4 (*A rule based system*) The system contains rules of the form

If t_1 *is* A_1 **AND** t_2 *is* A_2 **AND** $\ldots$ t_n *is* A_n **THEN ...**

$x_1, x_2, \ldots, x_n$ denote the degrees of satisfaction of the rule predicates t_1 is A_1, t_2 is A_2, etc. The goal is to calculate the overall degree of satisfaction of the combined predicate of the rule antecedent $f(x_1, x_2, \ldots, x_n)$.[2]

The input value 0 is interpreted as no membership, no preference, no evidence, no satisfaction, etc., and naturally, an aggregation of n zeros should yield 0. Similarly, the value 1 is interpreted as full membership (strongest preference, evidence), and an aggregation of ones should naturally yield 1. This implies a fundamental property of aggregation functions, the preservation of the bounds

$$f(\underbrace{0, 0, \ldots, 0}_{n-times}) = 0 \quad \text{and} \quad f(\underbrace{1, 1, \ldots, 1}_{n-times}) = 1. \tag{1.1}$$

The second fundamental property is the monotonicity condition. Consider aggregation of two inputs $\mathbf{x}$ and $\mathbf{y}$, such that $x_1 < y_1$ and $x_j = y_j$ for all $j = 2, \ldots, n$, e.g., $\mathbf{x} = (a, b, b, b), \mathbf{y} = (c, b, b, b), a < c$. Think of the j-th argument of f as the degree of preference with respect to the j-th criterion, and $\mathbf{x}$ and $\mathbf{y}$ as vectors representing two alternatives A and B. Thus B is preferred to A with respect to the first criterion, and we equally prefer the two alternatives with respect to all other criteria. Then it is not reasonable to prefer A to B. Of course the numbering of the criteria is not important, so monotonicity holds not only for the first but for any argument x_i.

For example, consider buying an item in a grocery store. There are two grocery shops close by (the two alternatives are whether to buy the item in one or the other shop), and the item costs less in shop A. The two criteria are the price and distance to the shop. We equally prefer the two alternatives with respect to the second criterion, and prefer shop A with respect to the price. After combining the two criteria, we prefer buying in shop A and not shop B.

---

[2]As a specific example, consider the rules in a fuzzy controller of an air conditioner: If temperature is HIGH AND humidity is MEDIUM THEN...

Mathematically (increasing) monotonicity in all arguments is expressed as

$$x_i \leq y_i \text{ for all } i \in \{1, \ldots, n\} \text{ implies } f(x_1, \ldots, x_n) \leq f(y_1, \ldots, y_n). \quad (1.2)$$

We will frequently use vector inequality $\mathbf{x} \leq \mathbf{y}$, which means that each component of $\mathbf{x}$ is no greater than the corresponding component of $\mathbf{y}$. Thus, increasing monotonicity can be expressed as $\mathbf{x} \leq \mathbf{y}$ implies $f(\mathbf{x}) \leq f(\mathbf{y})$. Condition (1.2) is equivalent to the condition that each *univariate* function $f_{\mathbf{x}}(t) = f(\mathbf{x})$ with $t = x_i$ and the rest of the components of $\mathbf{x}$ being fixed, is monotone increasing in t. Note that we use the term increasing not in the strict sense, that is it could be that $f(\mathbf{x}) = f(\mathbf{y})$ for $\mathbf{x} \neq \mathbf{y}$.[3]

The monotonicity in all arguments and preservation of the bounds are the two fundamental properties that characterize general aggregation functions. If any of these properties fails, we cannot consider function f as an aggregation function, because it will provide inconsistent output when used, say, in a decision support system.[4] All the other properties discussed in this book define specific classes of aggregation functions.

We should also note that a function f can be represented in several ways:

(a) as an algebraic formula (say, $f(\mathbf{x}) = x_1 + x_2 - x_3$),
(b) as a graph of the function (e.g., as 2D, 3D plot, or contour plot),
(c) verbally, as a sequence of steps (e.g., take the average of components of $\mathbf{x}$), or more formally, as an algorithm,
(d) as a lookup table,
(e) as a solution to some equation (algebraic, differential, or functional),
(f) as a computer subroutine that returns a value y for any specified $\mathbf{x}$ (so called oracle).

Finally, aggregation functions are also called *aggregation operators* in the literature. Generally the term *operator* is reserved (in Mathematics) for functions $f : X \to Y$, whose domain X and co-domain Y consist of more complicated objects than sets of real numbers. Typically both X and Y are sets of functions. Differentiation and integration operators are typical examples, see, e.g., [Wei02]. Therefore, we shall use the term *aggregation function* throughout this book.

While our focus in this book is on the *averaging* functions, we shall present a brief overview of other types of aggregation and give several definitions that are applicable to aggregation functions in general. A deeper discussion of aggregation functions is presented in [BPC07, GMMP09].

Definition 1.5 (*Aggregation function*) An aggregation function is a function of $n > 1$ arguments that maps the (n-dimensional) cube onto an interval $\mathbb{I} = [a, b]$, $f : \mathbb{I}^n \to \mathbb{I}$, with the properties

---

[3]In some works like [BPC07, GMMP09] the authors use the term non-decreasing. Since there might be a confusion with functions that are, for example, oscillating, we prefer to use the terms increasing and strictly increasing, see Definition 1.22.

[4]However it is quite feasible to consider non-monotone averages when some arguments are possibly corrupted, and therefore extend the notion of aggregation functions. We shall discuss this in Chap. 7.

(i) $f(\underbrace{a, a, \ldots, a}_{n-times}) = a$ and $f(\underbrace{b, b, \ldots, b}_{n-times}) = b.$

(ii) $\mathbf{x} \leq \mathbf{y}$ implies $f(\mathbf{x}) \leq f(\mathbf{y})$ for all $\mathbf{x}, \mathbf{y} \in \mathbb{I}^n$.

As we mentioned, typically in fuzzy systems $\mathbb{I} = [0, 1]$, but most of the averaging functions we present are defined on $\mathbb{R}_+ = [0, \infty)$ or even on the real line $\mathbb{R}$, hence we shall not unnecessarily restrict the domain and range of these functions.

It is often the case that aggregation of inputs of various sizes has to be considered in the same framework. In some applications, input vectors may have a varying number of components (for instance, some values can be missing). In theoretical studies, it is also often appropriate to consider a family of functions of $n = 2, 3, \ldots$ arguments with the same underlying property. The following mathematical construction of an *extended aggregation function* [MC97] allows one to define and work with such families of functions of any number of arguments.

Definition 1.6 (*Extended aggregation function*) An extended aggregation function is a mapping

$$F : \bigcup_{n\in\{1,2,\ldots\}} \mathbb{I}^n \to \mathbb{I},$$

such that the restriction of this mapping to the domain $\mathbb{I}^n$ for a fixed n is an n-ary aggregation function f, with the convention $F(x) = x$ for $n = 1$.

Thus, in simpler terms, an extended aggregation function[5] is a family of $2-$, $3-, \ldots$ variate aggregation functions, with the convention $F(x) = x$ for the special case $n = 1$. We shall use the notation f_n when we want to emphasize that an aggregation function has n arguments. In general, two members of such a family for distinct input sizes m and n need not be related. However, we shall see that in the most interesting cases they are related, and sometimes can be computed using one generic formula.

In the next subsection we study some generic properties of aggregation functions and extended aggregation functions. Generally a given property holds for an extended aggregation function F if and only if it holds for every member of the family f_n.

Examples of aggregation functions

Arithmetic mean $f_n(\mathbf{x}) = \dfrac{1}{n}(x_1 + x_2 + \cdots + x_n).$

Geometric mean $f_n(\mathbf{x}) = \sqrt[n]{x_1 x_2 \ldots x_n}.$

Harmonic mean $f_n(\mathbf{x}) = \dfrac{n}{\frac{1}{x_1} + \frac{1}{x_2} + \cdots + \frac{1}{x_n}}.$

Minimum $\min(\mathbf{x}) = \min\{x_1, \ldots, x_n\}.$

---

[5]Sometimes extended aggregation functions are also referred to as aggregation operators, see footnote 1.1

$$\text{Maximum} \qquad \max(\mathbf{x}) = \max\{x_1, \ldots, x_n\}.$$

$$\text{Product} \qquad f_n(\mathbf{x}) = x_1 x_2 \ldots x_n = \prod_{i=1}^{n} x_i.$$

$$\text{Bounded sum} \qquad f_n(\mathbf{x}) = \min\{1, \sum_{i=1}^{n} x_i\}.$$

Note that in all mentioned examples we have extended aggregation functions, since the above generic formulae are valid for any $n > 1$.

1.2 Applications of Aggregation Functions

Consider the following prototypical situations. We have several criteria with respect to which we assess different options (or objects), and every option fulfills each criterion only partially (and has a score on $[a, b]$ scale). Our aim is to evaluate the combined score for each option (possibly to rank the options).

We may decide to average all the scores. This is a typical approach in sports competitions (like artistic skating) where the criteria are scores given by judges. The total score is the arithmetic mean.

$$f(x_1, \ldots, x_n) = \frac{1}{n}(x_1 + x_2 + \cdots + x_n).$$

We may decide to take a different approach: low scores pull the overall score down. The total score will be no greater than the minimum individual criterion score. This is an example of conjunctive behavior. Conjunctive aggregation functions are suitable to model conjunctions like

If x *is* A **AND** y *is* B **AND** z *is* C **THEN** ...

where A, B and C are the criteria against which the parameters x, y, z are assessed. For example, this is how they choose astronauts: the candidates must fulfill this and that criteria, and having an imperfect score in just one criterion moves the total to that imperfect score, and further down if there is more than one imperfect score.

Yet another approach is to let high scores push each other up. We start with the largest individual score, and then each other nonzero score pushes the total up. The overall score will be no smaller than the maximum of all individual scores. This is an example of disjunctive behavior. Such aggregation functions model disjunctions in logical rules like

If x *is* A **OR** y *is* B **OR** z *is* C **THEN** ...

For example, consider collecting evidence supporting some hypothesis. Having more than one piece of supporting evidence makes the total support stronger than the support due to any single piece of evidence. A simple example may be: if you have fever you may have a cold. If you cough and sneeze, you may have a cold. But if you have fever, cough and sneeze at the same time, you are almost certain to have a cold.

It is also possible to think of an aggregation scheme where low scores pull each other down and high scores push each other up. We need to set some threshold to distinguish low and high scores, say 0.5. Then an aggregation function will have conjunctive behavior when all $x_i \leq 0.5$, disjunctive behavior when all $x_i \geq 0.5$, and either conjunctive, disjunctive or averaging behavior when there are some low and some high scores.

One typical use of such aggregation functions is when low scores are interpreted as "negative" information, and high scores as "positive" information [RCM12]. Sometimes scientists also use a bipolar scale ($\mathbb{I} = [-1, 1]$ instead of $[0, 1]$, in which case the threshold is 0).

Multiple attribute decision making

In problems of multiple attribute decision making (sometimes interchangeably called multiple criteria decision making[6]) [CH92, HM79, HY81], an alternative (a decision) has to be chosen based on several, usually conflicting criteria. The alternatives are evaluated by using attributes, or features, which may be expressed numerically.[7] For example, when buying a car, the attributes are usually the price, quality, fuel consumption, size, power, brand (a nominal attribute), etc. In order to choose the best alternative, one needs to combine the values of the attributes in some way. One popular approach is called *Multi-Attribute Utility Theory* (MAUT). It assigns a numerical score to each alternative, called its *utility* $u(a_1, \ldots, a_n)$. Values a_i denote the numerical scores of each attribute i. Note that a bigger value of a_i does not imply "better", for example when choosing a car, one may prefer medium-sized cars.

The basic assumption of MAUT is that the total utility is a function of individual utilities $x_i = u_i(a_i)$, i.e., $u(a) = u(u_1(a_1), \ldots, u_n(a_n)) = u(\mathbf{x})$. The individual utilities depend *only* on the corresponding attribute a_i, and not on the other attributes, although the attributes can be correlated of course. This is a simplifying assumption, but it allows one to model quite accurately many practical problems.

The rational decision-making axiom implies that one cannot prefer an alternative which differs from another alternative in that it is inferior with respect to some individual utilities, but not superior with respect to the other ones. Mathematically, this means that the function u is monotone non-decreasing with respect to all arguments. If we scale the utilities to $[0, 1]$, and add the boundary conditions $u(0, \ldots, 0) = 0$, $u(1, \ldots, 1) = 1$, we obtain that u is an aggregation function on $[0, 1]^n$.

---

[6]Multiple criteria decision making, besides multiple attribute decision making, also involves multiple objective decision making [HM79].

[7]Of course, sometimes attributes are qualitative, or ordinal. They may be converted to a numerical scale, e.g., using utility functions.

The most common aggregation functions used in MAUT are the additive and multiplicative utilities, which are the weighted arithmetic and geometric means (see p. 25). However disjunctive and conjunctive methods are also popular [CH92]. Compensation is an important notion in MAUT, expressing the concept of trade-off. It implies that the decrease of the total utility u due to some attributes can be compensated by the increase due to other attributes. For example, when buying an item, the increase in price can be compensated by the increase in quality.

Group decision making

Consider a group of n experts who evaluate one (or more) alternatives. Each expert expresses his/her evaluation on a numerical scale (which is the strength of this expert's preference). The goal is to combine all experts' evaluations into a single score [DP00, DP04, HL87]. By scaling the preferences to [0, 1] we obtain a vector of inputs $\mathbf{x} = (x_1, \ldots, x_n)$, where x_i is the degree of preference of the i-th expert. The overall evaluation will be some value $y = f(\mathbf{x})$.

The most commonly used aggregation method in group decision making is the weighted arithmetic mean. If experts have different standing, then their scores are assigned different weights. However, experts' opinions may be correlated, and one has to model their interactions and groupings. Further, one has to model such concepts as dictatorship, veto, oligarchies, etc.

From the mathematical point of view, the function f needs to be monotone non-decreasing (the increase in one expert's score cannot lead to the decrease of the overall score) and satisfy the boundary conditions. Thus f is an aggregation function.

Fuzzy logic and rule based systems

In fuzzy set theory [Zad65], membership of objects in fuzzy sets are numbers from [0, 1]. Fuzzy sets allow one to model vagueness and uncertainty which are very often present in natural languages. For example the set "ripe bananas" is fuzzy, as there are obviously different degrees of ripeness. Similarly, the sets of "small numbers", "tall people", "high blood pressure" are fuzzy, since there is no clear cutoff which discriminates objects between those that are in the set and those that are not. An object may simultaneously belong to a fuzzy set and its complement. A fuzzy set A defined on a set of objects X is represented by a membership function $\mu_A : X \rightarrow [0, 1]$, in such a way that for any object $x \in X$ the value $\mu_A(x)$ measures the degree of membership of x in the fuzzy set A.

The classical operations of fuzzy sets, intersection and union, are based on the minimum and maximum, i.e., $\mu_{A \cap B} = \min\{\mu_A, \mu_B\}$, $\mu_{A \cup B} = \max\{\mu_A, \mu_B\}$. However other aggregation functions, such as the product, have been considered almost since the inception of fuzzy set theory [BZ70]. Nowadays a large class of conjunctive and disjunctive functions, the triangular norms and conorms, and overlap and grouping functions are used to model fuzzy set intersection and union.

Fuzzy set theory has proved to be extremely useful for solving many real world problems, in which the data are imprecise, e.g., [DP80, DP00, YF94, Zad78, Zad85, Zim96]. Fuzzy control in industrial systems and consumer electronics is one notable example of the practical applications of fuzzy logic.

Rule based systems, especially fuzzy rule based systems [Zad85], involve aggregation of various numerical scores, which correspond to degrees of satisfaction of rule antecedents. A rule can be a statement like

If x *is* A **AND** y *is* B **AND** z *is* C **THEN** some action

The antecedents are usually membership values of x in A, y in B, etc. The strength of "firing" the rule is determined by an aggregation function that combines membership values $f(\mu_A(x), \mu_B(y), \mu_C(z))$.

Image processing

Another interesting application of aggregation functions comes from image processing where the intensity values of pixels need to be aggregated into a representative value for the purposes of filtering or some transformation of an image. Here we present the problem of image reduction as a specific example.

In image reduction the aim is to reduce the resolution of an image to either fit a particular display, for printing purposes, or to perform image recognition, retrieval or segmentation tasks much faster [BBP12, PFBMB15]. For example, the cameras of modern smartphones capture images with much higher resolution than the display of such device can reproduce, and therefore the image must be scaled down. Similarly, processing of a large number of high resolution images (e.g. in a video stream) with the aim of recognising some objects may be prohibitively expensive on a limited power autonomous device, whereas the same objects could be recognised at a lower resolution much faster.

Local averaging is a computationally efficient way of image reduction. Here one divides the image into (possibly overlapping) rectangular blocks and computes some sort of an average, or representative value of each block of pixels in an appropriate color space. This way the blocks of pixels are reduced to representative values make up the pixels of the reduced image.

Different aggregation functions could be used for aggregating pixel intensities in a block [BLVW15, PFBMB15, PJB12]. Such aggregation functions should be sufficiently robust to the noise in the image. Since a block of pixels with the same intensities should be reduced to a representative pixel with the same intensity, it makes sense to use averaging functions as they are all idempotent. At the same time uniform changes in intensity of the input image (image lightening or darkening) should not lead to the opposite changes in intensity of the output image. This restriction does not imply monotonicity but its weaker version which we discuss in Chap. 7.

More generally one can think of *data reduction* as a tool for accelerating subsequent processing of large data sets. Nowadays data is captured by billions of digital devices worldwide, and only a fraction can be analysed for any particular purpose. By reducing the data so that its interesting aspects are preserved while the inevitable noise is filtered out, one can reduce the computational cost of data analysis algorithms while maintaining their accuracy. Sophisticated yet computationally inexpensive aggregation functions will play a significant role in this area.

There are many other uses of aggregation functions, detailed in the references at the end of this chapter.

1.3 Classification and General Properties

1.3.1 Main Classes

As we have seen in the previous section, there are various semantics of aggregation, and the main classes are determined according to these semantics. In some cases we require that high and low inputs average each other, in other cases aggregation functions model logical connectives (disjunction and conjunction), so that the inputs reinforce each other, and sometimes the behavior of aggregation functions depends on the inputs. The four main classes of aggregation functions are [CKKM02, DP80, DP85, DP00, DP04]

- Averaging,
- Conjunctive,
- Disjunctive,
- Mixed.

Definition 1.7 (*Averaging aggregation*) An aggregation function f has averaging behavior (or is averaging) if for every $\mathbf{x} \in \mathbb{I}^n$ it is bounded by

$$\min(\mathbf{x}) \leq f(\mathbf{x}) \leq \max(\mathbf{x}).$$

Definition 1.8 (*Conjunctive aggregation*) An aggregation function f has conjunctive behavior (or is conjunctive) if for every $\mathbf{x}$ it is bounded by

$$f(\mathbf{x}) \leq \min(\mathbf{x}) = \min(x_1, x_2, \ldots, x_n).$$

Definition 1.9 (*Disjunctive aggregation*) An aggregation function f has disjunctive behavior (or is disjunctive) if for every $\mathbf{x}$ it is bounded by

$$f(\mathbf{x}) \geq \max(\mathbf{x}) = \max(x_1, x_2, \ldots, x_n).$$

Definition 1.10 (*Mixed aggregation*) An aggregation function f is mixed if it does not belong to any of the above classes, i.e., it exhibits different types of behavior on different parts of the domain.

1.3.2 Main Properties

Definition 1.11 (*Idempotency*) An aggregation function f is called idempotent if for every input $\mathbf{x} = (t, t, \ldots, t), t \in \mathbb{I}$ the output is $f(t, t \ldots, t) = t$.

Note 1.12 Because of monotonicity of f, idempotency is equivalent to averaging behavior.[8]

The aggregation functions *minimum* and *maximum* are the only two functions that are at the same time conjunctive (disjunctive) and averaging, and hence idempotent.

Example 1.13 The arithmetic mean is an averaging (idempotent) aggregation function

$$f(\mathbf{x}) = \frac{1}{n}(x_1 + x_2 + \cdots + x_n).$$

Example 1.14 The geometric mean is also an averaging (idempotent) aggregation function

$$f(\mathbf{x}) = \sqrt[n]{x_1 x_2 \ldots x_n}.$$

Example 1.15 The product is a conjunctive aggregation function

$$f(\mathbf{x}) = \prod_{i=1}^{n} x_i = x_1 x_2 \ldots x_n.$$

Examples of disjunctive and mixed functions will follow.

The diagonal and inverse diagonal of an aggregation function play an important role for some aggregation functions.

Definition 1.16 (*Diagonal section*) Let $f : \mathbb{I}^n \to \mathbb{I}$ be a function. The diagonal of f, $d_f : \mathbb{I} \to \mathbb{I}$ is the function $d_f(t) = f(t, t, \ldots, t)$. The inverse of the diagonal, if it exists, is the function $d_f^{-1} : \mathbb{I} \to \mathbb{I}$ which satisfies $d_f^{-1}(f(t, t, \ldots, t)) = t$.

If $f(t, \ldots, t)$ is continuous and strictly increasing, the diagonal is invertible.

Definition 1.17 (*Symmetry*) An aggregation function f is called symmetric, if its value does not depend on the permutation of the arguments, i.e.,

$$f(x_1, x_2, \ldots, x_n) = f(x_{P(1)}, x_{P(2)}, \ldots, x_{P(n)}),$$

for every $\mathbf{x}$ and every permutation $P = (P(1), P(2), \ldots, P(n))$ of $(1, 2 \ldots, n)$.

---

[8]Proof: Take any $\mathbf{x} \in \mathbb{I}^n$, and denote by $p = \min(\mathbf{x})$, $q = \max(\mathbf{x})$. By monotonicity, $p = f(p, p, \ldots, p) \le f(\mathbf{x}) \le f(q, q, \ldots, q) = q$. Hence $\min(\mathbf{x}) \le f(\mathbf{x}) \le \max(\mathbf{x})$. The converse: let $\min(\mathbf{x}) \le f(\mathbf{x}) \le \max(\mathbf{x})$. By taking $\mathbf{x} = (t, t, \ldots, t)$, $\min(\mathbf{x}) = \max(\mathbf{x}) = f(\mathbf{x}) = t$, hence idempotency.

The semantical interpretation of symmetry is anonymity, or equality. For example, equality of judges in sports competitions: all inputs are treated equally, and the output does not change if the judges swap seats.[9] On the other hand, in shareholders meetings the votes are not symmetric as they depend on the number of shares each voter has.

Example 1.18 The arithmetic and geometric means and the product in Examples 1.13–1.15 are symmetric aggregation functions. A weighted arithmetic mean with non-equal weights $w_1, w_2, \ldots w_n$, that are non-negative and add to one is not symmetric,

$$f(\mathbf{x}) = \sum_{i=1}^{n} w_i x_i = w_1 x_1 + w_2 x_2 + \cdots + w_n x_n.$$

Permutation of arguments is very important in aggregation, as it helps express symmetry, as well as to define other concepts. A permutation of $(1, 2 \ldots, 5)$ is just a tuple like $(5, 3, 2, 1, 4)$. There are $n! = 1 \times 2 \times 3 \times \cdots \times n$ possible permutations of $(1, 2, \ldots, n)$.

We will denote a vector whose components are arranged in the order given by a permutation P by $\mathbf{x}_P = (x_{P(1)}, x_{P(2)}, \ldots, x_{P(n)})$. In our example, $\mathbf{x}_P = (x_5, x_3, x_2, x_1, x_4)$. We will frequently use the following special permutations of the components of $\mathbf{x}$.

Definition 1.19 We denote by $\mathbf{x}_\nearrow$ the vector obtained from $\mathbf{x}$ by arranging its components in non-decreasing order, that is, $\mathbf{x}_\nearrow = \mathbf{x}_P$ where P is the permutation such that $x_{P(1)} \leq x_{P(2)} \leq \cdots \leq x_{P(n)}$.

Similarly, we denote by $\mathbf{x}_\searrow$ the vector obtained from $\mathbf{x}$ by arranging its components in non-increasing order, that is $\mathbf{x}_\searrow = \mathbf{x}_P$ where P is the permutation such that $x_{P(1)} \geq x_{P(2)} \geq \cdots \geq x_{P(n)}$.

Note 1.20 In fuzzy sets literature, the notation $\mathbf{x}_{()} = (x_{(1)}, \ldots, x_{(n)})$ is often used to denote both $\mathbf{x}_\nearrow$ and $\mathbf{x}_\searrow$, depending on the context.

Note 1.21 We can express the symmetry property by an equivalent statement that for every input vector $\mathbf{x}$

$$f(\mathbf{x}) = f(\mathbf{x}_\nearrow) \quad (\text{or} f(\mathbf{x}) = f(\mathbf{x}_\searrow)),$$

rather than $f(\mathbf{x}) = f(\mathbf{x}_P)$ for *every* permutation. This gives us a shortcut for calculating the value of a symmetric aggregation function for a given $\mathbf{x}$ by using sort operation.

Definition 1.22 (*Strict monotonicity*) An aggregation function f is strictly monotone increasing if

$$\mathbf{x} \leq \mathbf{y} \text{ but } \mathbf{x} \neq \mathbf{y} \text{ implies } f(\mathbf{x}) < f(\mathbf{y}) \text{ for every } \mathbf{x}, \mathbf{y} \in \mathbb{I}^n. \tag{1.3}$$

---

[9]It is frequently interpreted as anonymity criterion: anonymous ballot papers can be counted in any order.

Note 1.23 Notice the difference between $[\mathbf{x} \le \mathbf{y}, \mathbf{x} \ne \mathbf{y}]$ and $\mathbf{x} < \mathbf{y}$. The latter implies that for *all* components of $\mathbf{x}$ and $\mathbf{y}$ we have $x_i < y_i$, whereas the former means that at least one component of $\mathbf{y}$ is greater than that of $\mathbf{x}$, i.e., $\exists i$ such that $x_i < y_i$ and $\forall j : x_j \le y_j$.

Strict monotonicity is a rather restrictive property. Note that there are no strictly monotone conjunctive or disjunctive aggregation functions. This is because every conjunctive function coincides with $\min(\mathbf{x})$ for those $\mathbf{x}$ that have at least one zero component, and min is not strictly monotone (similarly, disjunctive aggregation functions coincide with $\max(\mathbf{x})$ for those $\mathbf{x}$ that have at least one component $x_i = 1$). However, strict monotonicity on the semi-open set $]a, b]^n$ (respectively $[a, b[^n)$ is often considered for conjunctive (disjunctive) aggregation functions. Of course, there are plenty of strictly increasing averaging aggregation functions, such as arithmetic means.

It is often the case that when an input vector contains a specific value, this value can be omitted. For example, consider a conjunctive aggregation function f to model the rule for buying bananas

If *price is LOW* **AND** *banana is RIPE* **THEN** buy it.

Conjunction means that we only want cheap and ripe bananas, but both are matters of degree, expressed on [0, 1] scale. If one of the arguments $x_i = 1$, e.g., we find perfectly ripe bananas, then the outcome of the rule is equal to the degree of our satisfaction with the price.

Definition 1.24 (*Neutral element*) An aggregation function f has a neutral element $e \in \mathbb{I}$, if for every $t \in \mathbb{I}$ in any position it holds

$$f(e, \ldots, e, t, e, \ldots, e) = t.$$

For extended aggregation functions, we have a stronger version of this property, which relates aggregation functions with a different number of arguments.

Definition 1.25 (*Strong neutral element*) An extended aggregation function F has a neutral element $e \in \mathbb{I}$, if for every $\mathbf{x}$ with $x_i = e$, for some $1 \le i \le n$, and every $n \ge 2$,

$$f_n(x_1, \ldots, x_{i-1}, e, x_{i+1}, \ldots, x_n) = f_{n-1}(x_1, \ldots, x_{i-1}, x_{i+1}, \ldots, x_n),$$

When $n = 2$, we have $f(t, e) = f(e, t) = t$. Then by iterating this property we obtain as a consequence that every member f_n of the family has the neutral element e, i.e.,

$$f_n(e, \ldots, e, t, e, \ldots, e) = t,$$

for t in any position.

Note 1.26 A neutral element, if it exists, is unique.[10] It can be any number from $\mathbb{I}$.

Note 1.27 Observe that if an aggregation function f on $[a, b]^n$ has neutral element $e = b$ (respectively $e = a$) then f is necessarily conjunctive (respectively disjunctive). Indeed, if f has neutral element $e = 1$, then by monotonicity it is $f(x_1, \ldots, x_n) \leq f(1, \ldots, 1, x_i, 1, \ldots, 1) = x_i$ for any $i \in \{1, \ldots, n\}$, and this implies $f \leq \min$ (the proof for the case $e = 0$ is analogous).

Note 1.28 The concept of a neutral element has been extended to that of neutral tuples, see [BCP07b].

Example 1.29 The product function $f(\mathbf{x}) = \prod x_i$ has neutral element $e = 1$. Similarly, min function has neutral element $e = b$ and max function has neutral element $e = a$. The arithmetic mean does not have a neutral element. We shall see later on that any triangular norm has $e = 1$, and any triangular conorm has $e = 0$.

It may also be the case that one specific value a of any argument yields the output a. For example, if we use conjunction for aggregation, then if any input is 0, then the output must be 0 as well. In the banana buying rule above, if the banana is green, we do not buy it at any price.

Definition 1.30 *(Absorbing element (annihilator))* An aggregation function f has an absorbing element $a \in \mathbb{I}$ if

$$f(x_1, \ldots, x_{i-1}, a, x_{i+1}, \ldots, x_n) = a,$$

for every $\mathbf{x}$ such that $x_i = a$ with a in any position.

Note 1.31 An absorbing element, if it exists, is unique. It can be any number from $\mathbb{I}$.

Example 1.32 Any conjunctive aggregation function has absorbing element $a = 0$. Any disjunctive aggregation function has absorbing element $a = 1$. This is a simple consequence of the Definitions 1.8 and 1.9. Some averaging functions also have an absorbing element, for example the geometric mean

$$f(\mathbf{x}) = \left(\prod_{i=1}^{n} x_i \right)^{1/n}$$

has the absorbing element $a = 0$.

---

[10]Proof: Assume f has two neutral elements e and u. Then $u = f(e, u) = e$, therefore $e = u$. For n variables, assume $e < u$. By monotonicity, $e = f(u, \ldots, u, \ldots, u) \geq f(e, e, \ldots, e, u, e \ldots, e) = u$, hence we have a contradiction. The case $e > u$ leads to a similar contradiction.

Note 1.33 An aggregation function with an annihilator in $]0, 1[$ cannot have a neutral element.[11] But it may have a neutral element if $a = 0$ or $a = 1$.

Note 1.34 The concept of an absorbing element has been extended to that of absorbing tuples, see [BCP07a].

Let us fix $\mathbb{I} = [0, 1]$ for a moment.

Definition 1.35 (*Zero divisor*) An element $a \in]0, 1[$ is a zero divisor of an aggregation function f if for all $i \in \{1, \ldots, n\}$ there exists some $\mathbf{x} \in]0, 1]^n$ such that its i-th component is $x_i = a$, and it holds $f(\mathbf{x}) = 0$, i.e., the equality

$$f(x_1, \ldots, x_{i-1}, a, x_{i+1}, \ldots, x_n) = 0,$$

can hold for some $\mathbf{x} > \mathbf{0}$ with a at any position.

Note 1.36 Because of monotonicity of f, if a is a zero divisor, then all values $b \in]0, a]$ are also zero divisors.

The interpretation of zero divisors is straightforward: if one of the inputs takes the value a, or a smaller value, then the aggregated value could be zero, for some $\mathbf{x}$. So it is possible to have the aggregated value zero, even if all the inputs are positive. The largest value a (or rather an upper bound on a) plays the role of a threshold, the lower bound on all the inputs which guarantees a non-zero output. That is, if b is *not* a zero divisor, then $f(\mathbf{x}) > 0$, if all $x_i \geq b$.

Example 1.37 Averaging aggregation functions do not have zero divisors. But the function $f(x_1, x_2) = \max\{0, x_1 + x_2 - 1\}$ has a zero divisor $a = 0.999$, which means that the output can be zero even if any of the components x_1 or x_2 is as large as 0.999, provided that the other component is sufficiently small. However, 1 is not a zero divisor.

Zero divisors exist for aggregation functions that exhibit conjunctive behavior, at least on parts of their domain, i.e., conjunctive and mixed aggregation functions. For disjunctive aggregation functions we have an analogous definition.

Definition 1.38 (*One divisor*) An element $a \in]0, 1[$ is a one divisor of an aggregation function f if for all $i = 1, \ldots, n$ there exists some $\mathbf{x} \in [0, 1[^n$ such that its i-th component is $x_i = a$ and it holds $f(\mathbf{x}) = 1$, i.e., the equality

$$f(x_1, \ldots, x_{i-1}, a, x_{i+1}, \ldots, x_n) = 1,$$

can hold for some $\mathbf{x} < \mathbf{1}$ with a at any position.

---

[11]Proof: Suppose $a \in]0, 1[$ is the absorbing element and $e \in [0, 1]$ is the neutral element. Then if $a \leq e$, we get the contradiction $a = 0$, since it is $a = f(a, \ldots, a, 0) \leq f(e, \ldots, e, 0) = 0$. Similarly, if $a > e$ then $a = f(a, \ldots, a, 1) \geq f(e, \ldots, e, 1) = 1$.

The interpretation is similar: the value of any inputs larger than a can make the output $f(\mathbf{x}) = 1$, even if none of the inputs is actually 1. On the other hand, if b is not a one divisor, then the output cannot be one if all the inputs are no larger than b.

The following property is useful for construction of n-ary aggregation functions from a single two-variable function.

Definition 1.39 (*Associativity*)   A two-argument function f is associative if $f(f(x_1, x_2), x_3) = f(x_1, f(x_2, x_3))$ holds for all x_1, x_2, x_3 in its domain.

Consequently, the n-ary aggregation function can be constructed in a unique way by iteratively applying f_2 as

$$f_n(x_1, \ldots, x_n) = f_2(f_2(\ldots f_2(x_1, x_2), x_3), \ldots, x_n).$$

Thus bivariate associative aggregation functions univocally define extended aggregation functions.

Example 1.40 The product, minimum and maximum are associative aggregation functions. The arithmetic mean is not associative.

Associativity simplifies calculation of aggregation functions, and it effectively allows one to easily aggregate any number of inputs. It is not the only way of doing this (for instance the arithmetic or geometric means are also easily computed for any number of inputs).

Another construction gives what are called recursive extended aggregation functions by Montero [DAMM01]. It involves a family of two-variate functions f_2^n, $n = 2, 3, \ldots$.

Definition 1.41 (*Recursive extended aggregation function*) An extended aggregation function F is recursive by Montero if the members f_n are defined from a family of two-variate aggregation functions f_2^n recursively as

$$f_n(x_1, \ldots, x_n) = f_2^n(f_{n-1}(x_1, \ldots, x_{n-1}), x_n),$$

starting with $f_2 = f_2^2$.

Each extended aggregation function built from an associative bivariate aggregation function is recursive by Montero, but the converse is not true.

Example 1.42 Define $f_2^n(t_1, t_2) = \frac{(n-1)t_1 + t_2}{n}$. Then $f_n(\mathbf{x}) = \frac{1}{n} \sum_{i=1}^{n} x_i$, the arithmetic mean (which is not associative).

Definition 1.43 (*Decomposable extended aggregation function*)   An extended aggregation function F is decomposable if for all $m, n = 1, 2, \ldots$ and for all $\mathbf{x} \in \mathbb{I}^m$, $\mathbf{y} \in \mathbb{I}^n$:

Table 1.1 The table of scores to be aggregated by m jurymen with respect to n criteria

Juror\criterion	1	2	3	...	n	Total
1	x_{11}	x_{12}	x_{13}	...	x_{1n}	y_1
2	x_{21}	x_{22}	x_{23}	...	x_{2n}	y_2
3	x_{31}	x_{32}	x_{33}	...	x_{3n}	y_3
4	x_{41}	x_{42}	x_{43}	...	x_{4n}	y_4
$\vdots$	$\vdots$	$\vdots$	$\vdots$	$\vdots$	$\vdots$	$\vdots$
m	x_{m1}	x_{m2}	x_{m3}	...	x_{mn}	y_m
Total	$\tilde{y}_1$	$\tilde{y}_2$	$\tilde{y}_3$	...	$\tilde{y}_n$	$\tilde{z}\backslash z$

$$f_{m+n}(x_1, \ldots, x_m, y_1, \ldots, y_n) \tag{1.4}$$
$$= f_{m+n}(\underbrace{f_m(x_1, \ldots, x_m), \ldots, f_m(x_1, \ldots, x_m)}_{m \ times}, y_1, \ldots, y_n).$$

A continuous decomposable extended aggregation function is always idempotent.

Another useful property, which generalizes both symmetry and associativity, and is applicable to extended aggregation functions, is called bisymmetry. Consider the situation in which m jurymen evaluate an alternative with respect to n criteria. Let $x_{ij}, i = 1, \ldots, m, j = 1, \ldots, n$ denote the score given by the i-th juryman with respect to the j-th criterion. To compute the global score $f_{mn}(x_{11}, \ldots, x_{1n}, \ldots, x_{mn})$ we can either evaluate the scores given by the i-th juryman, $y_i = f_n(x_{i1}, \ldots, x_{in})$, and then aggregate them as $z = f_m(y_1, \ldots, y_m)$, or, alternatively, aggregate scores of all jurymen with respect to each individual criterion j, i.e., compute $\tilde{y}_j = f_m(x_{1j}, \ldots, x_{mj})$, and then aggregate these scores as $\tilde{z} = f_n(\tilde{y}_1, \ldots, \tilde{y}_n)$. The third alternative is to aggregate all the scores by an aggregation function $f_{mn}(\mathbf{x})$.

This is illustrated in Table 1.1. We can either aggregate scores in each row, and then aggregate the totals in the last column of this table, or we can aggregate scores in each column, and then aggregate the totals in the last row, or aggregate all scores at once. The bisymmetry property simply means that all three methods lead to the same answer.

Definition 1.44 (*Bisymmetry*) An extended aggregation function F is bisymmetric if for all $m, n = 1, 2, \ldots$ and for all $\mathbf{x} \in \mathbb{I}^{mn}$:

$$f_{mn}(\mathbf{x}) = f_m(f_n(x_{11}, \ldots, x_{1n}), \ldots, f_n(x_{m1}, \ldots, x_{mn})) \tag{1.5}$$
$$= f_n(f_m(x_{11}, \ldots, x_{m1}), \ldots, f_m(x_{1n}, \ldots, x_{mn})).$$

Note 1.45 A symmetric associative extended aggregation function is bisymmetric. However there are symmetric and bisymmetric non-associative extended aggregation functions, e.g., the arithmetic and geometric means. The extended aggregation function defined by $f(\mathbf{x}) = x_1$ (projection to the first coordinate) is bisymmetric and associative but not symmetric. The extended aggregation function $f(\mathbf{x}) = \left(\sum_{i=1}^{n} \frac{x_i}{n}\right)^2$

(square of the arithmetic mean) is symmetric but neither associative nor bisymmetric. Every continuous associative extended aggregation function is bisymmetric, but not necessarily symmetric.

Let us finally mention two properties describing the stability of aggregation functions with respect to some changes of the scale:

Definition 1.46 (*Shift-invariance*) An aggregation function $f : \mathbb{I}^n \to \mathbb{I}$ is shift-invariant (or stable for translations) if for all λ and for all $(x_1, \ldots, x_n) \in \mathbb{I}^n$ it is

$$f(x_1 + \lambda, \ldots, x_n + \lambda) = f(x_1, \ldots, x_n) + \lambda$$

whenever $(x_1 + \lambda, \ldots, x_n + \lambda) \in \mathbb{I}^n$.

Definition 1.47 (*Homogeneity*) An aggregation function $f : \mathbb{I}^n \to \mathbb{I}$ is homogeneous of order 1 if for all λ and for all $(x_1, \ldots, x_n) \in \mathbb{I}^n$ it is

$$f(\lambda x_1, \ldots, \lambda x_n) = \lambda f(x_1, \ldots, x_n).$$

Aggregation functions which are both shift-invariant and homogeneous are known as *linear aggregation functions*.

Note that, due to the boundary conditions $f(a, \ldots, a) = a$ and $f(b, \ldots, b) = b$, either shift-invariant, homogeneous or linear aggregation functions are necessarily idempotent, and thus (see Note 1.12) they can only be found among averaging functions. A prototypical example of a linear aggregation function is the arithmetic mean.

An aggregation function is said to be *locally internal* when it always provides as the output the value of one of its arguments:

Definition 1.48 (*Local internality*) An aggregation function f is called *locally internal* if for all $x_1, \ldots, x_n \in \mathbb{I}^n$, $f(x_1, \ldots, x_n) \in \{x_1, \ldots, x_n\}$.

Evidently, any locally internal aggregation function is idempotent (and then, see Note 1.12, it is an averaging function), but not vice-versa. Projections and order statistics (see Chap. 2), along with the minimum and maximum, are trivial instances of locally internal aggregation functions. Other functions in this class are the left- and right-continuous idempotent uninorms characterized in [DB99]. For details on this topic, we refer the reader to [MMT03], where bivariate locally internal aggregation functions have been characterized and studied in detail in conjunction with additional properties, such as symmetry, associativity and existence of a neutral element.

In [YR97] Yager introduced the so-called *self-identity* property, applicable to extended aggregation functions, and defined as follows.

Definition 1.49 (*Self-identity*) An extended aggregation function F has the *self-identity property* if for all $n \geq 1$ and for all $x_1, \ldots, x_n \in \mathbb{I}^n$,

$$F(x_1, \ldots, x_n, F(x_1, \ldots, x_n)) = F(x_1, \ldots, x_n).$$

Note that extended aggregation functions that satisfy the self-identity property are necessarily idempotent (and hence averaging), but the converse is not true. The arithmetic mean, the a-medians (see Chap. 2) and the functions min and max are examples of extended idempotent functions with the self-identity property. A subclass of weighted means that possess this property was characterized in [YR97].

Extended aggregation functions with the self-identity property verify, in addition, the following two inequalities:

$$F(x_1, \ldots, x_n, k) \geq F(x_1, \ldots, x_n) \quad \text{if} \quad k \geq F(x_1, \ldots, x_n),$$
$$F(x_1, \ldots, x_n, k) \leq F(x_1, \ldots, x_n) \quad \text{if} \quad k \leq F(x_1, \ldots, x_n).$$

1.3.3 Duality

It is often useful to draw a parallel between conjunctive and disjunctive aggregation functions, as they often satisfy very similar properties, just viewed from a different angle. The concept of a dual aggregation function helps with mapping most properties of conjunctive aggregation functions to disjunctive ones. So essentially one studies conjunctive functions, and obtains the corresponding results for disjunctive functions by duality. There are also aggregation functions that are self-dual.

First we need the concept of negation.

Definition 1.50 (*Strict negation*) A univariate function N defined on $[0, 1]$ is called a strict negation, if its range is also $[0, 1]$ and it is strictly monotone decreasing.[12]

Definition 1.51 (*Strong negation*) A univariate function N defined on $[0, 1]$ is called a strong negation, if it is strictly decreasing and involutive (i.e., $N(N(t)) = t$ for all $t \in [0, 1]$).

Example 1.52 The most commonly used strong negation is the standard negation $N(t) = 1 - t$. We will use it throughout this book. Another example of negation is $N(t) = 1 - t^2$, which is strict but not strong.

Note 1.53 A strictly monotone bijection is always continuous. Hence strict and strong negations are continuous.

Despite its simplicity, the standard negation plays a fundamental role in the construction of strong negations, since any strong negation can be built from the standard negation using an automorphism[13] of the unit interval [Tri79]:

---

[12]A frequently used term is *bijection*: a bijection is a function $f : A \rightarrow B$, such that for every $y \in B$ there is exactly one $x \in A$, such that $y = f(x)$, i.e., it defines a one-to-one correspondence between A and B. Because N is strictly monotone, it is a one-to-one function. Its range is $[0,1]$, hence it is an *onto* mapping, and therefore a bijection.

[13] *Automorphism* is another useful term: An automorphism is a strictly increasing bijection of an interval onto itself $[a, b] \rightarrow [a, b]$.

Theorem 1.54 *A function* $N : [0, 1] \to [0, 1]$ *is a strong negation if and only if there exists an automorphism* $\varphi : [0, 1] \to [0, 1]$ *such that* $N = N_\varphi = \varphi^{-1} \circ (1 - Id) \circ \varphi$, *i.e.* $N(t) = N_\varphi(t) = \varphi^{-1}(1 - \varphi(t))$ *for any* $t \in [0, 1]$.

Example 1.55 Let us construct some strong negations:

- With $\varphi(a) = a^\lambda$ ($\lambda > 0$):

$$N_\varphi(t) = (1 - t^\lambda)^{1/\lambda}$$

(note that the standard negation is recovered with $\lambda = 1$);

- With $\varphi(a) = 1 - (1 - a)^\lambda$ ($\lambda > 0$):

$$N_\varphi(t) = 1 - [1 - (1 - t)^\lambda]^{1/\lambda}$$

(the standard negation is recovered with $\lambda = 1$);

- With $\varphi(a) = \frac{a}{\lambda + (1 - \lambda)a}$ ($\lambda > 0$):

$$N_\varphi(t) = \frac{\lambda^2(1 - t)}{t + \lambda^2(1 - t)}$$

(again, the standard negation is obtained with $\lambda = 1$);

- With $\varphi(a) = \frac{\ln(1 + \lambda a^\alpha)}{\ln(1 + \lambda)}$, ($\lambda > -1, \alpha > 0$):

$$N_\varphi(t) = \left(\frac{1 - t^\alpha}{1 + \lambda t^\alpha} \right)^{1/\alpha}.$$

Note that taking $\alpha = 1$ we get the family $N_\varphi(t) = \frac{1-t}{1+\lambda t}$, which is known as the Sugeno's family of strong negations (which includes, when $\lambda = 0$, the standard negation) (Fig. 1.1).

Note 1.56 The characterization given in Theorem 1.54 allows one to easily show that any strong negation N has a unique *fixed point*, i.e., there exists one and only one value in $[0, 1]$, which we will denote t_N, verifying $N(t_N) = t_N$. Indeed, since $N = N_\varphi$ for some automorphism φ, the equation $N(t_N) = t_N$ is equivalent to $\varphi^{-1}(1 - \varphi(t_N)) = t_N$, whose unique solution is given by $t_N = \varphi^{-1}(1/2)$. Note that, obviously, it is always $t_N \neq 0$ and $t_N \neq 1$.

Definition 1.57 (*Dual aggregation function*) Let $N : [0, 1] \to [0, 1]$ be a strong negation and $f : [0, 1]^n \to [0, 1]$ an aggregation function. Then the aggregation function f_d given by

$$f_d(x_1, \ldots, x_n) = N(f(N(x_1), N(x_2), \ldots, N(x_n)))$$

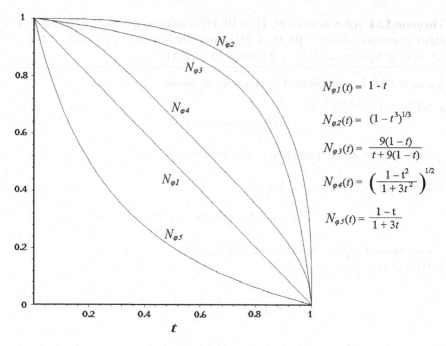

$N_{\varphi 1}(t) = 1 - t$

$N_{\varphi 2}(t) = (1 - t^3)^{1/3}$

$N_{\varphi 3}(t) = \dfrac{9(1 - t)}{t + 9(1 - t)}$

$N_{\varphi 4}(t) = \left(\dfrac{1 - t^2}{1 + 3t^2}\right)^{1/2}$

$N_{\varphi 5}(t) = \dfrac{1 - t}{1 + 3t}$

Fig. 1.1 Graphs of some strong negations in Example 1.55 with a fixed parameter λ

is called the dual of f with respect to N, or, for short, the N-dual of f. When using the standard negation, f_d is given by

$$f_d(x_1, \ldots, x_n) = 1 - f(1 - x_1, \ldots, 1 - x_n)$$

and we will simply say that f_d is the dual of f.

It is evident that the dual of a conjunctive aggregation function is disjunctive, and vice versa, regardless of what strong negation is used. Some functions are *self-dual*.

Definition 1.58 (*Self-dual aggregation function*) Given a strong negation N, an aggregation function f is self-dual with respect to N (for short, N-self-dual or N-invariant), if

$$f(\mathbf{x}) = N(f(N(\mathbf{x}))),$$

where $N(\mathbf{x}) = (N(x_1), \ldots, N(x_n))$. For the standard negation we have

$$f(\mathbf{x}) = 1 - f(1 - \mathbf{x}),$$

and it is simply said that f is self-dual.

For example, the arithmetic mean is self-dual. It is worth noting that there are no N-self-dual conjunctive or disjunctive aggregation functions.

1.3.4 Comparability

Sometimes it is possible to compare different aggregation functions and establish a certain order among them. We shall compare aggregation functions pointwise, i.e., for every $\mathbf{x} \in \mathbb{I}^n$.

Definition 1.59 An aggregation function f is stronger than another aggregation function of the same number of arguments g, if for all $\mathbf{x} \in \mathbb{I}^n$:

$$g(\mathbf{x}) \leq f(\mathbf{x}).$$

It is expressed as $g \leq f$. When f is stronger that g, it is equivalently said that g is weaker than f.

Not all aggregation functions are comparable . It may happen that f is stronger than g only on some part of the domain, and the opposite is true on the rest of the domain. In this case we say that f and g are *incomparable*.

Example 1.60 The strongest conjunctive aggregation function is the *minimum*, and the weakest disjunctive aggregation function is the *maximum* (see Definitions 1.8 and 1.9). Any disjunctive aggregation function is stronger than an averaging function, and any averaging function is stronger than a conjunctive one.

1.3.5 Continuity and Stability

We will be mostly interested in *continuous* aggregation functions, which intuitively are such functions that a small change in the input results in a small change in the output.[14] There are some interesting aggregation functions that are discontinuous, but from the practical point of view continuity is very important for producing a stable output.

The next definition is an even stronger continuity requirement. The reason is that simple, or even uniform continuity is not sufficient to distinguish functions that

---

[14] A real function of n arguments is continuous if for any sequences $\{x_{ij}\}, i = 1, \ldots, n$ such that $\lim_{j \to \infty} x_{ij} = y_i$ it holds $\lim_{j \to \infty} f(x_{1j}, \ldots x_{nj}) = f(y_1, \ldots, y_n)$. Because the domain $\mathbb{I}^n$ is a compact set, continuity is equivalent to its stronger version, uniform continuity. For monotone functions we have a stronger result: an aggregation function is uniformly continuous if and only if it is continuous in each argument (i.e., we can check continuity by fixing all variables but one, and checking continuity of each univariate function. However, general non-monotone functions can be continuous in each variable without being continuous).

produce a "small" change in value due to a small change of the argument.[15] The
following definition puts a bound on the actual change in value due to changes
in the input.

Definition 1.61 (*Lipschitz continuity*) An aggregation function f is called Lipschitz
continuous if there is a positive number M, such that for any two vectors $\mathbf{x}, \mathbf{y}$ in the
domain of definition of f:

$$|f(\mathbf{x}) - f(\mathbf{y})| \leq M d(\mathbf{x}, \mathbf{y}), \qquad (1.6)$$

where $d(\mathbf{x}, \mathbf{y})$ is a distance between $\mathbf{x}$ and $\mathbf{y}$.[16] The smallest such number M is called
the Lipschitz constant of f (in the distance d).

Typically the distance is the Euclidean distance between vectors,

$$d(\mathbf{x}, \mathbf{y}) = \sqrt{(x_1 - y_1)^2 + (x_2 - y_2)^2 + \cdots + (x_n - y_n)^2},$$

but it can be chosen as any norm $d(\mathbf{x}, \mathbf{y}) = ||\mathbf{x} - \mathbf{y}||$[17]; typically it is chosen as a
p-norm. A p-norm, $p \geq 1$ is a function $||\mathbf{x}||_p = \left(\sum_{i=1}^{n} |x_i|^p \right)^{1/p}$, for finite p, and
$||\mathbf{x}||_\infty = \max_{i=1\ldots,n} |x_i|$.

Thus, if the change in the input is $\delta = ||\mathbf{x} - \mathbf{y}||$, then the output will change by
at most $M\delta$. Hence M can be interpreted as the upper bound on the rate of change
of a function. If a function f is differentiable, then M is simply the upper bound
on the norm of its gradient. All differentiable functions are necessarily Lipschitz-
continuous, but not vice versa. However, any Lipschitz function is differentiable
"almost" everywhere.[18]

We pay attention to the rate of change of a function because of the ever present
input inaccuracies. If the aggregation function receives an inaccurate input $\tilde{\mathbf{x}} =
(x_1 + \delta_1, \ldots, x_n + \delta_n)$, contaminated with some noise $(\delta_1, \ldots, \delta_n)$, we do not expect
the output $f(\tilde{\mathbf{x}})$ to be substantially different from $f(\mathbf{x})$. The Lipschitz constant M
bounds the factor by which the noise is magnified.

---

[15]Think of this: a discontinuous (but integrable) function can be approximated arbitrarily well by
some continuous function (e.g., a polynomial). Thus based on their values, or graphs, we cannot
distinguish between continuous and discontinuous integrable functions, as the values of both functions coincide up to a tiny difference (which we can make as small as we want). A computer will
not see any difference between the two types of functions. Mathematically speaking, the subset of
continuous functions $C(\Omega)$ is *dense* in the set of integrable functions $L^1(\Omega)$ on a compact set.

[16]A *distance* between objects from a set S is a function defined on $S \times S$, whose values are non-
negative real numbers, with the properties: (1) $d(x, y) = 0$ if and only if $x = y$, (2) $d(x, y) =
d(y, x)$, and (3) $d(x, z) \leq d(x, y) + d(y, z)$ (triangular inequality). Such distance is called a *metric*.

[17]A *norm* is a function f on a vector space with the properties: (1) $f(\mathbf{x}) > 0$ for all nonzero $\mathbf{x}$ and
$f(\mathbf{0}) = 0$, (2) $f(a\mathbf{x}) = |a| f(\mathbf{x})$, and (3) $f(\mathbf{x} + \mathbf{y}) \leq f(\mathbf{x}) + f(\mathbf{y})$..

[18]I.e., it is differentiable on its entire domain, except for a subset of *measure zero*.

Note 1.62 Since $f(\mathbf{a}) = a$ and $f(\mathbf{b}) = b$, the Lipschitz constant of any aggregation function is $M \geq 1/||\mathbf{1}|| = |b-a|/||\mathbf{b}-\mathbf{a}||$. For p-norms we have $||\mathbf{1}|| = \sqrt[p]{n \cdot 1} \leq 1$, that is $M \geq n^{-1/p}$, so in principle M can be smaller than 1.

Definition 1.63 (*p-stable aggregation functions*) Given $p \geq 1$, an aggregation function is called *p*-stable if its Lipschitz constant in the *p*-norm $|| \cdot ||_p$ is 1. An extended aggregation function is *p*-stable if it can be represented as a family of *p*-stable aggregation functions.

Evidently, *p*-stable aggregation functions do not enhance input inaccuracies, as $|f(\tilde{\mathbf{x}}) - f(\mathbf{x})| \leq ||\tilde{\mathbf{x}} - \mathbf{x}||_p = ||||_p$.

Definition 1.64 (*1-Lipschitz aggregation functions*) An aggregation function f is called 1-Lipschitz if it is *p*-stable with $p = 1$, i.e., for all $\mathbf{x}, \mathbf{y}$:

$$|f(\mathbf{x}) - f(\mathbf{y})| \leq |x_1 - y_1| + |x_2 - y_2| + \cdots + |x_n - y_n|.$$

Definition 1.65 (*Kernel aggregation functions*) An aggregation function f is called kernel if it is *p*-stable with $p = \infty$, i.e., for all $\mathbf{x}, \mathbf{y}$:

$$|f(\mathbf{x}) - f(\mathbf{y})| \leq \max_{i=1,\ldots,n} |x_i - y_i|.$$

For kernel aggregation functions, the error in the output cannot exceed the largest error in the input vector.

Note 1.66 If an aggregation function is *p*-stable for a given $p > 1$, then it is also *q*-stable for any $1 \leq q < p$. This is because $||\mathbf{x}||_p \leq ||\mathbf{x}||_q$ for all $\mathbf{x}$.

Note 1.67 Aggregation functions which have Lipschitz constant k in 1-norm are called k-Lipschitz. Some properties and constructions are discussed in [BC09, BCJ10b].

Example 1.68 The product, minimum and maximum are *p*-stable extended aggregation functions for any p. The arithmetic mean is also *p*-stable for any p. The geometric mean is not Lipschitz, although it is continuous.[19]

---

[19]Take $f(x_1, x_2) = \sqrt{x_1 x_2}$, which is continuous for $x_1, x_2 \geq 0$, and let $x_2 = 1$. $f(t, 1) = \sqrt{t}$ is continuous but not Lipschitz. To see this, let $t = 0$ and $u > 0$. Then $|\sqrt{0} - \sqrt{u}| = \sqrt{u} > Mu = M|0 - u|$, or $u^{-\frac{1}{2}} > M$, for whatever choice of M, if we make u sufficiently small. Hence the Lipschitz condition fails.

1.4 Main Families and Prototypical Examples

1.4.1 Min and Max

The minimum and maximum functions are the two main aggregation functions that are used in fuzzy set theory and fuzzy logic. This is partly due to the fact that they are the only two operations consistent with a number of set-theoretical properties, and in particular mutual distributivity [BG73]. These connectives model fuzzy set intersection and union (or conjunction and disjunction).

They are defined for any number of arguments as

$$\min(\mathbf{x}) = \min_{i=1,\ldots,n} x_i, \tag{1.7}$$

$$\max(\mathbf{x}) = \max_{i=1,\ldots,n} x_i. \tag{1.8}$$

The minimum and maximum are conjunctive and disjunctive extended aggregation functions respectively, and simultaneously limiting cases of averaging aggregation functions.

Both minimum and maximum are symmetric and associative, and Lipschitz-continuous (in fact kernel aggregation functions). The min function (on $\mathbb{I} = [0, 1]$) has the neutral element $e = 1$ and the absorbing element $a = 0$, and the max function has the neutral element $e = 0$ and the absorbing element $a = 1$. They are dual to each other with respect to the standard negation $N(t) = 1 - t$ (and in fact, any strong negation N)

$$\max(\mathbf{x}) = 1 - \min(\mathbf{1} - \mathbf{x}) = 1 - \min_{i=1,\ldots,n} (1 - x_i),$$

$$\max(\mathbf{x}) = N(\min(N(\mathbf{x}))) = N(\min_{i=1,\ldots,n} (N(x_i))),$$

Most classes and parametric families of aggregation functions include maximum and minimum as members or as the limiting cases.

1.4.2 Means

Means are averaging aggregation functions. Formally, a mean is simply a function f with the property [Bul03]

$$\min(\mathbf{x}) \le f(\mathbf{x}) \le \max(\mathbf{x}).$$

Still there are other properties that define one or another family of means. We discuss them in Chap. 2.

Definition 1.69 (*Arithmetic mean*) The arithmetic mean is the function

$$M(\mathbf{x}) = \frac{1}{n}(x_1 + x_2 + \cdots + x_n) = \frac{1}{n}\sum_{i=1}^{n} x_i.$$

Definition 1.70 (*Weighting vector*) A vector $\mathbf{w} = (w_1, \ldots, w_n)$ is called a weighting vector if $w_i \in [0, 1]$ and $\sum_{i=1}^{n} w_i = 1$.

Definition 1.71 (*Weighted arithmetic mean (WAM)*) Given a weighting vector $\mathbf{w}$, the weighted arithmetic mean is the function

$$M_{\mathbf{w}}(\mathbf{x}) = w_1 x_1 + w_2 x_2 + \cdots + w_n x_n = \sum_{i=1}^{n} w_i x_i.$$

Definition 1.72 (*Geometric mean*) The geometric mean is the function

$$G(\mathbf{x}) = \sqrt[n]{x_1 x_2 \ldots x_n} = \left(\prod_{i=1}^{n} x_i\right)^{1/n}.$$

Definition 1.73 (*Harmonic mean*) The harmonic mean is the function

$$H(\mathbf{x}) = n \left(\sum_{i=1}^{n} \frac{1}{x_i}\right)^{-1}.$$

We discuss weighted versions of the above means, as well as many other means in Chap. 2.

1.4.3 Medians

Definition 1.74 (*Median*) The median is the function

$$Med(\mathbf{x}) = \begin{cases} \frac{1}{2}(x_{(k)} + x_{(k+1)}), & \text{if } n = 2k \text{ is even} \\ x_{(k)}, & \text{if } n = 2k - 1 \text{ is odd,} \end{cases}$$

where $x_{(k)}$ is the k-th largest (or smallest) component of $\mathbf{x}$.

Definition 1.75 (*a-Median*) Given a value $a \in \mathbb{I}$, the a-median is the function

$$Med_a(\mathbf{x}) = Med(x_1, \ldots, x_n, \overbrace{a, \ldots, a}^{n-1 \text{ times}}).$$

1.4.4 Ordered Weighted Averaging

Ordered weighted averaging functions (OWA) are also averaging aggregation functions, which associate weights not with a particular input, but rather with its value. They have been introduced by Yager [Yag88] and have become very popular in the fuzzy sets community.

Let $\mathbf{x}_{\searrow}$ be the vector obtained from $\mathbf{x}$ by arranging its components in non-increasing order $x_{(1)} \geq x_{(2)} \geq \cdots \geq x_{(n)}$.

Definition 1.76 (*OWA*) Given a weighting vector $\mathbf{w}$, the OWA function is

$$OWA_{\mathbf{w}}(\mathbf{x}) = \sum_{i=1}^{n} w_i x_{(i)} = \langle \mathbf{w}, \mathbf{x}_{\searrow} \rangle.$$

Note that calculation of the value of the OWA function can be done by using a `sort()` operation. If all weights are equal, OWA becomes the arithmetic mean. The vector of weights $\mathbf{w} = (1, 0, \ldots, 0)$ yields the maximum and $\mathbf{w} = (0, \ldots, 0, 1)$ yields the minimum function.

1.4.5 Choquet and Sugeno Integrals

These are two classes of averaging aggregation functions defined with respect to a fuzzy measure. They are useful to model interactions between the variables x_i.

Definition 1.77 (*Fuzzy measure*) Let $\mathcal{N} = \{1, 2, \ldots, n\}$. A discrete fuzzy measure is a set function[20] $v : 2^{\mathcal{N}} \to [0, 1]$ which is monotonic (i.e. $v(S) \leq v(T)$ whenever $S \subseteq T$) and satisfies $v(\emptyset) = 0$, $v(\mathcal{N}) = 1$.

Definition 1.78 (*Choquet integral*) The discrete Choquet integral with respect to a fuzzy measure v is given by

$$C_v(\mathbf{x}) = \sum_{i=1}^{n} x_{(i)} [v(\{j | x_j \geq x_{(i)}\}) - v(\{j | x_j \geq x_{(i+1)}\})], \qquad (1.9)$$

where $\mathbf{x}_{\nearrow} = (x_{(1)}, x_{(2)}, \ldots, x_{(n)})$ is a non-decreasing permutation of the input $\mathbf{x}$, and $x_{(n+1)} = \infty$ by convention.

---

[20] A set function is a function whose domain consists of all possible subsets of $\mathcal{N}$. For example, for $n = 3$, a set function is specified by $2^3 = 8$ values at $v(\emptyset)$, $v(\{1\})$, $v(\{2\})$, $v(\{3\})$, $v(\{1, 2\})$, $v(\{1, 3\})$, $v(\{2, 3\})$, $v(\{1, 2, 3\})$.

By rearranging the terms of the sum, Eq. (1.9) can also be written as

$$C_v(\mathbf{x}) = \sum_{i=1}^{n} \left[x_{(i)} - x_{(i-1)}\right] v(H_i). \qquad (1.10)$$

where $x_{(0)} = 0$ by convention, and $H_i = \{(i), \ldots, (n)\}$ is the subset of indices of $n - i + 1$ largest components of $\mathbf{x}$.

The class of Choquet integrals includes weighted arithmetic means and OWA functions as special cases, when the fuzzy measure is respectively additive and symmetric (i.e., when $v(H_i) = v(|H_i|)$ depends only on the cardinality of the set H_i). The Choquet integral is a piecewise linear idempotent function, uniquely defined by its values at the vertices of the unit cube $[0, 1]^n$, i.e., at the points $\mathbf{x}$, whose coordinates $x_i \in \{0, 1\}$. Note that there are 2^n such points, the same as the number of values that determine the fuzzy measure v. We consider these functions in detail in Chap. 2.

Definition 1.79 (*Sugeno integral*) The Sugeno integral with respect to a fuzzy measure v is given by

$$S_v(\mathbf{x}) = \max_{i=1,\ldots,n} \min\{x_{(i)}, v(H_i)\}, \qquad (1.11)$$

where $\mathbf{x}_\nearrow = (x_{(1)}, x_{(2)}, \ldots, x_{(n)})$ is a non-decreasing permutation of the input $\mathbf{x}$, and $H_i = \{(i), \ldots, (n)\}$.

In the special case of a symmetric fuzzy measure Sugeno integral becomes the median $S_v(\mathbf{x}) = Med(x_1, \ldots, x_n, 1, v(n - 1), v(n - 2), \ldots, v(1))$.

1.4.6 Conjunctive and Disjunctive Functions

The prototypical examples of conjunctive and disjunctive aggregation functions are so-called triangular norms and conorms respectively (t-norms and t-conorms). They are treated in detail in [BPC07, GMMP09, KMP00], and below are just a few typical examples. All functions in Examples 1.80–1.83 are symmetric and associative. T-norms and t-conorms are defined on $[0, 1]^2$ and by associativity, on $[0, 1]^n$.

Example 1.80 The product is a conjunctive extended aggregation function (it is a t-norm)

$$T_P(\mathbf{x}) = \prod_{i=1}^{n} x_i.$$

Example 1.81 The dual product, also called probabilistic sum, is a disjunctive extended aggregation function (it is a t-conorm)

$$S_P(\mathbf{x}) = 1 - \prod_{i=1}^{n}(1 - x_i).$$

Example 1.82 Łukasiewicz triangular norm and conorm are conjunctive and disjunctive extended aggregation functions

$$T_L(\mathbf{x}) = \max(0, \sum_{i=1}^{n} x_i - (n - 1)),$$

$$S_L(\mathbf{x}) = \min(1, \sum_{i=1}^{n} x_i).$$

Example 1.83 Einstein sum is a disjunctive aggregation function (it is a t-conorm). Its bivariate form is given by

$$f(x_1, x_2) = \frac{x_1 + x_2}{1 + x_1 x_2}.$$

Example 1.84 The function

$$f(x_1, x_2) = x_1 x_2^2$$

is a conjunctive ($x_1 x_2^2 \le x_1 x_2 \le \min(x_1, x_2)$), asymmetric aggregation function. It is not a t-norm.

1.4.7 Mixed Aggregation

In some situations, high input values are required to reinforce each other whereas low values pull the output down. Thus the aggregation function has to be disjunctive for high values, conjunctive for low values, and perhaps averaging if some values are high and some are low. This is typically the case when high values are interpreted as "positive" information, and low values as "negative" information. The classical expert systems MYCIN and PROSPECTOR [BS84, DHN76] use precisely this type of aggregation (on $[-1, 1]$ interval).

A different behavior may also be needed: aggregation of both high and low values moves the output towards some intermediate value. Thus certain aggregation functions need to be conjunctive, disjunctive or averaging in different parts of their domain.

Uninorms and nullnorms (see [BPC07]) are typical examples of such aggregation functions, but there are many others.

Example 1.85 The $3 - \Pi$ function [YR96] is

$$f(\mathbf{x}) = \frac{\displaystyle\prod_{i=1}^{n} x_i}{\displaystyle\prod_{i=1}^{n} x_i + \prod_{i=1}^{n}(1 - x_i)},$$

with the convention $\frac{0}{0} = 0$. It is conjunctive on $[0, \frac{1}{2}]^n$, disjunctive on $[\frac{1}{2}, 1]^n$ and averaging elsewhere. It is associative, with the neutral element $e = \frac{1}{2}$, and discontinuous on the boundaries of $[0, 1]^n$. It is a uninorm.

1.5 Composition and Transformation of Aggregation Functions

We have examined several prototypical examples of aggregation functions from different classes. Of course, this is a very limited number of functions, and they may not be sufficient to model a specific problem. The question arises as to how we can construct new aggregation functions from the existing ones. Which properties will be conserved, and which properties will be lost?

We consider two simple techniques for constructing new aggregation functions. The first technique is based on the monotonic transformation of the inputs and the second is based on iterative application of aggregation functions.

Let us consider univariate strictly increasing bijections (hence continuous) φ_1, $\varphi_2, \ldots, \varphi_n$ and ψ; $\varphi_i, \psi : \mathbb{I} \to \mathbb{I}$.

Proposition 1.86 *Let* $\varphi_1, \ldots, \varphi_n, \psi : \mathbb{I} \to \mathbb{I}$ *be strictly increasing bijections. For any aggregation function* f, *the function*

$$g(\mathbf{x}) = \psi(f(\varphi_1(x_1), \varphi_2(x_2), \ldots, \varphi_n(x_n)))$$

is an aggregation function.

Note 1.87 The functions φ_i, ψ may also be strictly decreasing (but all at the same time), we already saw in Sect. 1.3.3 that if we choose each φ_i and ψ as a strong negation, then we obtain a dual aggregation function g.

Of course, nothing can be said about the properties of g. However in some special cases we can establish which properties remain intact.

Proposition 1.88 [BB87] *Let* f *be an averaging aggregation function on* $\mathbb{I}^n$, *and* h *be a continuous strictly monotone function* $X \to \mathbb{I}$, *called scaling (or generating) function, and* $X \subseteq \bar{\mathbb{R}}$. *Then* $f_h(\mathbf{x}) = h^{-1}(f(h(\mathbf{x})))$ *is also an averaging aggregation function on* X^n.

Proposition 1.89 *Let f be an aggregation function and let g be defined as in Proposition 1.86. Then*

- *If f is continuous, so is g.*
- *If f is symmetric and $\varphi_1 = \varphi_2 = \cdots = \varphi_n$, then g is symmetric.*
- *If f is associative and $\psi^{-1} = \varphi_1 = \cdots = \varphi_n$ then g is associative.*

Next, take $\varphi_1(t) = \varphi_2(t) = \cdots = \varphi_n(t) = t$. That is, consider a composition of functions $g(\mathbf{x}) = (\psi \circ f)(\mathbf{x}) = \psi(f(\mathbf{x}))$. We examine how the choice of ψ can affect the behavior of the aggregation function f.

It is clear that ψ needs to be monotone non-decreasing. Depending on its properties, it can modify the type of the aggregation function.

Proposition 1.90 *Let f be an aggregation function and let $\psi : \mathbb{I} \to \mathbb{I}$ be a non-decreasing function satisfying $\psi(a) = a$ and $\psi(b) = b$. If $\psi(f(b, \ldots, b, t, b, \ldots, b)) \leq t$ for all $t \in [a, b]$ and at any position, then $g = \psi \circ f$ is a conjunctive aggregation function.*

Proof The proof is simple: For any fixed position i, and any $\mathbf{x}$ we have $g(\mathbf{x}) \leq \max\limits_{x_j \in [a,b], j \neq i} g(\mathbf{x}) = g(b, \ldots, b, x_i, b \ldots, b) \leq x_i$. This holds for every i, therefore $g(\mathbf{x}) \leq \min(\mathbf{x})$. By applying Proposition 1.86 we complete the proof. $\qquad\square$

Proposition 1.90 will be mainly used when choosing f as an averaging aggregation function. Not every averaging aggregation function can be converted to a conjunctive function using Proposition 1.90: the value $f(\mathbf{x})$ must be distinct from b for $\mathbf{x} \neq (b, b, \ldots, b)$. Furthermore, if $\max\limits_{i=1,\ldots,n} f(b, \ldots, b, a, b, \ldots, b) = c < b$ with a in the i-th position, then necessarily $\psi(t) = a$ for $t \leq c$. If ψ is a bijection, then f must have $c = a$ as absorbing element. The main point of Proposition 1.90 is that one can construct conjunctive aggregation functions from many types of averaging functions (discussed in Chap. 2) by a simple transformation, and that its condition involves single variate functions $\psi(f(b, \ldots, b, t, b, \ldots, b))$, which is not difficult to verify.

Note 1.91 A similar construction also transforms averaging functions with the absorbing element $a = 1$, if ψ is a bijection, to disjunctive functions (by using duality). However it does not work the other way around, i.e., to construct averaging functions from either conjunctive or disjunctive functions. This can be achieved by using the idempotization method, see [CKKM02], p. 28.

Example 1.92 Take the geometric mean $f(\mathbf{x}) = \sqrt{x_1 x_2}$, which is an averaging function with the absorbing element $a = 0$. Take $\psi(t) = t^2$. Composition $(\psi \circ f)(\mathbf{x}) = x_1 x_2$, yields the product function, which is conjunctive.

Example 1.93 Take the harmonic mean $f(\mathbf{x}) = 2(\frac{1}{x_1} + \frac{1}{x_2})^{-1} = \frac{2x_1 x_2}{x_1 + x_2}$, which also has the absorbing element $a = 0$. Take again $\psi(t) = t^2$. Composition $g(\mathbf{x}) = (\psi \circ f)(\mathbf{x}) = \frac{(2x_1 x_2)^2}{(x_1 + x_2)^2}$ is a conjunctive aggregation function (we can check that $g(x_1, 1) = \frac{4x_1^2}{(1+x_1)^2} \leq x_1$). Now take $\psi(t) = \frac{t}{2-t}$. A simple computation yields

$g(x_1, 1) = x_1$ and $g(x_1, x_2) = \frac{x_1 x_2}{x_1 + x_2 - x_1 x_2}$, a Hamacher triangular norm T_0^H, which is conjunctive .

Let us now consider an iterative application of aggregation functions. Consider three aggregation functions $f, g : \mathbb{I}^n \to \mathbb{I}$ and $h : \mathbb{I}^2 \to \mathbb{I}$, i.e., h is a bivariate function. Then the combination

$$H(\mathbf{x}) = h(f(\mathbf{x}), g(\mathbf{x}))$$

is also an aggregation function. It is continuous if f, g and h are. Depending on the properties of these functions, the resulting aggregation function may also possess certain properties.

Proposition 1.94 *Let f and g be n-ary aggregation functions, h be a bivariate aggregation function, and let H be defined as $H(\mathbf{x}) = h(f(\mathbf{x}), g(\mathbf{x}))$. Then*

- *If f and g are symmetric then H is also symmetric.*
- *If f, g and h are averaging functions, then H is averaging.*
- *If f, g and h are associative, H is not necessarily associative.*
- *If any or all f, g and h have a neutral element, H does not necessarily have a neutral element.*
- *If f, g and h are conjunctive (disjunctive), H is also conjunctive (disjunctive).*

Previously we mentioned that in certain applications the use of [0, 1] scale is not very intuitive. One situation is when we aggregate pieces of "positive" and "negative" information, for instance evidence that confirms and disconfirms a hypothesis. It may be more natural to use a bipolar [−1, 1] scale, in which negative values refer to negative evidence and positive values refer to positive evidence. In some early expert systems (MYCIN [BS84] and PROSPECTOR [DHN76]) the [−1, 1] scale was used.

The question is whether the use of a different scale brings anything new to the mathematics of aggregation. The answer is negative, the aggregation functions on two different closed intervals are isomorphic, i.e., any aggregation function on the scale [a, b] can be obtained by a simple linear transformation from an aggregation function on [0, 1]. Thus, the choice of the scale is a question of interpretability, not of the type of aggregation.

Transformation from one scale to another is straightforward, and it can be done in many different ways. The most common formulas are the following. Let $f^{[a,b]}$ be an aggregation function on the interval [a, b], and let $f^{[0,1]}$ be the corresponding aggregation function on [0, 1]. Then

$$f^{[a,b]}(x_1, \ldots, x_n) = (b-a) f^{[0,1]} \left(\frac{x_1 - a}{b - a}, \ldots, \frac{x_n - a}{b - a} \right) + a, \qquad (1.12)$$

$$f^{[0,1]}(x_1, \ldots, x_n) = \frac{f^{[a,b]} \left((b-a)x_1 + a, \ldots, (b-a)x_n + a \right) - a}{b - a}, \quad (1.13)$$

or in vector form

$$f^{[a,b]}(\mathbf{x}) = (b-a)f^{[0,1]}\left(\frac{\mathbf{x}-\mathbf{a}}{b-a}\right) + a,$$

$$f^{[0,1]}(\mathbf{x}) = \frac{f^{[a,b]}\left((b-a)(\mathbf{x}+\mathbf{a})\right) - a}{b-a}.$$

Thus for transformation to and from a bipolar scale we use

$$f^{[-1,1]}(x_1, \ldots, x_n) = 2f^{[0,1]}\left(\frac{x_1+1}{2}, \ldots, \frac{x_n+1}{2}\right) - 1, \qquad (1.14)$$

$$f^{[0,1]}(x_1, \ldots, x_n) = \frac{f^{[-1,1]}\left(2x_1-1, \ldots, 2x_n-1\right)+1}{2}. \qquad (1.15)$$

1.6 How to Choose an Aggregation Function

There are infinitely many aggregation functions. They are grouped in various families, such as means, triangular norms and conorms, Choquet and Sugeno integrals, and many others. The question is how to choose the most suitable aggregation function for a specific application. Is one aggregation function enough, or should different aggregation functions be used in different parts of the application?

There are two components to the answer. First of all, the selected aggregation function must be consistent with the semantics of the aggregation procedure. That is, if one models a conjunction, averaging or disjunctive aggregation functions are not suitable. Should the aggregation function be symmetric, have a neutral or absorbing element, or be idempotent? Is the number of inputs always the same? What is the interpretation of input values? Answering these questions should result in a number of mathematical properties, based on which a suitable class or family can be chosen.

The second issue is to choose the appropriate member of that class or family, which does what it is supposed to do—produces adequate outputs for given inputs. It is expected that the developer of a system has some rough idea of what the appropriate outputs are for some prototype inputs. Thus we arrive at the issue of fitting the data.

The data may come from different sources and in different forms. First, it could be the result of some mental experiment: let us take the input values $(1, 0, 0)$. What output do we expect?

Second, the developer of an application could ask the domain experts to provide their opinion on the desired outputs for selected inputs. This can be done by presenting the experts some prototypical cases (either the input vectors, or domain specific situations before they are translated into the inputs). If there is more than one expert, their outputs could be either averaged, or translated into the range of possible output values, or the experts could be brought together to find a consensus.

Third, the data could be collected in an experiment, by asking a group of lay people or experts about their input and output values, but without associating these values with some aggregation rule. For example, an interesting experiment reported in [Zim96, ZZ80] consisted in asking a group of people about the membership values they would use for different objects in the fuzzy sets "metallic", "container", and then in the combined set "metallic container". The goal was to determine a model for intersection of two sets. The subjects were asked the questions about membership values on three separate days, to discourage them from building some inner model for aggregation.

Fourth, the data can be collected automatically by observing the responses of subjects to various stimuli. For example, by presenting a user of a computer system with some information and recording their actions or decisions.

In the most typical case, the data comes in pairs $(\mathbf{x}, y)$, where $\mathbf{x} \in [0, 1]^n$ is the input vector and $y \in [0, 1]$ is the desired output. There are several pairs, which will be denoted by a subscript k: $(\mathbf{x}_k, y_k), k = 1, \ldots, K$. However there are variations of the data set: (a) some components of vectors $\mathbf{x}_k$ may be missing, (b) vectors $\mathbf{x}_k$ may have varying dimension by construction, and (c) the outputs y_k could be specified as a range of values (i.e., the interval $[\underline{y}_k, \overline{y}_k]$).

In fitting an aggregation function to the data, we will distinguish interpolation and approximation problems. In the case of interpolation, our aim is to fit the specified output values exactly. For instance, the pairs $((0, 0, \ldots, 0), 0)$ and $((1, 1, \ldots, 1), 1)$ should always be interpolated. On the other hand, when the data comes from an experiment, it will normally contain some errors, and therefore it is pointless to interpolate the inaccurate values y_k. In this case our aim is to stay close to the desired outputs without actually matching them. This is the approximation problem.

There are of course other issues to take into account when choosing an aggregation function, such as simplicity, numerical efficiency, easiness of interpretation, and so on [Zim96]. There are no general rules here, and it is up to the system developer to make an educated choice. In what follows, we concentrate on the first two criteria: to be consistent with semantically important properties of the aggregation procedure, and to fit the desired data.

We now formalize the selection problem.

Problem 1.95 *(Selection of an aggregation function)* Let us have a number of mathematical properties $\mathcal{P}_1, \mathcal{P}_2, \ldots$ and the data $\mathcal{D} = \{(\mathbf{x}_k, y_k)\}_{k=1}^{K}$. Choose an aggregation function f consistent with $\mathcal{P}_1, \mathcal{P}_2, \ldots$, and satisfying $f(\mathbf{x}_k) \approx y_k$, $k = 1, \ldots, K$.

Of course the approximate equalities may be satisfied exactly, if the properties $\mathcal{P}_1, \mathcal{P}_2, \ldots$ allow this. We shall also consider a variation of the selection problem when y_k are given as intervals, in which case we require $f(\mathbf{x}_k) \in [\underline{y}_k, \overline{y}_k]$, or even approximately satisfy this condition.

The satisfaction of approximate equalities $f(\mathbf{x}_k) \approx y_k$ is usually translated into the following minimization problem.

$$\text{minimize } ||\mathbf{r}|| \qquad\qquad (1.16)$$
$$\text{subject to } f \text{ satisfies } \mathcal{P}_1, \mathcal{P}_2, \ldots,$$

where $||\mathbf{r}||$ is the norm of the residuals, i.e., $\mathbf{r} \in R^K$ is the vector of the differences between the predicted and observed values $r_k = f(\mathbf{x}_k) - y_k$. There are many ways to choose the norm, and the most popular are the p-norms: the least squares norm ($p = 2$)

$$||\mathbf{r}||_2 = \left(\sum_{k=1}^{K} r_k^2 \right)^{1/2},$$

the least absolute deviation norm ($p = 1$)

$$||\mathbf{r}||_1 = \sum_{k=1}^{K} |r_k|,$$

the Chebyshev norm ($p = \infty$)

$$||\mathbf{r}||_\infty = \max_{k=1,\ldots,K} |r_k|,$$

or their weighted analogues, like

$$||\mathbf{r}|| = \left(\sum_{k=1}^{K} u_k r_k^2 \right)^{1/2},$$

where the weight $u_k \geq 0$ determines the relative importance to fit the k-th value y_k.[21]

Example 1.96 Consider choosing the weights of a weighted arithmetic mean consistent with the data set $\{(\mathbf{x}_k, y_k)\}_{k=1}^{K}$ using the least squares approach. We minimize the sum of squares

$$\text{minimize } \sum_{k=1}^{K} \left(\sum_{i=1}^{n} w_i x_{ik} - y_k \right)^2$$
$$\text{subject to } \sum_{i=1}^{n} w_i = 1,$$
$$w_1, \ldots, w_n \geq 0.$$

This is a quadratic programming problem (see Sect. 1.7.4), which is solved by a number of standard methods.

---

[21] Values y_k may have been recorded with different accuracies, or specified by experts of different standing.

The use of p-norms for $p > 1$ makes the fitting problem susceptible to large errors in the data, called outliers. Just a single outlier could pull the model in its own direction to make it very inaccurate. There are alternative fitting criteria that include using Huber-type functions, the least median of residuals, the least trimmed squares or maximum trimmed likelihood criteria [Hub03, MMY06, RL03, YB10]. These methods are robust to outliers (even if half of the data have gross errors) but are much more expensive computationally because of a huge number of locally optimal solutions.

In some studies [BJL11, BCJ10a, BJ11, KNL94] it was suggested that for decision making problems, the actual numerical value of the output $f(\mathbf{x}_k)$ was not as important as the ranking of the outputs. For instance, if $y_k \leq y_l$, then it should be $f(\mathbf{x}_k) \leq f(\mathbf{x}_l)$. Indeed, people are not really good at assigning consistent numerical scores to their preferences, but they are good at ranking the alternatives. Thus it is argued [KNL94] that a suitable choice of aggregation function should be consistent with the ranking of the outputs y_k rather than their numerical values. The use of the mentioned fitting criteria does not preserve the ranking of outputs, unless they are interpolated. Preservation of ranking of outputs can be done by imposing the constraints $f(\mathbf{x}_k) \leq f(\mathbf{x}_l)$ if $y_k \leq y_l$ for all pairs k, l.

1.7 Supplementary Material: Some Methods for Approximation and Optimization

In this section we temporarily step away from aggregation functions and outline some of the methods of numerical analysis, which are used later on as tools for constructing aggregation functions. Most of the material in this section can be found in standard numerical analysis textbooks [BF01, CK04]. A few recently developed methods will also be presented, and we will provide the references to the articles which discuss these methods in detail.

1.7.1 Univariate Approximation and Smoothing

Consider a set of data $(x_k, y_k), k = 1, \ldots, K, x_k, y_k \in \mathbb{R}$. The aim of interpolation is to define a function f, which can be used to calculate the values at x distinct from x_k. The interpolation conditions are specified as $f(x_k) = y_k$ for all $k = 1, \ldots, K$. We assume that the abscissae are ordered $x_k < x_{k+1}, k = 1, \ldots, K - 1$.

When the data contain inaccuracies, it is pointless to interpolate these data. Methods of approximation and smoothing are applied, which produce functions f that are regularized in some sense, and fit the data in the least squares, least absolute deviation or some other sense.

Let us consider some basis $\{B_1, \ldots, B_n\}$, so that approximation is sought in the form

$$f(x) = \sum_{i=1}^{n} a_i B_i(x). \tag{1.17}$$

The number of basis functions n is not the same as the number of data K. Regularization by restricting n to a small number $n \ll K$ is very typical. When B_i are polynomials, the method is called polynomial regression. The basis functions are usually chosen as a system of orthogonal polynomials (e.g., Legendre or Chebyshev polynomials) to ensure a well conditioned system of equations.

The functions B_i can be chosen as B-splines, with a small number of knots t_k fixed in the interval $[x_1, x_K]$. These splines are called regression splines. Regardless what are the basis functions, the coefficients are found by solving the over-determined linear system $\mathbf{B}\mathbf{a} = \mathbf{y}$, with

$$\mathbf{B} = \begin{bmatrix} B_1(x_1) & B_2(x_1) & \ldots & B_n(x_1) \\ B_1(x_2) & B_2(x_2) & \ldots & B_n(x_2) \\ \vdots & & & \vdots \\ \vdots & & & \vdots \\ B_1(x_K) & B_2(x_K) & \ldots & B_n(x_K) \end{bmatrix}.$$

Note that the matrix is rectangular, as $n \ll K$, and its rank is usually n. Since not all the equations can be fitted simultaneously, we shall talk about a system of approximate equalities $\mathbf{B}\mathbf{a} \approx \mathbf{y}$.

In the case of least squares approximation, one minimizes the Euclidean norm of the residuals $||\mathbf{B}\mathbf{a} - \mathbf{y}||_2$, or explicitly,

$$\min_{\mathbf{a} \in \mathbb{R}^n} \left(\sum_{k=1}^{K} (\sum_{i=1}^{n} a_i B_i(x_k) - y_k)^2 \right)^{1/2}. \tag{1.18}$$

One can solve the system $\mathbf{B}\mathbf{a} = \mathbf{y}$ directly using QR-factorization. The pseudo-solution will be precisely the vector $\mathbf{a}$ minimizing the Euclidean norm $||\mathbf{B}\mathbf{a} - \mathbf{y}||$.

An alternative to least squares approximation is the least absolute deviation (LAD) approximation [BS83]. Here one minimizes

$$\min_{\mathbf{a} \in \mathbb{R}^n} \sum_{k=1}^{K} | \sum_{i=1}^{n} a_i B_i(x_k) - y_k |, \tag{1.19}$$

possibly subject to some additional constraints discussed later. It is known that the LAD criterion is less sensitive to outliers in the data.

To solve the minimization problem (1.19) one uses the following trick to convert it to a linear programming (LP) problem. Let $r_k = f(x_k) - y_k$ be the k-th residual. We represent it as a difference of a positive and negative parts $r_k = r_k^+ - r_k^-, r_k^+, r_k^- \geq 0$. The absolute value is $|r_k| = r_k^+ + r_k^-$. Now the problem (1.19) is converted into an LP problem with respect to $\mathbf{a}, \mathbf{r}^+, \mathbf{r}^-$

$$\text{minimize} \quad \sum_{k=1}^{K} (r_k^+ + r_k^-), \tag{1.20}$$

$$\text{s.t.} \sum_{i=1}^{n} a_i B_i(x_k) - (r_k^+ - r_k^-) = y_k, \; k = 1, \ldots, K$$

$$r_k^+, r_k^- \geq 0.$$

The solution is performed by using the simplex method, or by a specialized version of the simplex method [BS83]. It is important to note that the system of linear constraints typically has a sparse structure, therefore for large K the use of special programming libraries that employ sparse matrix representation is needed.

1.7.2 Approximation with Constraints

Consider a linear least squares or least absolute deviation problem, in which together with the data, additional information is available. For example, there are known bounds on the function f, $L(x) \leq f(x) \leq U(x)$, or f is known to be monotone increasing, convex, either on the whole of its domain, or on given intervals. This information has to be taken into account when calculating the coefficients $\mathbf{a}$, otherwise the resulting function may fail to satisfy these requirements (even if the data are consistent with them). This is the problem of constrained approximation.

If the constraints are non-linear, then the problem becomes a nonlinear (and sometimes global) optimization problem. This is not desirable, as numerical solutions to such problems could be very expensive. If the problem could be formulated in such a way that the constraints are linear, then it can be solved by standard quadratic and linear programming methods.

Constraints on coefficients

One typical example is linear least squares or least absolute deviation approximation, with the constraints on coefficients $a_i \geq 0$, $\sum_{i=1}^{n} a_i = 1$. There are multiple instances of this problem in our book, when determining the weights of various inputs (in the multivariate setting). We have a system of linear constraints, and in the case of the least squares approximation, we have the problem

$$\text{minimize} \sum_{k=1}^{K} (\sum_{i=1}^{n} a_i B_i(x_k) - y_k)^2 \tag{1.21}$$

$$\text{s.t.} \quad \sum_{i=1}^{n} a_i = 1, a_i \geq 0.$$

It is easy to see that this is a quadratic programming problem, see Sect. 1.7.4. It is advisable to use standard QP algorithms, as they have proven convergence and are very efficient.

In the case of the least absolute deviation, it is quite easy to modify linear programming problem (1.20) to incorporate the new constraints on the variables a_i.

The mentioned constrained linear least squares or least absolute deviation problems are often stated as follows:

$$\text{Solve } \mathbf{Ax} \approx \mathbf{b} \tag{1.22}$$
$$\text{s.t. } \mathbf{Cx} = \mathbf{d}$$
$$\mathbf{Ex} \le \mathbf{f},$$

where $\mathbf{A}, \mathbf{C}, \mathbf{E}$ are matrices of size $k \times n$, $m \times n$ and $p \times n$, and $\mathbf{b}, \mathbf{d}, \mathbf{f}$ are vectors of length k, m, p respectively. The solution to the system of approximate inequalities is performed in the least squares or least absolute deviation sense [BS83, LH95]. Here vector $\mathbf{x}$ plays the role of the unknown coefficients $\mathbf{a}$. There are specially adapted versions of quadratic and linear programming algorithms, see Sect. 1.7.5.

Monotonicity constraints

Another example of constrained approximation problem is monotone (or isotone) approximation. Here the approximated function f is known to be monotone increasing (decreasing), perhaps on some interval, and this has to be incorporated into the approximation process. There are many methods devoted to univariate monotone approximation, most of which are based on spline functions [Bel00, Bel02, DB01, Die95, Sch81].

For regression splines, when using B-spline representation, the problem is exactly of the type (1.21) or (1.22), as monotonicity can be expressed as a set of linear inequalities on spline coefficients. It is solved by a simpler version of problem (1.22), called Non-Negative Least Squares (NNLS) [HH81, LH95].

Alternative methods for interpolating and smoothing splines are based on insertion of extra knots (besides the data x_i) and solving a convex nonlinear optimization problem. We mention the algorithms by Schumaker [Sch83] and McAllister and Roulier [MR81] for quadratic interpolating splines, and by Anderson and Elfving [AE91] for cubic smoothing splines.

1.7.3 Multivariate Approximation

Linear regression

Linear regression is probably the best known method of multivariate approximation. It consists in building a hyperplane which fits the data best in the least squares sense. Let the equation of the hyperplane be

$$f(\mathbf{x}) = a_0 + a_1 x_1 + a_2 x_2 + \cdots + a_n x_n.$$

Then the vector of coefficients **a** can be determined by solving the least squares problem

$$\text{minimize} \sum_{k=1}^{K} \left(a_0 + \sum_{i=1}^{n} a_i x_{ik} - y_k\right)^2,$$

where x_{ik} is the i-th component of the vector $\mathbf{x}_k$. Linear regression problem can be immediately generalized if we choose

$$f(\mathbf{x}) = a_0 + \sum_{i=1}^{n} a_i B_i(x_i),$$

where B_i are some given functions of the i-th component of **x**. In fact, one can define more than one function B_i for the i-th component (we will treat this case below). Then the vector of unknown coefficients can be determined by solving $\mathbf{Ba} \approx \mathbf{y}$ in the least squares sense, in the same way as for the univariate functions described on p. 36. The solution essentially can be obtained by using QR-factorization of **B**.

It is also possible to add some linear constraints on the coefficients **a**, for example, non-negativity. Then one obtains a constrained least squares problem, which is an instance of QP. By choosing to minimize the least absolute deviation instead of the least squares criterion, one obtains an LAD problem, which is converted to LP. Both cases can be stated as problem (1.22) on p. 38.

Tensor product schemata

Evidently, linear regression, even with different sets of basis functions B_i, is limited to relatively simple dependence of y on the arguments **x**. A more general method is to represent a multivariate function f as a tensor product of univariate functions

$$f_i(x_i) = \sum_{j=1}^{J_i} a_{ij} B_{ij}(x_i).$$

Thus each univariate function f_i is written in the form (1.17).

Now, take a product $f(\mathbf{x}) = f_1(x_1) f_2(x_2) \ldots f_n(x_n)$. It can be written as

$$f(\mathbf{x}) = \sum_{m=1}^{J} b_m B_m(\mathbf{x}),$$

where $b_m = a_{1j_1} a_{2j_2} \ldots a_{nj_n}$, $B_m(\mathbf{x}) = B_{1j_1}(x_1) B_{2j_2}(x_2) \ldots B_{nj_n}(x_n)$ and $J = J_1 J_2 \ldots J_n$. In this way we clearly see that the vector of unknown coefficients (of length J) can be found by solving a least squares (or LAD) problem (1.18), once we write down the components of the matrix **B**, namely $\mathbf{B}_{km} = B_m(\mathbf{x}_k)$.

In addition, we can add restrictions on the coefficients, and obtain a constrained LS or LAD problem (1.22). So in principle one can apply the same method of solution as in the univariate case. The problem with the tensor product approach is the sheer number of basis functions and coefficients. For example, if one uses tensor product splines (i.e., B_{ij} are univariate B-splines), say $J_1 = J_2 = \ldots J_n = 5$ and works with the data in $\mathbb{R}^5$, there are $5^5 = 3125$ unknown coefficients. So the size of the matrix $\mathbf{B}$ will be $K \times 3125$. Typically one needs a large number of data $K > J$, otherwise the system is ill-conditioned. Furthermore, depending on the choice of B_{ij}, these data need to be appropriately distributed over the domain (otherwise we may get entire zero columns of $\mathbf{B}$). The problem quickly becomes worse in higher dimensions—a manifestation of the so-called curse of dimensionality.

Inclusion of monotonicity and convexity constraints for tensor product regression splines is possible. In a suitably chosen basis [Bel00] monotonicity (with respect to each variable) is written as a set of linear inequalities. Then the LS or LAD approximation becomes the problem (1.22), which is solved by either quadratic or linear programming techniques (see Sect. 1.7.5). These methods handle well degeneracy of the matrices (e.g., when $K < J$), but one should be aware of their limitations, and general applicability of tensor product schemata to small dimension.

Lipschitz approximation

This is a scattered data interpolation and approximation technique based on the Lipschitz properties of the function f [Bel05, Bel06]. Since the Lipschitz condition is expressed as

$$|f(\mathbf{x}) - f(\mathbf{y})| \leq M\|\mathbf{x} - \mathbf{y}\|,$$

with $\| \cdot \|$ being some norm, it translates into the tight lower and upper bounds on any Lipschitz function with Lipschitz constant M that can interpolate the data,

$$\sigma_u(\mathbf{x}) = \min_{k=1,\ldots,K} \{y_k + M\|\mathbf{x} - \mathbf{x}_k\|\},$$
$$\sigma_l(\mathbf{x}) = \max_{k=1,\ldots,K} \{y_k - M\|\mathbf{x}_k - \mathbf{x}\|\}.$$

Then the best possible interpolant in the worst case scenario is given by the arithmetic mean of these bounds. Calculation of the optinal interpolant is straightforward, and no solution to any system of equations (or any training) is required.

The method also works with monotonicity constraints, by using the bounds

$$\sigma_u(\mathbf{x}) = \min_k\{y_k + M\|(\mathbf{x} - \mathbf{x}_k)_+\|\},$$
$$\sigma_l(\mathbf{x}) = \max_k\{y_k - M\|(\mathbf{x}_k - \mathbf{x})_+\|\}, \tag{1.23}$$

where $\mathbf{z}_+$ denotes the positive part of vector $\mathbf{z}$: $\mathbf{z}_+ = (\bar{z}_1, \ldots, \bar{z}_n)$, with

$$\bar{z}_i = \max\{z_i, 0\}.$$

In fact many other types of constraints can be included as simple bounds on f, [Bel07, BC07, BC09, BCL07], see Chap. 7.

What is interesting about Lipschitz interpolant, is that it provides the best possible solution in the worst case scenario, i.e., it delivers a function which minimizes the largest distance from any Lipschitz function that interpolates the data.

If one is interested in smoothing, then the method of Lipschitz smoothing is applied [Bel05]. It consists in determining the smoothened values of y_k that are compatible with a chosen Lipschitz constant. This problem is set as either a QP or LP problem, depending whether we use the LS or LAD criterion.

This method has been generalized for locally Lipschitz functions, where the Lipschitz constant depends on the values of $\mathbf{x}$, and it works for monotone functions.

1.7.4 Convex and Non-convex Optimization

When fitting a function to the data, or determining the vector of weights, one has to solve an optimization problem. We have seen that methods of univariate and multivariate approximation require solving such problems, notably the quadratic and linear programming problems. In other cases, like ANN training, the optimization problem is nonlinear. There are several types of optimization problems that frequently arise, and below we outline some of the methods developed for each type. We consider continuous optimization problems, where the domain of the objective function f is $\mathbb{R}^n$ or a compact subset.

We distinguish unconstrained and constrained optimization. In the first case the domain is $\mathbb{R}^n$, in the second case the feasible domain is some subset $X \subset \mathbb{R}^n$, typically determined by a system of equalities and inequalities. We write

$$\text{minimize} \qquad f(\mathbf{x}) \qquad\qquad (1.24)$$
$$\text{subject to} \quad g_i(\mathbf{x}) = 0, i = 1, \dots k,$$
$$h_i(\mathbf{x}) \le 0, i = 1, \dots m.$$

A special case arises when the functions g_i, h_i are linear (or affine). The feasible domain is then a convex polytope, and when in addition to this the objective function is linear or convex quadratic, then special methods of linear and quadratic programming are applied.

We also distinguish convex and non-convex optimization. A convex function f satisfies the following condition,

$$f(\alpha \mathbf{x} + (1 - \alpha)\mathbf{y}) \le \alpha f(\mathbf{x}) + (1 - \alpha) f(\mathbf{y}),$$

for all $\alpha \in [0, 1]$ and all $\mathbf{x}, \mathbf{y} \in Dom(f)$. If the inequality is strict, it is called strictly convex. Convex functions have a unique minimum (possibly many minimizers), which is the global minimum of the function. Thus if one could check the necessary conditions for a minimum (called KKT (Karush-Kuhn-Tucker) conditions), then one can be certain that the global minimum has been found. Numerical minimization can be performed by any descent scheme, like a quasi-Newton method, steepest descent, coordinate descent, etc. [BGLS00, DS83, Fle00, NW99, PR02].

If the function is not convex, it may still have a unique minimum, although the use of descent methods is more problematic. A special class is that of log-convex (or T-convex) functions, which are functions f, such that $\tilde{f} = \exp(f)$ (or $\tilde{f} = T(f)$) is convex. They are also treated by descent methods (for instance, one can just minimize $\tilde{f}$ instead of f as the minimizers coincide).

General non-convex functions can have multiple local minima, and frequently their number grows exponentially with the dimension n. This number can easily reach 10^{20}–10^{60} for $n < 30$. While locating a local minimum can be done by using descent methods (called local search in this context), there is no guarantee whatsoever that the solution found is anywhere near the global minimum. With such a number of local minima, their enumeration is practically infeasible. This is the problem of global optimization, treated in Sect. 1.7.4.

Whenever it is possible to take advantage of convexity or its variants, one should always do this, as more general methods will waste time by chasing non-existent local minima. On the other hand, one should be aware of the implications of non-convexity, especially the multiple local minima problem, and apply proper global optimization algorithms.

We shall also mention the issue of non-differentiable (or non-smooth) optimization. Most local search methods, like quasi-Newton, steepest descent, conjugate gradient, etc., assume the existence of the derivatives (and sometimes all second order derivatives, the Hessian matrix). Not every objective function is differentiable, for example $f(x) = |x|$, or a maximum of differentiable functions. Calculation of descent direction at those points where f does not have a gradient is problematic. Generalizations of the notion of gradient (like Clarke's subdifferential, or quasi-differential [Cla83, DR86]) are applied. What exacerbates the problem is that the local/global minimizers are often those points where f is not differentiable. There are a number of derivative-free methods of non-smooth optimization [Bag02, BGLS00, Pow02, Pow04], and we particularly mention the Bundle methods [Bag02, Kiw85, Lem78].

Univariate optimization

The classical method of univariate optimization is the Newton's method. It works by choosing an initial approximation x_0 and then iterating the following step

$$x_{k+1} = x_k - \frac{f'(x_k)}{f''(x_k)}.$$

Of course, the objective function f needs to be twice differentiable and the derivatives known explicitly. It is possible to approximate the derivatives using finite difference approximation, which leads to the secant method, and more generally to various quasi-Newton schemata.

Newton's method converges to a local minimum of f, and only if certain conditions are satisfied, typically if the initial approximation is close to the minimum (which is unknown in the first place).

For non-differentiable functions the generalizations of the gradient (subgradient, quasi-gradient [Cla83, DR86]) are often used. The minimization scheme is similar to the Newton's method, but the derivative is replaced with an approximation of its generalized version. Golden section method (on an interval) is another classical method, which does not require approximation of the derivatives.

If the objective function is not convex, there could be multiple local minima. Locating the global minimum (on a bounded interval) should be done by using a global optimization method. Grid search is the simplest approach, but it is not the most efficient. It is often augmented with some local optimization method, like the Newton's method (i.e., local optimization method is called from the nodes of the grid as starting points).

If the objective function is known to be Lipschitz-continuous, and an estimate of its Lipschitz constant M is available, Pijavski-Shubert method [Pij72, Shu72] is an efficient way to find and confirm the global optimum. While this method is not as fast as the Newton's method, it guarantees the globally optimal solution in the case of multiextremal objective functions, and does not suffer from lack of convergence.

Multivariate constrained optimization

Linear programming

As we mentioned, a constrained optimization problem involves constraints on the variables, that may be linear or nonlinear. If the constraints are linear, and the objective function is also linear, then we have a special case called a linear programming problem (LP). It takes a typical form

$$\text{minimize} \quad \sum_{i=1}^{n} c_i x_i = \mathbf{c}^t \mathbf{x}$$
$$\text{s.t.} \quad \mathbf{Ax} = \mathbf{b}$$
$$\mathbf{Cx} \leq \mathbf{d}$$
$$x_i \geq 0, i = 1, \ldots, n.$$

Here $\mathbf{A}, \mathbf{C}$ are matrices of size $k \times n$, $m \times n$ and $\mathbf{b}, \mathbf{d}$ are vectors of size k and m respectively. Maximization problem is obtained by exchanging the signs of the coefficients c_i, and similarly "greater than" type inequalities are transformed into "smaller than". The condition of non-negativity of x_i can in principle be dropped (such x_i are called unrestricted) with the help of artificial variables, but it is stated as in the standard formulation of LP, and because most solution algorithms assume it by default.

Each LP problem has an associated dual problem (see any textbook on linear programming, e.g., [Chv83, Van01]), and the solution to the dual problem allows one to recover that of the primal and vice versa. In some cases solution to the dual problem is computationally less expensive than that to the primal, typically when k and m are large.

The two most used solution methods are the simplex method and the interior point method. In most practical problems, both types of algorithms have an equivalent running time, even though in the worst case scenario the simplex method is exponential and the interior point method is polynomial in complexity.

Quadratic programming

A typical quadratic programming problem is formulated as

$$\text{minimize} \quad \tfrac{1}{2}\mathbf{x}^t\mathbf{Q}\mathbf{x} + \mathbf{c}^t\mathbf{x}$$
$$\text{s.t.} \quad \mathbf{A}\mathbf{x} = \mathbf{b}$$
$$\mathbf{C}\mathbf{x} \le \mathbf{d}$$
$$x_i \ge 0, i = 1, \ldots, n.$$

Here $\mathbf{Q}$ is a symmetric positive semidefinite matrix (hence the objective function is convex), $\mathbf{A}, \mathbf{C}$ are matrices of constraints, $\mathbf{c}, \mathbf{b}, \mathbf{d}$ are vectors of size n, k, m respectively, and the factor $\tfrac{1}{2}$ is written for standardization (note that most programming libraries assume it!).

If $\mathbf{Q}$ is indefinite (meaning that the objective function is neither convex nor concave) the optimization problem is extremely complicated because of a very large number of local minima (an instance of an NP-hard problem). If $\mathbf{Q}$ is negative definite, this is the problem of concave programming, which is also NP-hard. Standard QP algorithms do not treat these cases, but specialized methods are available.

General constrained nonlinear programming

This is an optimization problem in which the objective function is not linear or quadratic, or constraints $h_i(\mathbf{x}) \le 0, i = 1, \ldots, m$ are nonlinear. There could be multiple minima, so that this is the problem of global optimization. If the objective function is convex, and the constraints define a convex feasible set, the minimum is unique. It should be noted that even a problem of finding a feasible $\mathbf{x}$ is already complicated.

The two main approaches to constrained optimization are the penalty function and the barrier function approach [NW99]. In the first case, an auxiliary objective function $\tilde{f}(\mathbf{x}) = f(\mathbf{x}) + \lambda P(\mathbf{x})$ is minimized, where P is the penalty term, a function which is zero in the feasible domain and non-zero elsewhere, increasing with the degree to which the constraints are violated. It can be smooth or non-smooth [PR02]. λ is a penalty parameter; it is often the case that a sequence of auxiliary objective functions is minimized, with decreasing values of λ. Minimization of $\tilde{f}$ is done by local search methods.

In the case of a barrier function, typical auxiliary functions are

$$\tilde{f}(\mathbf{x}) = f(\mathbf{x}) + \lambda \sum - \ln(-h_i(\mathbf{x})), \quad \tilde{f}(\mathbf{x}) = f(\mathbf{x}) + \lambda \sum (-h_i(\mathbf{x}))^{-r}$$

but now the penalty term is non-zero inside the feasible domain, and grows as $\mathbf{x}$ approaches the boundary.

Recently Sequential Quadratic Programming methods (SQP) have gained popularity for solving constrained nonlinear programming problems, especially those that arise in nonlinear approximation [Sch03]. In essence, this method is based on solving a sequence of QP subproblems at each iteration of the nonlinear optimization problem, by linearizing constraints and approximating the Lagrangian function of the problem (1.24)

$$L(\mathbf{x}, \boldsymbol{\lambda}) = f(\mathbf{x}) + \sum_{i=1}^{k} \lambda_i g_i(\mathbf{x}) + \sum_{i=m+1}^{k+m} \lambda_i h_{i-k}(\mathbf{x})$$

quadratically (variables λ_i are called the Lagrange multipliers). We refer the reader to [Sch03] for its detailed analysis.

Note that all mentioned methods converge to a locally optimal solution, if f or functions g_i are non-convex. There could be many local optima, and to find the global minimum, global optimization methods are needed.

Multilevel optimization

It is often the case that with respect to some variables the optimization problem is convex, linear or quadratic, and not with respect to the others. Of course one can treat it as a general NLP, but knowing that in most cases we will have a difficult global optimization problem, it makes sense to use the special structure for a subset of variables. This will reduce the complexity of the global optimization problem by reducing the number of variables.

Suppose that we have to minimize $f(\mathbf{x})$ and f is convex with respect to the variables x_i, $i \in \mathcal{I} \subset \{1, 2, \ldots, n\}$, and let $\tilde{\mathcal{I}} = \{1, \ldots, n\} \setminus \mathcal{I}$ denote the complement of this set. Then we have

$$\min_{\mathbf{x}} f(\mathbf{x}) = \min_{x_i : i \in \tilde{\mathcal{I}}} \min_{x_i : i \in \mathcal{I}} f(\mathbf{x}).$$

This is a bi-level optimization problem. At the inner level we treat the variables x_i, $i \in \tilde{\mathcal{I}}$ as constants, and perform minimization with respect to those whose indices are in $\mathcal{I}$. This is done by some efficient local optimization algorithm.

At the outer level we have the global optimization problem

$$\min_{x_i : i \in \tilde{\mathcal{I}}} \tilde{f}(\mathbf{x}),$$

where the function $\tilde{f}$ is the solution to the inner problem. In other words, each time we need a value of $\tilde{f}$, we solve the inner problem with respect to x_i, $i \in \mathcal{I}$.

Global optimization: stochastic methods

Global optimization methods are traditionally divided into two broad categories: sto-chastic and deterministic [HPT00, PR02, Pin96]. Stochastic methods do not guaran-tee the globally optimal solution but in probability (i.e., they converge to the global optimum with probability 1, as long as they are allowed to run indefinitively). Of course any algorithm has to be stopped at some time. It is argued that stochastic optimization methods return the global minimum in a finite number of steps with high probability.

Unfortunately there are no rules for how long a method should run to deliver the global optimum with the desired probability, as it depends on the objective function. In some problems this time is not that big, but in others stochastic methods converge extremely slowly, and never find the global solution after any reasonable running time. This is a manifestation of the so called curse of dimensionality, as the issue is aggravated when the number of variables is increased, in fact the complexity of the optimization problem grows exponentially with the dimension. Even if the class of the objective functions is limited to smooth or Lipschitz functions, the global optimization problem is NP-hard [HPT00, HP95, HT93, Pin96] (Fig. 1.2).

The methods in this category include pure random search (i.e., just evaluate and compare the values of f at randomly chosen points), multistart local search, heuris-tics like simulated annealing, genetic algorithms, tabu search, and many others, see [Mic94, PR02]. The choice of the method depends very much on the specific prob-lem, as some methods work faster for certain problem classes. All methods in this category are competitive with one another. There is no general rule for choosing any particular method, it comes down to trial and error.

Global optimization: deterministic methods

Deterministic methods guarantee a globally optimal solution for some classes of objective functions (e.g., Lipschitz functions), however their running time is very large. It also grows exponentially with the dimension, as the optimization problem is NP-hard.

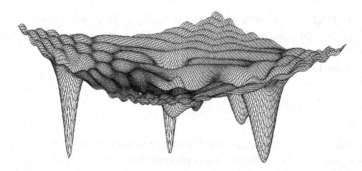

Fig. 1.2 An objective function with multiple local minima and stationary points

We mention in this category the grid search (i.e., systematic exploration of the domain, possible with the help of local optimization), Branch-and-Bound methods (especially the αBB method [Flo00]), space-filling curves (i.e., representing a multivariate function f through a special univariate function whose values coincide with those of f along an infinite curve which "fills" the domain, see [SS00]), and multivariate extensions of the Pijavski-Shubert method [Bel03, Bel04, Bel08, HJ95].

One such extension is known as the Cutting Angle methods [Rub00], and the algorithm for the Extended Cutting Angle Method (ECAM) is described in [BB02, Bel03, Bel04, Bel08]. In up to 10 variables this method is quite efficient numerically.

We wish to reiterate that there is no magic bullet in global optimization: it is NP-hard in the best case (when restricting the class of the objective functions). It is therefore very important to identify some of the variables, with respect to which the objective function is linear, convex or unimodal, and set up a multilevel optimization problem, as this would reduce the number of global variables, and improve the computational complexity.

1.7.5 Main Tools and Libraries

There are a number of commercial and free open source programming libraries that provide efficient and thoroughly tested implementations of the approximation and optimization algorithms discussed in this Appendix. Below is just a sample from an extensive collection of such tools, that in the authors' view are both reliable and sufficiently simple to be used by less experienced users. Valuable references are http://plato.asu.edu/sub/nonlsq.html, http://www-fp.mcs.anl.gov/hs/software/index.php, http://www2.informs.org/Resources/

An online textbook on optimization is available from, http://www.mpri.lsu.edu/textbook/TablCont.htm

Linear programming

A typical linear programming problem is formulated as

$$\text{minimize} \quad \sum_{i=1}^{n} c_i x_i = \mathbf{c}^t \mathbf{x}$$

$$\text{s.t.} \quad \mathbf{A}\mathbf{x} = \mathbf{b}$$

$$\mathbf{C}\mathbf{x} \leq \mathbf{d}$$

$$x_i \geq 0, i = 1, \ldots, n,$$

where $\mathbf{A}, \mathbf{C}$ are matrices of size $k \times n$, $m \times n$ and $\mathbf{b}, \mathbf{d}$ are vectors of size k and m respectively.

The two most used solution methods are the simplex method and the interior point method [Chv83, Van01]. There are a number of standard implementations of both methods. Typically, the user of a programming library is required to specify the entries of the arrays $\mathbf{c}, \mathbf{A}, \mathbf{b}, \mathbf{C}, \mathbf{d}$, point to the unrestricted variables, and sometimes specify the lower and upper bounds on x_i.[22] Most libraries use sparse matrix representation, but they also provide adequate conversion tools.

The packages GLPK (GNU Linear Programming Toolkit) and LPSOLVE http://www.gnu.org/software/glpk/,   http://www.lpsolve.sourceforge.net/5.5/   are both open source and very efficient and reliable. Both implement sparse matrix representation.

Commercial alternatives include CPLEX http://www.ilog.com, LINDO http://www.lindo.com, MINOS http://www.web.stanford.edu/group/SOL/download.html and many others. These packages also include quadratic and general nonlinear programming, as well as mixed integer programming modules.

Quadratic programming

A typical quadratic programming problem is formulated as

$$
\begin{aligned}
\text{minimize} \quad & \tfrac{1}{2}\mathbf{x}^t \mathbf{Q}\mathbf{x} + \mathbf{c}^t \mathbf{x} \\
\text{s.t.} \quad & \mathbf{A}\mathbf{x} = \mathbf{b} \\
& \mathbf{C}\mathbf{x} \leq \mathbf{d} \\
& x_i \geq 0, i = 1, \ldots, n.
\end{aligned}
$$

Here $\mathbf{Q}$ is a symmetric positive semidefinite matrix (hence the objective function is convex), $\mathbf{A}, \mathbf{C}$ are matrices of constraints, $\mathbf{c}, \mathbf{b}, \mathbf{d}$ are vectors of size n, k, m respectively. Note that most libraries assume the factor $\frac{1}{2}$!

An open source QP solver which supports sparse matrices is OOQP, http://www.cs.wisc.edu/~swright/ooqp/. It requires a separate module which should be downloaded from HSL Archive, module MA27, http://www.hsl.rl.ac.uk/archive/specs/ma27.pdf.

There are many alternative QP solvers, for example Algorithm 559, http://www.netlib.org/toms/559, as well as already mentioned commercial CPLEX, LINDO and MINOS packages, see the guides to optimization software,http://plato.asu.edu/sub/nonlsq.html, http://www-fp.mcs.anl.gov/hs/software/index.php.

Least absolute deviation problem

As we mentioned, the LAD problem is converted into an LP problem by using (1.20), hence any LP solver can be used. However there are specially designed versions of the simplex method suitable for LAD problem [BS83].

The LAD problem is formulated as follows. Solve the system of equations $\mathbf{A}\mathbf{x} \approx \mathbf{b}$ in the least absolute deviation sense, subject to constraints $\mathbf{C}\mathbf{x} = \mathbf{d}$ and $\mathbf{E}\mathbf{x} \leq \mathbf{f}$, where $\mathbf{A}, \mathbf{C}, \mathbf{E}$ and $\mathbf{b}, \mathbf{d}, \mathbf{f}$ are matrices and vectors defined by the user. The computer code can be found in *netlib* http://www.netlib.org/ as Algorithm 552 http://www.netlib.org/toms/552, see also Algorithm 615, Algorithm 478 and Algorithm 551 in the same library.

---

[22]Even though the bounds can be specified through the general constraints in $\mathbf{C}, \mathbf{d}$, in the algorithms the bounds are processed differently (and more efficiently).

For Chebyschev approximation code see Algorithm 495 in *netlib*, http://www.netlib.org/toms/495.

We should note that all the mentioned algorithms do not use sparse matrix representation, hence they work well with dense matrices of constraints or when the number of constraints is not large. For large sparse LADs, use the generic LP methods.

Constrained least squares

While general QP methods can be applied to this problem, specialized algorithms are available. The Algorithm 587 from *netlib* solves the following problem called LSEI (Least Squares with Equality and Inequality constraints).

Solve the system of equations $Ax \approx b$ in the least squares sense, subject to constraints $Cx = d$ and $Ex \leq f$, where A, C, E and b, d, f are matrices and vectors defined by the user. The algorithm handles well degeneracy in the systems of equations/constraints. The computer code (in FORTRAN) can be downloaded from *netlib* http://www.netlib.org/toms/587.

Nonlinear optimization

As a general reference we recommend the following repositories:
http://www-fp.mcs.anl.gov/hs/software/index.php
http://www2.informs.org/Resources/
http://www.gams.nist.gov/
http://www.mat.univie.ac.at/neum/glopt/software_l.html.

Univariate global optimization

For convex problems the golden section methods is very reliable, it is often combined with the Newton's method. There are multiple implementations, see the references to nonlinear optimization above.

For non-convex multiextremal objective functions, we recommend Pijavski-Shubert method, it is implemented in GANSO library as the special case of ECAM http://www.deakin.edu.au/~gleb/ganso.html.

Multivariate global optimization

There are a number of repositories and links at http://www.gams.nist.gov/, http://www.mat.univie.ac.at/neum/glopt/software_g.html.

GANSO library http://www.deakin.edu.au/~gleb/ganso.html implements a number of global methods, both deterministic (ECAM) and stochastic (multistart random search, heuristics) and also their combinations. It has C/C++, Fortran, Matlab and Maple interfaces.

Spline approximation

Various implementations of interpolating, smoothing and regression splines (univariate and bivariate) are available from *Netlib* and *TOMS* http://www.netlib.org/tom. Monotone univariate and tensor product regression splines are implemented in *tspline* package http://www.deakin.edu.au/~gleb/tspline.html.

Another implementation is FITPACK http://www.netlib.org/fitpack/.

Multivariate monotone approximation

tspline package http://www.deakin.edu.au/~gleb/tspline.html implements monotone tensor-product regression splines.

The method of monotone Lipschitz approximation is available from LibLip library http://www.deakin.edu.au/~gleb/lip.html, and also http://packages.debian.org/stable/libs/liblip2.

References

[AE91]   L.-E. Andersson, T. Elfving, Interpolation and approximation by monotone cubic splines. J. Approx. Theory **66**, 302–333 (1991)
[Bag02]   A. Bagirov, A method for minimization of quasidifferentiable functions. Optim. Methods Softw. **17**, 31–60 (2002)
[BB02]   L.M. Batten, G. Beliakov, Fast algorithm for the cutting angle method of global optimization. J. Global Optim. **24**, 149–161 (2002)
[Bel00]   G. Beliakov, Shape preserving approximation using least squares splines. Approximation Theory Appl. **16**, 80–98 (2000)
[Bel02]   G. Beliakov, Monotone approximation of aggregation operators using least squares splines. Int. J. Uncertainty Fuzziness Knowl. Based Syst. **10**, 659–676 (2002)
[Bel03]   G. Beliakov, Geometry and combinatorics of the cutting angle method. Optimization **52**, 379–394 (2003)
[Bel04]   G. Beliakov, The cutting angle method—a tool for constrained global optimization. Optim. Methods Softw. **19**, 137–151 (2004)
[Bel05]   G. Beliakov, Monotonicity preserving approximation of multivariate scattered data. BIT **45**, 653–677 (2005)
[Bel06]   G. Beliakov, Interpolation of Lipschitz functions. J. Comp. Appl. Math. **196**, 20–44 (2006)
[Bel07]   G. Beliakov, Construction of aggregation operators for automated decision making via optimal interpolation and global optimization. J. Ind. Manage. Optim. **3**, 193–208 (2007)
[Bel08]   G. Beliakov, Extended cutting angle method of global optimization. Pac. J. Optim. **4**, 153–176 (2008)
[BBP12]   G. Beliakov, H. Bustince, D. Paternain, Image reduction using means on discrete product lattices. IEEE Trans. Image Process. **21**, 1070–1083 (2012)
[BC07]   G. Beliakov, T. Calvo, Construction of aggregation operators with noble reinforcement. IEEE Trans. Fuzzy Syst. **15**, 1209–1218 (2007)
[BC09]   G. Beliakov, T. Calvo, Construction of k-Lipschitz triangular norms and conorms from empirical data. IEEE Trans. Fuzzy Syst. **17**, 1217–1220 (2009)
[BCJ10a]   G. Beliakov, T. Calvo, S. James, Aggregation of preferences in recommender systems, in *Recommender Systems Handbook*. ed. by P.B. Kantor, F. Ricci, L. Rokach, B. Shapira (Springer, 2010)

[BCJ10b]   G. Beliakov, T. Calvo, S. James, On Lipschitz properties of generated aggregation functions. Fuzzy Sets Syst. **161**, 1437–1447 (2010)

[BCL07]   G. Beliakov, T. Calvo, J. Lázaro, Pointwise construction of Lipschitz aggregation operators with specific properties. Int. J. Uncertainty Fuzziness Knowl. Based Syst. **15**, 193–223 (2007)

[BCP07a]   G. Beliakov, T. Calvo, A. Pradera, Absorbent tuples of aggregation operators. Fuzzy Sets Syst. **158**, 1675–1691 (2007)

[BCP07b]   G. Beliakov, T. Calvo, A. Pradera, Handling of neutral information by aggregation operators. Fuzzy Sets Syst. **158**, 861–880 (2007)

[BJ11]   G. Beliakov, S. James, Citation-based journal ranks: the use of fuzzy measures. Fuzzy Sets Syst. **167**, 101–119 (2011)

[BJL11]   G. Beliakov, S. James, G. Li, Learning Choquet-integralbased metrics for semisupervised clustering. IEEE Trans. Fuzzy Syst. **19**, 562–574 (2011)

[BLVW15]   G. Beliakov, G. Li, H.Q. Vu, T. Wilkin, Characterizing compactness of geometrical clusters using fuzzy measures. IEEE Trans. Fuzzy Syst. **23**, 1030–1043 (2015)

[BPC07]   G. Beliakov, A. Pradera, T. Calvo, *Aggregation Functions: A Guide for Practitioners*, vol. 221, Studies in Fuzziness and Soft Computing (Springer, Berlin, 2007)

[BG73]   R.E. Bellman, M. Giertz, On the analytic formalism of the theory of fuzzy sets. Inf. Sci. **5**, 149–156 (1973)

[BZ70]   R.E. Bellman, L. Zadeh, Decisionmaking in a fuzzy environment. Manage. Sci. **17**, 141–164 (1970)

[BS83]   P. Bloomfield, W.L. Steiger, *Least Absolute Deviations. Theory, Applications and Algorithms* (Stuttgart: Birkhauser, Boston, 1983)

[BGLS00]   J.F. Bonnans, J.C. Gilbert, C. Lemarechal, C.A. Sagastizabal, *Numerical Optimization. Theoretical and Practical Aspects* (Springer, Berlin, 2000)

[BB87]   J.M. Borwein, P.B. Borwein, *PI and the AGM: A Study in Analytic Number Theory and Computational Complexity* (Wiley, New York, 1987)

[BS84]   B. Buchanan, E. Shortliffe, *Rule-based Expert Systems. The MYCIN Experiments of the Stanford Heuristic Programming Project*. Reading, (MA: Addison-Wesley, 1984)

[Bul03]   P.S. Bullen, *Handbook of Means and Their Inequalities* (Kluwer, Dordrecht, 2003)

[BF01]   R.L. Burden, J.D. Faires, *Numerical Analysis* (Brooks/Cole, Pacific Grove, 2001)

[CKKM02]   T. Calvo, A. Kolesárová, M. Komorníková, R. Mesiar, in *Aggregation Operators: Properties, Classes And Construction Methods*, ed. by T. Calvo, G. Mayor, R. Mesiar. Aggregation Operators. New Trends and Applications (Physica-Verlag, Heidelberg, 2002), pp. 3–104

[CH92]   S.-J. Chen, C.-L. Hwang, *Fuzzy Multiple Attribute Decision Making: Methods and Applications* (Springer, Berlin, 1992)

[CK04]   E.W. Cheney, D. Kincaid, *Numerical Mathematics and Computing* (Brooks/Cole, Belmont, 2004)

[Chv83]   V. Chvatal, *Linear Programming* (W.H. Freeman, New York, 1983)

[Cla83]   F.H. Clarke, *Optimization and Nonsmooth Analysis* (Wiley, New York, 1983)

[DB99]   B. De Baets, Idempotent uninorms. Eur. J. Oper. Res. **118**, 631–642 (1999)

[DB01]   C. De Boor, *A Practical Guide to Splines* (Springer, New York, 2001)

[DAMM01]   J. Del Amo, J. Montero, E. Molina, Representation of recursive rules. Eur. J. Oper. Res. **130**, 29–53 (2001)

[DR86]   V.F. Demyanov, A. Rubinov, *Quasi-differential Calculus* (Optimization Software Inc, New York, 1986)

[DS83]   J.E. Dennis, R.B. Schnabel, *Numerical Methods for Unconstrained Optimization and Nonlinear Equations* (Prentice-Hall, Englewood Cliffs, 1983)

[Die95]   P. Dierckx, *Curve and Surface Fitting with Splines* (Clarendon press, Oxford, 1995)

[DP80]   D. Dubois, H. Prade, *Fuzzy Sets and Systems. Theory and Applications* (Academic Press, New York, 1980)

[DP85]   D. Dubois, H. Prade, A review of fuzzy set aggregation connectives. Inf. Sci. **36**, 85–121 (1985)

[DP00]   D. Dubois, H. Prade, *Fundamentals of Fuzzy Sets* (Kluwer, Boston, 2000)
[DP04]   D. Dubois, H. Prade, On the use of aggregation operations in information fusion processes. Fuzzy Sets Syst. **142**, 143–161 (2004)
[DHN76]   R. Duda, P. Hart, and N. Nilsson, Subjective Bayesian methods for rule-based inference systems. Proc. Nat. Comput. Conf. **45**, 1075–1082 (1976)
[Fle00]   R. Fletcher, *Practical Methods of Optimization*, 2nd edn. (Wiley, New York, 2000)
[Flo00]   C.A. Floudas, *Deterministic Global Optimization: Theory, Methods, and Applications* (Kluwer, Dordrecht, 2000)
[GMMP09]   M. Grabisch, J.-L. Marichal, R. Mesiar, E. Pap, *Aggregation Functions Encyclopedia of Mathematics and Its Foundations* (Cambridge University Press, 2009)
[HJ95]   P. Hansen, B. Jaumard, Lipschitz optimization, in *Handbook of Global Optimization*. ed. by R. Horst, P. Pardalos (Dordrecht: Kluwer, 1995), pp. 407–493
[HH81]   K.H. Haskell, R. Hanson, An algorithm for linear least squares problems with equality and nonnegativity constraints. Math. Program. **21**, 98–118 (1981)
[HPT00]   R. Horst, P. Pardalos, N.V. Thoai (eds.) *Introduction to Global Optimization*. 2nd edn. (Dordrecht: Kluwer, 2000)
[HP95]   R. Horst, P.M. Pardalos, *Handbook of Global Optimization* (Kluwer, Dordrecht, 1995)
[HT93]   R. Horst, H. Tuy, *Global Optimization: Deterministic Approaches*. 2nd rev. (Springer, Berlin, 1993)
[Hub03]   P.J. Huber, *Robust Statistics* (Wiley, New York, 2003)
[HL87]   C.-L. Hwang, M.J. Lin, *Group Decision Making under Multiple Criteria* (Springer, Berlin, 1987)
[HM79]   C.-L. Hwang, A.S.M. Masud, *Multiple Objective Decitions Making. Methods and Applications* (Springer, Berlin, 1979)
[HY81]   C.-L. Hwang, K. Yoon, *Multiple Attribute Decision Making. Methods and Applications* (Springer, Berlin, 1981)
[KNL94]   U. Kaymak, H.R. van Nauta Lemke, Selecting an aggregation operator for fuzzy decision making, in *3rd IEEE International Conference on Fuzzy Systems*, vol. 2. (1994), pp. 1418–1422
[Kiw85]   K.C. Kiwiel, *Methods of Descent of Nondifferentiable Optimization*, vol. 1133, Lecture Notes in Mathematics (Springer, Berlin, 1985)
[KMP00]   E.P. Klement, R. Mesiar, E. Pap, *Triangular Norms* (Kluwer, Dordrecht, 2000)
[LH95]   C. Lawson, R. Hanson, *Solving Least Squares Problems* (SIAM, Philadelphia, 1995)
[Lem78]   C. Lemarechal, Bundle methods in non-smooth optimization, in *Non-smooth Optimization*, ed. by C. Lemarechal, R. Mifflin (Pergamon Press, Oxford, 1978), pp. 78–102
[MMY06]   R. Maronna, R. Martin, V. Yohai, *Robust Statistics: Theory and Methods* (Wiley, New York, 2006)
[MMT03]   J. Martín, G. Mayor, J. Torrens, On locally internal monotonic operations. Fuzzy Sets Syst. **137**, 27–42 (2003)
[MC97]   G. Mayor, T. Calvo, Extended aggregation functions. In: IFSA'97. Vol. 1. Prague, 1997, pp. 281-285
[MR81]   D.F. McAllister, J.A. Roulier, An algorithm for computing a shape-preserving oscillatory quadratic spline. ACM Trans. Math. Softw. **7**, 331–347 (1981)
[Mic94]   Z. Michalewicz, *Genetic Algorithms + Data Structures = Evolution Programs*, 2nd, extended. (Springer, Berlin, 1994)
[NW99]   J. Nocedal, S. Wright, *Numerical Optimization* (Springer, New York, 1999)
[PR02]   P.M. Pardalos, G.C. Resende (eds.), *Handbook of Applied Optimization* (Oxford University Press, New York, 2002)
[PFBMB15]   D. Paternain, J. Fernandez, H. Bustince, R. Mesiar, G. Beliakov, Construction of image reduction operators using averaging aggregation functions. Fuzzy Sets Syst. **261**, 81–111 (2015)
[PJB12]   D. Paternain, A. Jurio, G. Beliakov, Color image reduction by minimizing penalty functions, in *Proceedings of the 9th IEEE International Conference on Fuzzy Systems (FUZZIEEE)*, Brisbane, Australia (2012)

[Pij72]   S.A. Pijavski, An algorithm for finding the absolute extremum of a function. USSR Comput. Math. Math. Phys. **2**, 57–67 (1972)

[Pin96]   J. Pinter, *Global Optimization in Action: Continuous and Lipschitz Optimization-Algorithms, Implementations, and Applications* (Kluwer, Dordrecht, 1996)

[Pow02]   M.J.D. Powell, UOBYQA: unconstrained optimization by quadratic approximation. Math. Program. **92**, 555–582 (2002)

[Pow04]   M.J.D. Powell. The NEWUOA software for unconstrained optimization without derivatives. Tech. rep. DAMTP NA2004/08, Department of Applied Mathematics and Theoretical Physics, Cambridge University, Cambridge CB3 9EW, UK, 2004

[RCM12]   J.T. Rodríguez, F. Camilo, F. Montero, On the semantics of bipolarity and fuzziness, in *Eurofuse 2011*, ed. by B. De Baets, et al. (Springer, Berlin, 2012), pp. 193–205

[RL03]   P.J. Rousseeuw, A.M. Leroy, *Robust Regression and Outlier Detection* (Wiley, New York, 2003)

[Rub00]   A.M. Rubinov, *Abstract Convexity and Global Optimization* (Kluwer, Dordrecht, 2000)

[Sch03]   K. Schittkowski, *Numerical Data Fitting in Dynamical Systems: a Practical Introduction with Applications and Software* (Kluwer, Dordrecht, 2003)

[Sch81]   L.L. Schumaker, *Spline Functions: Basic Theory* (Wiley, New York, 1981)

[Sch83]   L.L. Schumaker, On shape preserving quadratic interpolation. SIAM J. Numer. Anal. **20**, 854–864 (1983)

[Shu72]   B. Shubert, A sequential method seeking the global maximum of a function. SIAM J. Numer. Anal. **9**, 379–388 (1972)

[SS00]   R.G. Strongin, Y.D. Sergeyev, *Global Optimization with Non-convex Constraints: Sequential and Parallel Algorithms* (Kluwer, Dordrecht, 2000)

[Tri79]   E. Trillas. "Sobre funciones de negación en la teoría de los subconjuntos difusos". In: Stochastica III (1979), 47–59 (in Spanish) Reprinted (English version) in Advances of Fuzzy Logic, S. Barro et al. (eds), Universidad de Santiago de Compostela 31–43, 1998

[Van01]   R.J. Vanderbei, *Linear Programming: Foundations and Extensions* (Kluwer, Boston, 2001)

[Wei02]   E.W. Weisstein, *CRC Concise Encyclopedia of Mathematics* (CRC Press, Boca Raton, 2002)

[YR97]   R.R. Yager, A. Rybalov, Noncommutative self-identity aggregation. Fuzzy Sets Syst. **85**, 73–82 (1997)

[Yag88]   R.R. Yager, On ordered weighted averaging aggregation operators in multicriteria decision making. IEEE Trans. Syst. Man Cybern. **18**, 183–190 (1988)

[YB10]   R.R. Yager, G. Beliakov, OWA operators in regression problems. IEEE Trans. Fuzzy Syst. **18**, 106–113 (2010)

[YF94]   R.R. Yager, D. Filev, *Essentials of Fuzzy Modelling and Control* (Wiley, New York, 1994)

[YR96]   R.R. Yager, A. Rybalov, Uninorm aggregation operators. Fuzzy Sets Syst. **80**, 111–120 (1996)

[Zad65]   L. Zadeh, Fuzzy sets. Inf. Control **8**, 338–353 (1965)

[Zad78]   L. Zadeh, Fuzzy sets as a basis for a theory of possibility. Fuzzy Sets Syst. **1**, 3–28 (1978)

[Zad85]   L. Zadeh, The role of fuzzy logic in the management of uncertainty in expert systems, in *Approximate Reasoning in Expert Systems*, ed. by M.M. Gupta, A. Kandel, W. Bandler, J.B. Kiszka (Elsevier, North-Holland, 1985)

[Zim96]   H.-J. Zimmermann, *Fuzzy Set Theory—and its Applications* (Kluwer, Boston, 1996)

[ZZ80]   H.-J. Zimmermann, P. Zysno, Latent connectives in human decision making. Fuzzy Sets Syst. **4**, 37–51 (1980)

Chapter 2
Classical Averaging Functions

Abstract This chapter presents the classical means, starting with the weighted arithmetic and power means, and then continuing to the quasi-arithmetic means. The topics of generating functions, comparability and weights selection are covered. Several interesting classes of non-quasi-arithmetic means are presented, including Gini, Bonferroni, logarithmic and Bajraktarevic means. Methods of extension of symmetric bivariate means to the multivariate case are also discussed.

2.1 Semantics

Averaging is the most common way to combine the inputs. It is commonly used in voting, multicriteria and group decision making, constructing various performance scores, statistical analysis, etc. The basic rule is that the total score cannot be above or below any of the inputs. The aggregated value is seen as some sort of representative value of all the inputs.

We shall adopt the following generic definition [BPC07, Bul03, CKKM02, DP85, DP04].

Definition 2.1 (*Averaging aggregation*) An aggregation function f is averaging if for every $\mathbf{x}$ it is bounded by the minimum and the maximum

$$\min(\mathbf{x}) \leq f(\mathbf{x}) \leq \max(\mathbf{x}).$$

We remind that due to monotonicity of aggregation functions, averaging functions are idempotent, and vice versa, see Note 1.12, p. 10. That is, an aggregation function f is averaging if and only if $f(t, \ldots, t) = t$ for any $t \in \mathbb{I}$.

Formally, the minimum and maximum functions can be considered as averaging, however they are the limiting cases, right on the border with conjunctive and disjunctive functions. There are also some types of mixed aggregation functions, such as uninorms or nullnorms, that include averaging functions as particular cases.

© Springer International Publishing Switzerland 2016

G. Beliakov et al., *A Practical Guide to Averaging Functions*,
Studies in Fuzziness and Soft Computing 329,
DOI 10.1007/978-3-319-24753-3_2

The term *mean* is often used synonymously with the average to denote a measure of central tendency of the data.[1] A prototypical example of means is the arithmetic mean, often referred to as "the" mean.

However, the concepts of other means like the geometric and harmonic means, as well as the mode and midrange were already known to the Greeks [Eis71, Rub71]. Attempts to axiomatise the means were made in the 1920–1930s. Chisini [Chi29] gave a definition of a mean with respect to a function of n arguments f as a number M such that if every number x_i is replaced by M we get an equality

$$f(x_1, \ldots, x_n) = f(M, \ldots, M).$$

This definition was later criticised (e.g., by Gini [Gin58]) because it neither implies that the solution to the above equation exists, nor that M is in the range of the arguments, one of the fundamental requirements stipulated by Cauchy [Cau21]. Gini [Gin58] proposes to define a mean of several quantities as some value obtained as a result of a certain mathematical procedure with the given quantities, which either coincides with one of those quantities or is bounded by the largest and the smallest quantity.

Gini [Gin58] separates the means into analytical and positional. The analytical means are the ones obtained by application of a certain formula that acts on the values of the arguments. In contrast, positional means are computed by taking into account the relative position of the arguments. The arithmetic mean is an example of analytical means whereas the median is a positional mean. This classification is rather fuzzy, as Gini states himself, because of significant overlap.

Another distinction made by Gini is between hard means (where the result depends on all the inputs) and "moving" means (where changes to some inputs may not affect the value of the mean). In our terms the hard means correspond to strictly increasing averages. The representative examples of both classes are again the arithmetic mean and the median.

2.1.1 Measure of Orness

The *measure of orness*, also called the *degree of orness* or *attitudinal character*, is an important numerical characteristic of averaging aggregation functions. It was first defined in 1974 by Dujmovic [Duj73, Duj74], and then rediscovered several times, see [FR94, Yag88], mainly in the context of OWA functions (Sect. 3.1). It is applicable to any averaging function (and even to some other aggregation functions, like ST-OWA [FSM03]).

---

[1] In some languages there are no distinct terms that refer separately to the average and the mean, e.g. moyenne (Fr), medie (It.), promedio (or media) (Sp.), sredniaya (Ru.), Gemiddelde (Dut.), Gjen-nomsnitt (No.), whereas in others there are, e.g. Durchschnitt or Mittelwerte (Ger.). For etymology of the words *mean* and *average* see [Eis71]. We will use both terms synonymously.

Basically, the measure of orness measures how far a given averaging function is from the max function, which is the weakest disjunctive function. The measure of orness is computed for any averaging function [Duj74, FSM03, FR94] using

Definition 2.2 (*Measure of orness*) Let f be an averaging aggregation function. Then its measure of orness is

$$orness(f) = \frac{\int_{\mathbb{I}^n} f(\mathbf{x})d\mathbf{x} - \int_{\mathbb{I}^n} \min(\mathbf{x})d\mathbf{x}}{\int_{\mathbb{I}^n} \max(\mathbf{x})d\mathbf{x} - \int_{\mathbb{I}^n} \min(\mathbf{x})d\mathbf{x}}. \tag{2.1}$$

Clearly, $orness(\max) = 1$ and $orness(\min) = 0$, and for any f, $orness(f) \in [0, 1]$. The calculation of the integrals of max and min functions was performed in [Duj73] and results in simple equations

$$\int_{[0,1]^n} \max(\mathbf{x})d\mathbf{x} = \frac{n}{n+1} \text{ and } \int_{[0,1]^n} \min(\mathbf{x})d\mathbf{x} = \frac{1}{n+1}. \tag{2.2}$$

Hence (since both max and min are homogeneous functions) we get

$$\int_{[0,b]^n} \max(\mathbf{x})d\mathbf{x} = b^{n+1}\frac{n}{n+1} \text{ and } \int_{[0,b]^n} \min(\mathbf{x})d\mathbf{x} = b^{n+1}\frac{1}{n+1}, \tag{2.3}$$

and then for $a, b \geq 0$:

$$\int_{[a,b]^n} \max(\mathbf{x})d\mathbf{x} = a(b-a)^n + \frac{(b-a)^{n+1}n}{n+1} = (b-a)^n\frac{a+bn}{n+1} \text{ and}$$

$$\int_{[a,b]^n} \min(\mathbf{x})d\mathbf{x} = a(b-a)^n + \frac{(b-a)^{n+1}}{n+1} = (b-a)^n\frac{an+b}{n+1} \tag{2.4}$$

A different measure of orness, the average orness value, is proposed in [FSM03].

Definition 2.3 (*Average orness value*) Let f be an averaging aggregation function. Then its average orness value is

$$\overline{orness}(f) = \int_{\mathbb{I}^n} \frac{f(\mathbf{x}) - \min(\mathbf{x})}{\max(\mathbf{x}) - \min(\mathbf{x})}d\mathbf{x}. \tag{2.5}$$

Both the measure of orness and the average orness value are $\frac{1}{2}$ for weighted arithmetic means, and later we will see that both quantities coincide for OWA functions. However computation of the average orness value for other averaging functions is more involved (typically performed by numerical methods), therefore we will use mainly the measure of orness in Definition 2.2.

There are some alternative definitions of the measure of orness aimed at simplifying calculations (especially for quasi-arithmetic means, [Liu06]); we will discuss them in the relevant sections. Recently the concept of the orness measure has been revisited in [KSP14] in the context of OWA functions, we discuss the details in Sect. 3.2.1.

The measure of orness was used by Dujmovic [Duj07] to classify the averages into seven classes, including disjunction, hard and soft partial disjunction, neutral averaging, soft and hard partial conjunction, and conjunction. The mentioned classes are ordered according to the orness, starting from 1 (the max function), to some threshold value $\frac{1}{2} < \alpha < 1$, then between α and $\frac{1}{2}$, $\frac{1}{2}$ (neutrality), to some other threshold value $\frac{1}{2} > \beta > 0$, from β to 0, and then 0 (the min function).

The hard partial conjunctions model mandatory requirements (in the decision making domain), and soft partial conjunctions model desired but not mandatory requirements. The hard partial disjunction models sufficient requirements, and soft partial disjunction models desired but not sufficient requirements.

2.2 Classical Means

While means are treated synonymously with averaging functions, the classical treatment of means (see, e.g., [Bul03]) excludes certain types of averaging functions, which have been developed quite recently, in particular ordered weighted averaging and various integrals. On the other hand some classical means (e.g., some Gini means) lack monotonicity, and therefore are not aggregation functions. Some of these averages will be considered in Chap. 7. In this chapter we will concentrate on various classical means, and present other types of averaging, or mean-type functions in separate chapters.

2.2.1 Arithmetic Mean

The arithmetic mean is the most widely used averaging function.

Definition 2.4 (*Arithmetic mean*) The arithmetic mean, or the average of n values, is the function

$$M(\mathbf{x}) = \frac{1}{n}(x_1 + x_2 + \cdots + x_n) = \frac{1}{n}\sum_{i=1}^{n} x_i.$$

Since M is properly defined for any number of arguments, it is an extended aggregation function, see Definition 1.6.

Main properties

- The arithmetic mean M is a strictly increasing aggregation function;
- M is a symmetric function;

- M is an additive function, i.e., $M(\mathbf{x} + \mathbf{y}) = M(\mathbf{x}) + M(\mathbf{y})$ for all $\mathbf{x}, \mathbf{y} \in \mathbb{I}^n$ such that $\mathbf{x} + \mathbf{y} \in \mathbb{I}^n$;
- M is a homogeneous function, i.e., $M(\lambda\mathbf{x}) = \lambda M(\mathbf{x})$ for all $\mathbf{x} \in \mathbb{I}^n$ and for all $\lambda \in \mathbb{R}$;
- The orness measure $orness(M) = \frac{1}{2}$;
- M is a Lipschitz continuous function, with the Lipschitz constant in any $\| \cdot \|_p$ norm (see p. 22) $n^{-1/p}$, the smallest Lipschitz constant of all aggregation functions.

When the inputs are not symmetric, it is a common practice to associate each input with a weight, a number $w_i \in [0, 1]$ which reflects the relative contribution of this input to the total score. For example, in shareholders' meetings, the strength of each vote is associated with the number of shares this shareholder possesses. The votes are usually just added to each other, and after dividing by the total number of shares, we obtain a weighted arithmetic mean. Weights can also represent the reliability of an input or its importance.

Weights are not the only way to obtain asymmetric functions, we will study other methods in Sect. 2.4 and in Chap. 5. Recall from Chap. 1 the definition of a weighting vector:

Definition 2.5 (*Weighting vector*) A vector $\mathbf{w} = (w_1, \ldots, w_n)$ is called a weighting vector if $w_i \in [0, 1]$ and $\sum_{i=1}^{n} w_i = 1$.

Definition 2.6 (*Weighted arithmetic mean (WAM)*) Given a weighting vector $\mathbf{w}$, the weighted arithmetic mean is the function

$$M_{\mathbf{w}}(\mathbf{x}) = w_1 x_1 + w_2 x_2 + \cdots + w_n x_n = \sum_{i=1}^{n} w_i x_i = \langle \mathbf{w}, \mathbf{x} \rangle.$$

Main properties

- A weighted arithmetic mean $M_{\mathbf{w}}$ is a strictly increasing aggregation function, if all $w_i > 0$;
- $M_{\mathbf{w}}$ is an asymmetric (unless $w_i = 1/n$ for all $i \in \{1, \ldots, n\}$) idempotent function;
- $M_{\mathbf{w}}$ is an additive function, i.e., $M_{\mathbf{w}}(\mathbf{x} + \mathbf{y}) = M_{\mathbf{w}}(\mathbf{x}) + M_{\mathbf{w}}(\mathbf{y})$ for all $\mathbf{x}, \mathbf{y} \in \mathbb{I}^n$ such that $\mathbf{x} + \mathbf{y} \in \mathbb{I}^n$;
- $M_{\mathbf{w}}$ is a homogeneous function, i.e., $M_{\mathbf{w}}(\lambda\mathbf{x}) = \lambda M_{\mathbf{w}}(\mathbf{x})$ for all $\mathbf{x} \in \mathbb{I}^n$ and for all $\lambda \in \mathbb{R}$;
- Jensen's inequality: for any convex function[2] $g : \mathbb{I} \to [-\infty, \infty]$, $g(M_{\mathbf{w}}(\mathbf{x})) \leq M_{\mathbf{w}}(g(x_1), \ldots, g(x_n))$.
- $M_{\mathbf{w}}$ is a Lipschitz continuous function, in fact it is a kernel aggregation function (see p. 23);

---

[2]A function g is convex if and only if $g(\alpha t_1 + (1 - \alpha)t_2) \leq \alpha g(t_1) + (1 - \alpha)g(t_2)$ for all $t_1, t_2 \in Dom(g)$ and $\alpha \in [0, 1]$.

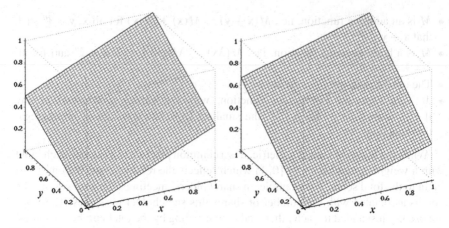

Fig. 2.1 3D plots of weighted arithmetic means $M_{(\frac{1}{2},\frac{1}{2})}$ and $M_{(\frac{1}{3},\frac{2}{3})}$

- $M_{\mathbf{w}}$ is a shift-invariant function (see p. 17);
- The orness measure $orness(M_{\mathbf{w}}) = \frac{1}{2}$;[3]
- $M_{\mathbf{w}}$ is a special case of the Choquet integral (see Sect. 4.1) with respect to an additive fuzzy measure (Fig. 2.1).

Geometric and harmonic means

Weighted arithmetic means are good for averaging inputs that can be added together. Frequently the inputs are not added but multiplied. For example, when averaging the rates of investment return over several years the use of the arithmetic mean is incorrect. This is because the rate of return (say 10 %) signifies that in one year the investment was multiplied by a factor 1.1. If the return is 20 % in the next year, then the total is multiplied by 1.2, which means that the original investment is multiplied by a factor of 1.1×1.2. The average return is calculated by using the geometric mean of 1.1 and 1.2, which gives ≈ 1.15.

Definition 2.7 (*Geometric mean*) The geometric mean is the function

---

[3]It is easy to check that

$$\int_{[0,1]^n} M(\mathbf{x})d\mathbf{x} = \frac{1}{n}\left(\int_0^1 x_1 dx_1 + \cdots + \int_0^1 x_n dx_n\right) = \frac{n}{n}\int_0^1 t\,dt = \frac{1}{2}.$$

Substituting the above value in (2.1) we obtain $orness(M) = \frac{1}{2}$. Following, for a weighted arithmetic mean we also obtain

$$\int_{[0,1]^n} M_{\mathbf{w}}(\mathbf{x})d\mathbf{x} = w_1\int_0^1 x_1 dx_1 + \cdots + w_n\int_0^1 x_n dx_n = \sum_{i=1}^n w_i\int_0^1 t\,dt = \frac{1}{2}.$$

.

$$G(\mathbf{x}) = \sqrt[n]{x_1 x_2 \ldots x_n} = \left(\prod_{i=1}^{n} x_i \right)^{1/n}.$$

Definition 2.8 (*Weighted geometric mean (WGM)*) Given a weighting vector $\mathbf{w}$, the weighted geometric mean is the function

$$G_{\mathbf{w}}(\mathbf{x}) = \prod_{i=1}^{n} x_i^{w_i}.$$

Definition 2.9 (*Harmonic mean*) The harmonic mean is the function

$$H(\mathbf{x}) = n \left(\sum_{i=1}^{n} \frac{1}{x_i} \right)^{-1}.$$

Definition 2.10 (*Weighted harmonic mean (WHM)*) Given a weighting vector $\mathbf{w}$, the weighted harmonic mean is the function

$$H_{\mathbf{w}}(\mathbf{x}) = \left(\sum_{i=1}^{n} \frac{w_i}{x_i} \right)^{-1}.$$

Note 2.11 If the weighting vector $\mathbf{w}$ is given without normalization, i.e., $W = \sum_{i=1}^{n} w_i \neq 1$, then one can either normalize it first by dividing each component by W, or use the alternative expressions for weighted geometric and harmonic means

$$G_{\mathbf{w}}(\mathbf{x}) = \left(\prod_{i=1}^{n} x_i^{w_i} \right)^{1/W},$$

$$H_{\mathbf{w}}(\mathbf{x}) = W \left(\sum_{i=1}^{n} \frac{w_i}{x_i} \right)^{-1}.$$

Geometric-Arithmetic Mean Inequality

The following result is an extended version of the well known geometric-arithmetic means inequality

$$H_{\mathbf{w}}(\mathbf{x}) \leq G_{\mathbf{w}}(\mathbf{x}) \leq M_{\mathbf{w}}(\mathbf{x}), \tag{2.6}$$

for any vector $\mathbf{x}$ and weighting vector $\mathbf{w}$, with equality if and only if $\mathbf{x} = (t, t, \ldots, t)$.

Another curious relation between these three means is that for $n = 2$ we have $G(x, y) = \sqrt{M(x, y) \cdot H(x, y)}$ (Figs. 2.2 and 2.3).

Power means

A further generalization of the arithmetic mean is a family called power means (also called *root-power means*). This family is defined by

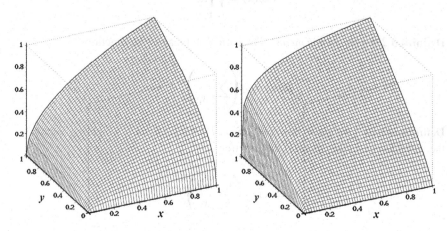

Fig. 2.2 3D plots of weighted geometric means $G_{(\frac{1}{2}, \frac{1}{2})}$ and $G_{(\frac{1}{5}, \frac{4}{5})}$

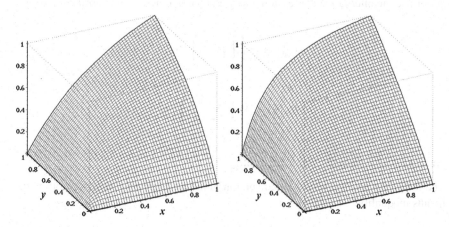

Fig. 2.3 3D plots of weighted harmonic means $H_{(\frac{1}{2}, \frac{1}{2})}$ and $H_{(\frac{1}{5}, \frac{4}{5})}$

Definition 2.12 (*Power mean*) For $r \in \mathbb{R}$, the power mean is the function

$$M_{[r]}(\mathbf{x}) = \left(\frac{1}{n} \sum_{i=1}^{n} x_i^r\right)^{1/r},$$

if $r \neq 0$, and $M_{[0]}(\mathbf{x}) = G(\mathbf{x})$.[4]

Definition 2.13 (*Weighted power mean*) Given a weighting vector $\mathbf{w}$ and $r \in \mathbb{R}$, the weighted power mean is the function

$$M_{\mathbf{w},[r]}(\mathbf{x}) = \left(\sum_{i=1}^{n} w_i x_i^r\right)^{1/r},$$

if $r \neq 0$, and $M_{\mathbf{w},[0]}(\mathbf{x}) = G_{\mathbf{w}}(\mathbf{x})$.

Note 2.14 The family of weighted power means is *augmented* to $r = -\infty$ and $r = \infty$ by using the limiting cases

$$M_{\mathbf{w},[-\infty]}(\mathbf{x}) = \lim_{r \to -\infty} M_{\mathbf{w},[r]}(\mathbf{x}) = \min(\mathbf{x}),$$

$$M_{\mathbf{w},[\infty]}(\mathbf{x}) = \lim_{r \to \infty} M_{\mathbf{w},[r]}(\mathbf{x}) = \max(\mathbf{x}).$$

However min and max are not themselves power means.

The limiting case of the weighted geometric mean is also obtained as

$$M_{\mathbf{w},[0]}(\mathbf{x}) = \lim_{r \to 0} M_{\mathbf{w},[r]}(\mathbf{x}) = G_{\mathbf{w}}(\mathbf{x}).$$

Of course, the family of weighted power means includes the special cases $M_{\mathbf{w},[1]}(\mathbf{x}) = M_{\mathbf{w}}(\mathbf{x})$, and $M_{\mathbf{w},[-1]}(\mathbf{x}) = H_{\mathbf{w}}(\mathbf{x})$. Another special case is the weighted *quadratic mean*

$$M_{\mathbf{w},[2]}(\mathbf{x}) = Q_{\mathbf{w}}(\mathbf{x}) = \sqrt{\sum_{i=1}^{n} w_i x_i^2}.$$

Main properties

- The weighted power mean $M_{\mathbf{w},[r]}$ is a strictly increasing aggregation function, if all $w_i > 0$ and $0 < r < \infty$;
- $M_{\mathbf{w},[r]}$ is a continuous function on $[0, \infty)^n$;

---

[4]We shall use square brackets in the notation $M_{[r]}$ for power means to distinguish them from quasi-arithmetic means M_g (see Sect. 2.3), where parameter g denotes a generating function rather than a real number. The same applies to the weighted power means.

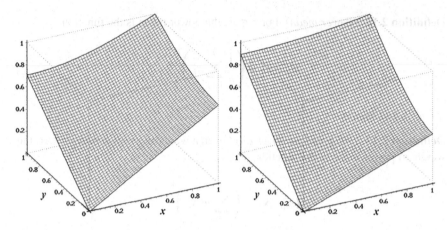

Fig. 2.4 3D plots of weighted quadratic mean $Q_{(\frac{1}{2},\frac{1}{2})}$ and $Q_{(\frac{1}{5},\frac{4}{5})}$

- $M_{\mathbf{w},[r]}$ is an asymmetric idempotent function (symmetric if all $w_i = \frac{1}{n}$);
- $M_{\mathbf{w},[r]}$ is a homogeneous function, i.e., $M_{\mathbf{w},[r]}(\lambda\mathbf{x}) = \lambda M_{\mathbf{w},[r]}(\mathbf{x})$ for all $\mathbf{x} \in [0,\infty)^n$ and for all $\lambda \in \mathbb{R}$; it is the only homogeneous weighted quasi-arithmetic mean (this class is introduced in Sect. 2.3);
- Weighted power means are comparable: $M_{\mathbf{w},[r]}(\mathbf{x}) \leq M_{\mathbf{w},[s]}(\mathbf{x})$ if $r \leq s$; this implies the geometric-arithmetic mean inequality;
- $M_{\mathbf{w},[r]}$ has absorbing element (always $a = 0$) if and only if $r \leq 0$ (and all weights w_i are positive);
- $M_{\mathbf{w},[r]}$ does not have a neutral element[5] (Figs. 2.4 and 2.5).

Measure of orness

Calculations for the geometric mean yields

$$orness(G) = -\frac{1}{n-1} + \frac{n+1}{n-1}\left(\frac{n}{n+1}\right)^n,$$

but for other means explicit formulas are known only for special cases, e.g., $n = 2$

$$\int_{[0,1]^2} Q(\mathbf{x})d\mathbf{x} = \frac{1}{3}\left(1 + \frac{1}{\sqrt{2}}\ln(1+\sqrt{2})\right) \approx 0.541075,$$

$$\int_{[0,1]^2} H(\mathbf{x})d\mathbf{x} = \frac{4}{3}(1 - \ln(2)), \quad \text{and}$$

---

[5]The limiting cases min ($r = -\infty$) and max ($r = \infty$) which have neutral elements $e = 1$ and $e = 0$ respectively, are not themselves power means.

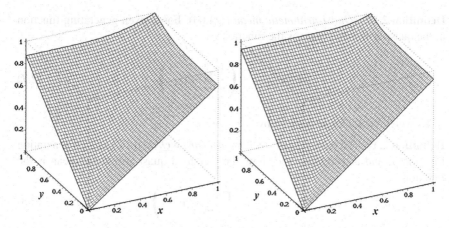

Fig. 2.5 3D plots of power means $M_{[5]}$ and $M_{[10]}$

$$\int_{[0,1]^2} M_{[-2]}(\mathbf{x})d\mathbf{x} = \frac{2}{3}(2 - \sqrt{2}),$$

from which, when $n = 2$, $orness(Q) \approx 0.623225$, $orness(H) \approx 0.22741$, and $orness(M_{[-2]}) = 3 - 2\sqrt{2}$.

Definition 2.15 (*Dual weighted power mean*) Let $M_{\mathbf{w},[r]}$ be a weighted power mean on [0, 1]. The function

$$\bar{M}_{\mathbf{w},[r]}(\mathbf{x}) = 1 - M_{\mathbf{w},[r]}(\mathbf{1} - \mathbf{x})$$

is called the dual weighted power mean.

Note 2.16 The dual weighted power mean is obviously a mean (the class of means is closed under duality). The absorbent element, if any, becomes $a = 1$. The extensions of weighted power means satisfy $\bar{M}_{\mathbf{w},[\infty]}(\mathbf{x}) = M_{\mathbf{w},[-\infty]}(\mathbf{x})$ and $\bar{M}_{\mathbf{w},[-\infty]}(\mathbf{x}) = M_{\mathbf{w},[\infty]}(\mathbf{x})$. The weighted arithmetic mean $M_{\mathbf{w}}$ is self-dual.

2.3 Weighted Quasi-arithmetic Means

2.3.1 Definitions

Quasi-arithmetic means generalize power means. Consider a univariate continuous strictly monotone function $g : \mathbb{I} \to [-\infty, \infty]$, which we call a *generating function*. Of course, g is invertible, but it is not necessarily a bijection (i.e., its range may be $Ran(g) \subset [-\infty, \infty]$).

Definition 2.17 (*Quasi-arithmetic mean (QAM)*) For a given generating function g, the quasi-arithmetic mean is the function

$$M_g(\mathbf{x}) = g^{-1}\left(\frac{1}{n}\sum_{i=1}^{n} g(x_i)\right). \tag{2.7}$$

Its weighted analogue is given by

Definition 2.18 (*Weighted quasi-arithmetic mean (WQAM)*) For a given generating function g, and a weighting vector $\mathbf{w}$, the weighted quasi-arithmetic mean is the function

$$M_{\mathbf{w},g}(\mathbf{x}) = g^{-1}\left(\sum_{i=1}^{n} w_i g(x_i)\right). \tag{2.8}$$

The weighted power means are a subclass of weighted quasi-arithmetic means with the generating function

$$g(t) = \begin{cases} t^r, & \text{if } r \neq 0, \\ \log(t), & \text{if } r = 0. \end{cases}$$

Note 2.19 Observe that if $Ran(g) = [-\infty, \infty]$, then we have the summation $-\infty + \infty$ or $+\infty - \infty$ if $x_i = 0$ and $x_j = 1$ for some $i \neq j$. When this occurs, a convention, such as $-\infty + \infty = +\infty - \infty = -\infty$, is adopted, and continuity of $M_{\mathbf{w},g}$ is lost.

Note 2.20 If the weighting vector $\mathbf{w}$ is not normalized, i.e., $W = \sum_{i=1}^{n} w_i \neq 1$, then weighted quasi-arithmetic means are expressed as

$$M_{\mathbf{w},g}(\mathbf{x}) = g^{-1}\left(\frac{1}{W}\sum_{i=1}^{n} w_i g(x_i)\right).$$

2.3.2 Main Properties

- Weighted quasi-arithmetic means are continuous if and only if $Ran(g) \neq [-\infty, \infty]$ [Kom01];
- Weighted quasi-arithmetic means on $[a, b]$ with strictly positive weights are strictly monotone increasing on $]a, b[^n$;
- The class of weighted quasi-arithmetic means is closed under duality. That is, given a strong negation N, the N-dual of a weighted quasi-arithmetic mean (on $[0, 1]$) $M_{\mathbf{w},g}$ is in turn a weighted quasi-arithmetic mean, given by $M_{\mathbf{w},g\circ N}$. For the standard negation, the dual of a weighted quasi-arithmetic mean is characterized by the generating function $h(t) = g(1 - t)$;
- The following result regarding self-duality holds (see, e.g., [PTC02]): Given a strong negation N, a weighted quasi-arithmetic mean $M_{\mathbf{w},g}$ is N-self-dual if and

only if N is the strong negation generated by g, i.e., if $N(t) = g^{-1}(g(0) + g(1) - g(t))$ for any $t \in [0, 1]$. This implies, in particular:

- Weighted quasi-arithmetic means, such that $g(0) = \pm\infty$ or $g(1) = \pm\infty$ are never N-self-dual (in fact, they are dual to each other);
- Weighted arithmetic means are always self-dual (i.e., N-self-dual with respect to the standard negation $N(t) = 1 - t$);

- The generating function is not defined uniquely, but up to an arbitrary linear transformation, i.e., if $g(t)$ is a generating function of some weighted quasi-arithmetic mean, then $ag(t) + b$, $a, b \in \mathbb{R}$, $a \neq 0$ is also a generating function of *the same mean*,[6] provided $Ran(g) \neq [-\infty, \infty]$;
- There are incomparable quasi-arithmetic means. Two quasi-arithmetic means M_g and M_h satisfy $M_g \leq M_h$ if and only if either the composite $g \circ h^{-1}$ is convex and g is decreasing, or $g \circ h^{-1}$ is concave and g increasing;
- Consequently, $M_g \leq M_{Id}$ if and only if g^{-1} is a convex function. By using duality, $M_{Id} \leq M_g$ if and only if g^{-1} is concave.
- The only homogeneous weighted quasi-arithmetic means are weighted power means;
- Weighted quasi-arithmetic means do not have a neutral element.[7] They may have an absorbing element only when all the weights are strictly positive and $g(a) = \pm\infty$ or $g(b) = \pm\infty$, and in such cases the corresponding absorbing elements are, respectively, a and b.
- Quasi-arithmetic means may or may not be Lipschitz continuous. Their Lipschitz properties are discussed in detail in [BCJ10].

Convexity and concavity of the generator

By comparing a quasi-arithmetic mean with the arithmetic mean $M = M_{Id}$ we can establish the convexity or concavity of its generator, namely the inverse g^{-1} is convex if and only if $M_g \leq M$. This is the case of a partial conjunction in the sense of Dujmovic [Duj07], i.e., $orness(M_g) \leq orness(M) = \frac{1}{2}$. For concave g^{-1} we have a partial disjunction $orness(M_g) \geq \frac{1}{2}$.

However the convexity (concavity) of the inverse g^{-1} does not imply concavity (convexity) of the mean M_g, although this is the case for many means, such as power and exponential means. As a counterexample consider a convex (on $[0,1]$) generator $g(t) = e^{-\frac{2}{t}}$, and the concave inverse $g^{-1}(t) = -\frac{2}{\ln(t)}$. The resulting mean $M_g \geq M$ but is neither convex nor concave as can be appreciated from Fig. 2.6. For example, $M_g(x, 0) = \frac{2}{\frac{2}{x} + \ln(2)}$ is concave whereas $M_g(x, 1) = \frac{2}{\ln(2) - \ln(e^{-\frac{2}{x}} + e^{-2})}$ is convex.

Recently the convexity (concavity) of the generator was connected to the supermodularity (submodularity) of the function M_g [CM09]. A function f is called supermodular if $f(\mathbf{x} \vee \mathbf{y}) + f(\mathbf{x} \wedge \mathbf{y}) \geq f(\mathbf{x}) + f(\mathbf{y})$, and submodular if the

---

[6]For this reason, one can assume that g is monotone increasing, as otherwise we can simply take $-g$.
[7]Observe that the limiting cases min and max are not quasi-arithmetic means.

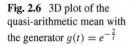

Fig. 2.6 3D plot of the quasi-arithmetic mean with the generator $g(t) = e^{-\frac{2}{t}}$

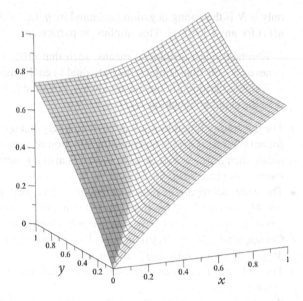

inequality is reversed.[8] The operations $\vee$, $\wedge$ are the usual componentwise maximum and minimum. Supermodularity is related to the property of increasing differences: a function f is supermodular if and only if it has increasing differences (which in the bi-variate case reads $f(x + a, y + b) - f(x, y + b) \geq f(x + a, y) - f(x, y)$ for positive a, b). A twice differentiable function f is supermodular if and only if its mixed partial derivatives $\frac{\partial^2 f}{\partial x_i \partial x_j} \geq 0$ for all distinct i, j. Then M_g is supermodular if and only if g^{-1} is convex [CM09], and hence if and only if $M_g \leq M$ is a partial conjunction. The case of partial disjunction equivalent to submodular functions M_g and concave g^{-1} is dealt with by using duality. The result also applies to weighted quasi-arithmetic means.

Orness measure

Calculation of the orness measure using (2.1) is technically difficult as the closed form expression for the integrals is available only in special cases. Some alternative definitions of the orness measure were presented, such as the one in [Duj05]

$$orness(M_g) = \frac{g(b) - \int_a^b g(t)dt}{g(b) - g(a)},$$

or the one in [Liu06]

$$orness(M_g) = \frac{g^{-1}\left(\frac{\int_a^b g(t)dt}{b-a}\right) - a}{b - a}.$$

---

[8]This concept of supermodularity of a function on $\mathbb{I}^n$ is different from supermodularity and submodularity of fuzzy measures in Definition 4.14.

2.3.3 Examples

Example 2.21 (*Weighted power means*) Weighted power means are a special case of weighted quasi-arithmetic means, with $g(t) = t^r$, $r \neq 0$ and $g(t) = \log(t)$ if $r = 0$. Note that the generating function $g(t) = \frac{t^r - 1}{r}$ defines exactly the same power mean (as a particular case of a linear transformation of g).

Example 2.22 (*Harmonic and geometric means*) These classical means, defined on p. 60, 61, are special cases of the power means, obtained when $g(t) = t^r$, $r = -1$ and $r = 0$ respectively.

Example 2.23 Let $g(t) = \log \frac{t}{1-t}$. The corresponding quasi-arithmetic mean M_g is given by

$$
M_g(\mathbf{x}) = \begin{cases} \dfrac{\sqrt[n]{\prod\limits_{i=1}^{n} x_i}}{\sqrt[n]{\prod\limits_{i=1}^{n} x_i} + \sqrt[n]{\prod\limits_{i=1}^{n} (1-x_i)}}, & \text{if } \{0, 1\} \not\subseteq \{x_1, \ldots, x_n\}s \\[4mm] 0, & \text{otherwise,} \end{cases}
$$

that is, $M_g = \frac{G}{G+1-G_d}$ and with the convention $\frac{0}{0} = 0$.

Example 2.24 (*Weighted trigonometric means*) Let $g_1(t) = \sin(\frac{\pi}{2}t)$, $g_2(t) = \cos(\frac{\pi}{2}t)$, and $g_3(t) = \tan(\frac{\pi}{2}t)$ be the generating functions. The weighted trigonometric means are the functions

$$
SM_{\mathbf{w}}(\mathbf{x}) = \frac{2}{\pi} \arcsin\left(\sum_{i=1}^{n} w_i \sin(\frac{\pi}{2}x_i)\right),
$$

$$
CM_{\mathbf{w}}(\mathbf{x}) = \frac{2}{\pi} \arccos\left(\sum_{i=1}^{n} w_i \cos(\frac{\pi}{2}x_i)\right) \text{ and}
$$

$$
TM_{\mathbf{w}}(\mathbf{x}) = \frac{2}{\pi} \arctan\left(\sum_{i=1}^{n} w_i \tan(\frac{\pi}{2}x_i)\right).
$$

Their 3D plots are presented on Figs. 2.7 and 2.8.

Example 2.25 (*Weighted exponential means*) Let the generating function be

$$
g(t) = \begin{cases} \gamma^t, & \text{if } \gamma \neq 1, \\ t, & \text{if } \gamma = 1. \end{cases}
$$

The weighted exponential mean is the function

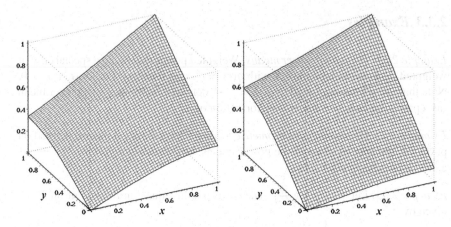

Fig. 2.7 3D plots of weighted trigonometric means $SM_{(\frac{1}{2},\frac{1}{2})}$ and $SM_{(\frac{1}{3},\frac{4}{3})}$

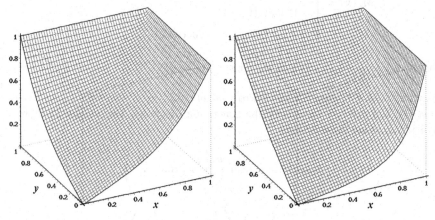

Fig. 2.8 3D plots of weighted trigonometric means $TM_{(\frac{1}{2},\frac{1}{2})}$ and $TM_{(\frac{1}{3},\frac{4}{3})}$

$$EM_{\mathbf{w},\gamma}(\mathbf{x}) = \begin{cases} \log_\gamma \left(\sum_{i=1}^n w_i \gamma^{x_i} \right), & \text{if } \gamma \neq 1, \\ M_{\mathbf{w}}(\mathbf{x}), & \text{if } \gamma = 1. \end{cases}$$

3D plots of some weighted exponential means are presented on Figs. 2.9 and 2.10.

Example 2.26   There is another mean also known as exponential [Bul03], given for $\mathbf{x} \geq \mathbf{1}$ by

$$f(\mathbf{x}) = \exp\left(\left(\prod_{i=1}^n \log(x_i) \right)^{1/n} \right).$$

It is a quasi-arithmetic mean with a generating function $g(t) = \log(\log(t))$, and its inverse $g^{-1}(t) = \exp(\exp(t))$.

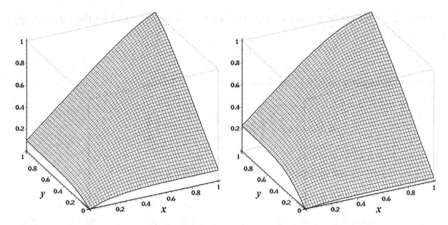

Fig. 2.9 3D plots of weighted exponential means $EM_{(\frac{1}{2},\frac{1}{2}),0.001}$ and $EM_{(\frac{1}{5},\frac{4}{5}),0.001}$

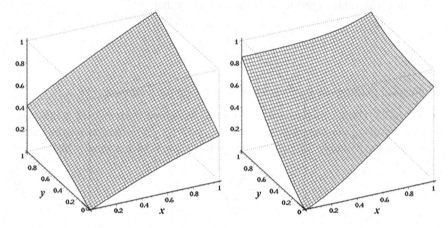

Fig. 2.10 3D plots of exponential means $EM_{(\frac{1}{2},\frac{1}{2}),0.5}$ and $EM_{(\frac{1}{2},\frac{1}{2}),100}$

In the domain $[0,1]^n$ one can use a generating function $g(t) = \log(-\log(t))$, so that its inverse is $g^{-1}(t) = \exp(-\exp(t))$. This mean is discontinuous, since $Ran(g) = [-\infty, \infty]$. We obtain the expression

$$f(\mathbf{x}) = \exp\left(-\prod_{i=1}^{n}\left(-\log(x_i)\right)^{1/n}\right).$$

Example 2.27 (*Weighted radical means*) Let $\gamma > 0$, $\gamma \neq 1$, and let the generating function be

$$g(t) = \gamma^{1/t}.$$

The weighted radical mean is the function

$$RM_{\mathbf{w},\gamma}(\mathbf{x}) = \left(\log_\gamma \left(\sum_{i=1}^n w_i \gamma^{1/x_i}\right)\right)^{-1}.$$

3D plots of some radical means are presented on Fig. 2.11.

Example 2.28 (*Weighted basis-exponential means*) Weighted basis-exponential means are obtained by using the generating function $g(t) = t^t$ and $t \geq \frac{1}{e}$ (this generating function is decreasing on $[0, \frac{1}{e}[$ and increasing on $]\frac{1}{e}, \infty]$, hence the restriction). The value of this mean is such a value y that

$$y^y = \sum_{i=1}^n w_i x_i^{x_i}.$$

For practical purposes this equation has to be solved for y numerically.

Example 2.29 (*Weighted basis-radical means*) Weighted basis-radical means are obtained by using the generator $g(t) = t^{1/t}$ and $t \geq \frac{1}{e}$ (restriction for the same reason as in the Example 2.28). The value of this mean is such a value y that

$$y^{1/y} = \sum_{i=1}^n w_i x_i^{1/x_i}.$$

For practical purposes this equation has to be solved for y numerically.

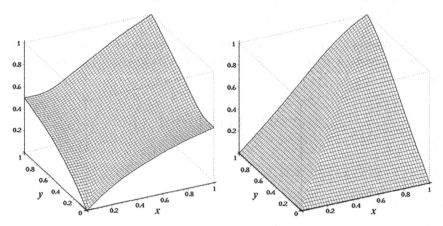

Fig. 2.11 3D plots of radical means $RM_{(\frac{1}{2},\frac{1}{2}),0.5}$ and $RM_{(\frac{1}{2},\frac{1}{2}),100}$

Example 2.30 (*Tsallis q-exponential means*) Another interesting generator was recently proposed in [RA14]. The authors used the q-exponentials e_q^t that are solutions to the differential equation $F' = F^q$ with the initial condition $F(0) = 1$. These are given by

$$F(t) = e_q^t = \lim_{r \to 1-q} (1 + rt)^{\frac{1}{r}} = (1 + (1 - q)t)^{\frac{1}{1-q}},$$

when $q \neq 1$ and e^t otherwise. The generators g themselves are derived as $g_1(t) = e_q^t - 1$ and $g_2(t) = e_q^{-(1-t)}$.

2.3.4 Calculation

Generating functions offer a nice way of calculating the values of weighted quasi-arithmetic means. Note that we can write

$$M_{\mathbf{w},g}(\mathbf{x}) = g^{-1}\left(M_{\mathbf{w}}(g(\mathbf{x}))\right),$$

where $g(\mathbf{x}) = (g(x_1), \ldots, g(x_n))$. Thus calculation can be performed in three steps:

1. Transform all the inputs by calculating vector $g(\mathbf{x})$;
2. Calculate the weighted arithmetic mean of the transformed inputs;
3. Calculate the inverse g^{-1} of the computed mean.

However one needs to be careful with the limiting cases, for example when $g(x_i)$ becomes infinite. Typically this is an indication of existence of an absorbing element, this needs to be picked up by the computer subroutine. Similarly, special cases like $M_{[r]}(\mathbf{x})$, $r \to \pm\infty$ have to be accommodated (in these cases the subroutine has to return the minimum or the maximum).

2.3.5 Weighting Triangles

When we are interested in using weighted quasi-arithmetic means as extended aggregation functions, we need to have a clear rule as to how the weighting vectors are calculated for each dimension $n = 2, 3, \ldots$. For symmetric quasi-arithmetic means we have a simple rule: for each n the weighting vector $\mathbf{w}^n = (\frac{1}{n}, \ldots, \frac{1}{n})$. For weighted means we need the concept of a weighting triangle [Cal+00, MC97].

Definition 2.31 (*Weighting triangle*) A *weighting triangle* or *triangle of weights* is a set of numbers $w_i^n \in [0, 1]$, for $i = 1, \ldots, n$ and $n \geq 1$, such that: $\sum_{i=1}^{n} w_i^n = 1$, for all $n \geq 1$. It will be represented in the following form

$$1$$
$$w_1^2 \quad w_2^2$$
$$w_1^3 \quad w_2^3 \quad w_3^3$$
$$w_1^4 \quad w_2^4 \quad w_3^4 \quad w_4^4$$
$$\cdots$$

Weighting triangles will be denoted by $\triangle w_i^n$.

Example 2.32 A basic example is the "normalized" Pascal triangle

$$1$$
$$1/2 \qquad 1/2$$
$$1/4 \qquad 2/4 \qquad 1/4$$
$$1/8 \qquad 3/8 \qquad 3/8 \qquad 1/8$$
$$1/16 \quad 4/16 \quad 6/16 \quad 4/16 \quad 1/16$$
$$\cdots \cdots$$

The generic formula for the weighting vector of dimension n in this weighting triangle is [9]

$$\mathbf{w}^n = \frac{1}{2^{n-1}} \left(\binom{n-1}{0}, \binom{n-1}{1}, \ldots, \binom{n-1}{n-1} \right)$$

for each $n \geq 1$.

It is possible to generate weighting triangles in different ways [CM99]:

Proposition 2.33 *The following methods generate weighting triangles:*

1. *Let $\lambda_1, \lambda_2, \ldots \geq 0$ be a sequence of non-negative real numbers such that $\lambda_1 > 0$. Define the weights using*

$$w_i^n = \frac{\lambda_{n-i+1}}{\lambda_1 + \cdots + \lambda_n},$$

   for all $i = 1, \ldots, n$ and $n \geq 1$;
2. *Let N be a strong negation.[10] Generate the weights using N by*

$$w_i^n = N\left(\frac{i-1}{n}\right) - N\left(\frac{i}{n}\right),$$

   for all $i = 1, \ldots, n$ and $n \geq 1$;
3. *Let Q be a monotone non-decreasing function $Q : [0, 1] \to [0, 1]$ such that[11] $Q(0) = 0$ and $Q(1) = 1$. Generate the weights using function Q by [Yag91]*

---

[9]Recall $\binom{n}{m} = \frac{n!}{m!(n-m)!}$.

[10]See Definition 1.51 on p. 18.

[11]This is a so-called quantifier, see p. 120.

$$w_i^n = Q\left(\frac{i}{n}\right) - Q\left(\frac{i-1}{n}\right),$$

for all $i = 1, \ldots, n$ and $n \geq 1$.

Another way to construct weighting triangles is by using fractal structures exemplified below. Such weighting triangles cannot be generated by any of the methods in Proposition 2.33.

Example 2.34 The following two triangles belong to the Sierpinski family [CM99]

$$1$$
$$1 \cdot \tfrac{1}{4} \quad 3 \cdot \tfrac{1}{4}$$
$$1 \cdot \tfrac{1}{4} \quad 3 \cdot \tfrac{1}{4^2} \quad 3^2 \cdot \tfrac{1}{4^2}$$
$$1 \cdot \tfrac{1}{4} \quad 3 \cdot \tfrac{1}{4^2} \quad 3^2 \cdot \tfrac{1}{4^3} \quad 3^3 \cdot \tfrac{1}{4^3}$$
$$\cdots \cdots$$

and

$$1$$
$$1 \cdot \tfrac{1}{9} \quad 8 \cdot \tfrac{1}{9}$$
$$1 \cdot \tfrac{1}{9} \quad 8 \cdot \tfrac{1}{9^2} \quad 8^2 \cdot \tfrac{1}{9^2}$$
$$1 \cdot \tfrac{1}{9} \quad 8 \cdot \tfrac{1}{9^2} \quad 8^2 \cdot \tfrac{1}{9^3} \quad 8^3 \cdot \tfrac{1}{9^3}$$
$$\cdots \cdots$$

In general, given two real numbers a, b such that $a < b$ and $\frac{a}{b} \neq \frac{b-a}{b}$, it is possible to define a weighting triangle by [CM99]

$$1$$
$$\tfrac{a}{b} \quad \tfrac{b-a}{b}$$
$$\tfrac{a}{b} \quad (\tfrac{b-a}{b})(\tfrac{a}{b}) \quad (\tfrac{b-a}{b})^2$$
$$\tfrac{a}{b} \quad (\tfrac{b-a}{b})(\tfrac{a}{b}) \quad (\tfrac{b-a}{b})^2(\tfrac{a}{b}) \quad (\tfrac{b-a}{b})^3$$
$$\cdots$$

This weighting triangle also belongs to the Sierpinski family, and a generic formula for the weights is

$$w_i^n = \frac{a}{b} \cdot \left(\frac{b-a}{b}\right)^{i-1}, \quad i = 1, \ldots, n-1 \quad \text{and} \quad w_n^n = \left(\frac{b-a}{b}\right)^{n-1}.$$

The question of stability of weighted quasi-arithmetic means with respect to the number of arguments was recently addressed in [BJ13, GRMRB14, RGMR13], see Sect. 6.5. This method connects the weighting vectors of various dimensions through

$$w_i^n = (1 - w_n^n)w_i^{n-1}, i = 1, 2, \ldots, n - 1,$$

as shown in Corollary 6.89.

Let us now mention two characterization theorems, which relate continuity, strict monotonicity and the properties of decomposability and bisymmetry to the class of weighted quasi-arithmetic means.

Theorem 2.35 (Kolmogorov-Nagumo) *An extended aggregation function F is continuous, decomposable,[12] and strictly monotone if and only if there is a monotone bijection $g : \mathbb{I} \to \mathbb{I}$, such that for each $n > 1$, f_n is a quasi-arithmetic mean M_g.*

The next result is a generalized version of Kolmogorov and Nagumo characterization, due to Aczél [Acz48].

Theorem 2.36 *An extended aggregation function F is continuous, bisymmetric,[13] idempotent, and strictly monotone if and only if there is a monotone bijection $g : \mathbb{I} \to \mathbb{I}$, and a weighting triangle $\triangle w_i^n$ with all positive weights, so that for each $n > 1$, f_n is a weighted quasi-arithmetic mean $M_{\mathbf{w}^n, g}$ (i.e., $f_n = M_{\mathbf{w}^n, g}$).*

Note 2.37 If we omit the strict monotonicity of F, we recover the class of non–strict means introduced by Fodor and Marichal [FM97].

2.3.6 Weights Dispersion

An important quantity associated with weighting vectors is their dispersion, also called entropy.

Definition 2.38 (*Weights dispersion (entropy)*) For a given weighting vector $\mathbf{w}$ its measure of dispersion (entropy) is

$$Disp(\mathbf{w}) = -\sum_{i=1}^{n} w_i \log w_i, \tag{2.9}$$

with the convention $0 \cdot \log 0 = 0$.

The weights dispersion measures the degree to which a weighted aggregation function f takes into account all inputs. For example, in the case of weighted means, among the two weighting vectors $\mathbf{w}_1 = (0, 1)$ and $\mathbf{w}_2 = (0.5, 0.5)$ the second one may be preferable, since the corresponding weighted mean uses information from two sources rather than a single source, and is consequently less sensitive to input inaccuracies.

---

[12]See Definition 1.43. Continuity and decomposability imply idempotency.
[13]See Definition 1.44.

A useful normalization of this measure is

$$-\frac{1}{\log n} \sum_{i=1}^{n} w_i \log w_i.$$

Along with the orness value (p. 57), the weights entropy is an important parameter in choosing weighting vectors of both quasi-arithmetic means and OWA functions (see p. 106).

There are also other entropy measures (e.g., Rényi entropy) frequently used in studies of weighted aggregation functions, e.g., [Yag95].[14]

2.3.7 How to Choose Weights

Choosing weights of weighted arithmetic means

In each application the weighting vector of the weighted arithmetic mean will be different. We examine the problem of choosing the weighting vector which fits best some empirical data, the pairs $(\mathbf{x}_k, y_k)$, $k = 1, \ldots, K$. Our goal is to determine the best weighted arithmetic mean that minimizes the norm of the differences between the predicted $(f(\mathbf{x}_k))$ and observed (y_k) values. We will use the least squares or least absolute deviation criterion, as discussed on p. 34. In the first case we have the following optimization problem

$$\min \quad \sum_{k=1}^{K} \left(\sum_{i=1}^{n} w_i x_{ik} - y_k \right)^2 \qquad (2.10)$$

$$\text{s.t.} \quad \sum_{i=1}^{n} w_i = 1, w_i \geq 0, i = 1, \ldots, n.$$

It is easy to recognize a standard quadratic programming problem (QP), with a convex objective function. There are plenty of standard methods for its solution.

We mentioned on p. 34 that one can use a different fitting criterion, such as the least absolute deviation (LAD) criterion, which translates into a different optimization problem

$$\min \quad \sum_{k=1}^{K} \left| \sum_{i=1}^{n} w_i x_{ik} - y_k \right| \qquad (2.11)$$

$$\text{s.t.} \quad \sum_{i=1}^{n} w_i = 1, w_i \geq 0, i = 1, \ldots, n.$$

This problem is subsequently converted into a linear programming problem (LP).

---

[14]These measures of entropy can be obtained by relaxing the subadditivity condition which characterizes Shannon entropy [TY05].

Particular attention is needed for the case when the quadratic (resp. linear) programming problems have singular matrices. Such cases appear when there are few data, or when the input values are linearly dependent. While modern quadratic and linear programming methods accommodate for such cases, the minimization problem will typically have multiple solutions. An additional criterion is then used to select one of these solutions, and typically this criterion relates to the dispersion of weights, or the entropy [Tor02], as defined in Definition 2.38. Torra [Tor02] proposes to solve an auxiliary univariate optimization problem to maximize weights dispersion, subject to a given value of (2.10).

Specifically, one solves the problem

$$\min \qquad \sum_{i=1}^{n} w_i \log w_i \qquad\qquad (2.12)$$

$$\text{s.t.} \ \sum_{i=1}^{n} w_i = 1, w_i \geq 0, i = 1, \ldots, n,$$

$$\sum_{k=1}^{K} \left(\sum_{i=1}^{n} w_i x_{ik} - y_k \right)^2 = A,$$

where A is the value of the solution of problem (2.10). It turns out that if Problem (2.10) has multiple solutions, they are expressed in parametric form as linear combinations of one another. Further, the objective function in (2.12) is convex. Therefore problem (2.12) is a convex programming problem subject to linear constraints, and it can be solved by standard methods, see [Tor02].

A different additional criterion is the so-called measure of orness (discussed in Sect. 2.1), which measures how far a given averaging function is from the max function, which is the weakest disjunctive function. It is applicable to any averaging function, and is frequently used as an additional constraint or criterion when constructing these functions. However, for any weighted arithmetic mean, the measure of orness is always $\frac{1}{2}$, therefore this parameter does not discriminate between arithmetic means with different weighting vectors.

Preservation of ordering of the outputs

We recall from Sect. 1.6, p. 32, that sometimes one not only has to fit an aggregation function to the numerical data, but also preserve the ordering of the outputs. That is, if $y_j \leq y_k$ then we expect $f(\mathbf{x}_j) \leq f(\mathbf{x}_k)$.

First, arrange the data, so that the outputs are in non-decreasing order, i.e., $y_k \leq y_{k+1}, k = 1, \ldots, K - 1$. Define the additional linear constraints

$$\langle \mathbf{x}_{k+1} - \mathbf{x}_k, \mathbf{w} \rangle = \sum_{i=1}^{n} w_i (x_{i,k+1} - x_{ik}) \geq 0,$$

$k = 1, \ldots, K - 1$. We add the above constraints to problem (2.10) or (2.11) and solve it. The addition of an extra $K - 1$ constraints neither changes the structure of the optimization problem, nor drastically affects its complexity.

Choosing weights of weighted quasi-arithmetic means

Consider the case of weighted quasi-arithmetic means, when a given generating function g is given. As before, we have a data set $(\mathbf{x}_k, y_k)$, $k = 1, \ldots, K$, and we are interested in finding the weighting vector $\mathbf{w}$ that fits the data best. When we use the least squares, as discussed on p. 34, we have the following optimization problem

$$\min \sum_{k=1}^{K} \left(g^{-1} \left(\sum_{i=1}^{n} w_i g(x_{ik}) \right) - y_k \right)^2 \tag{2.13}$$

$$\text{s.t. } \sum_{i=1}^{n} w_i = 1, w_i \geq 0, i = 1, \ldots, n.$$

This is a nonlinear optimization problem, but it can be reduced to quadratic programming by the following artifice. Let us apply g to y_k and the inner sum in (2.13). We obtain

$$\min \sum_{k=1}^{K} \left(\sum_{i=1}^{n} w_i g(x_{ik}) - g(y_k) \right)^2 \tag{2.14}$$

$$\text{s.t. } \sum_{i=1}^{n} w_i = 1, w_i \geq 0, i = 1, \ldots, n.$$

We recognize a standard quadratic programming problem (QP), with a convex objective function. This approach was discussed in detail in [Bel03, Bel05, BMV04, Tor02]. There are plenty of standard methods of solution.

If one uses the least absolute deviation (LAD) criterion (p. 34) we obtain a different optimization problem

$$\min \sum_{k=1}^{K} \left| \sum_{i=1}^{n} w_i g(x_{ik}) - g(y_k) \right| \tag{2.15}$$

$$\text{s.t. } \sum_{i=1}^{n} w_i = 1, w_i \geq 0, i = 1, \ldots, n.$$

This problem is subsequently converted into a linear programming problem (LP).

As in the case of weighted arithmetic means, in the presence of multiple optimal solutions, one can use an additional criterion of the dispersion of weights [Tor02].

One recent result relates the choice of the weighting vectors to stability of the aggregation functions with respect to changes to the number of arguments [BJ13, BJGRM15, GMRR12, GRMRB14]. Section 6.5 discusses this type of stability, and here we just mention that the notion of stability can be translated into the (nonlinear) constraints on the coefficients of the weighting vectors.

Preservation of ordering of the outputs

Similarly to what we did for weighted arithmetic means (see also Sect. 1.6, p. 32), we will require that the ordering of the outputs is preserved, i.e., if $y_j \leq y_k$ then we expect $f(\mathbf{x}_j) \leq f(\mathbf{x}_k)$. We arrange the data, so that the outputs are in non-decreasing order, $y_k \leq y_{k+1}, k = 1, \ldots, K - 1$. Then we define the additional linear constraints

$$\langle \mathbf{g}(x_{k+1}) - g(\mathbf{x}_k), \mathbf{w} \rangle = \sum_{i=1}^{n} w_i(g(x_{i,k+1}) - g(x_{ik})) \geq 0,$$

$k = 1, \ldots, K - 1$. We add the above constraints to problem (2.14) or (2.15) and solve it. The addition of extra $K - 1$ constraints does not change the structure of the optimization problem, nor drastically affects its complexity.

Choosing generating functions

Consider now the case when the generating function g is also unknown, and hence needs to be found based on the data. We study two cases: (a) when g is given algebraically, with one or more unknown parameters to estimate (e.g., $g_p(t) = t^p$, p unknown) and (b) when no specific algebraic form of g is given.

In the first case we solve the problem

$$\min_{p,\mathbf{w}} \sum_{k=1}^{K} \left(\sum_{i=1}^{n} w_i g_p(x_{ik}) - g_p(y_k) \right)^2 \tag{2.16}$$

$$\text{s.t.} \sum_{i=1}^{n} w_i = 1, w_i \geq 0, i = 1, \ldots, n,$$

$$\text{conditions on } p.$$

While this general optimization problem is non-convex and nonlinear (i.e., difficult to solve), we can convert it to a bi-level optimization problem

$$\min_{p} \left[\min_{\mathbf{w}} \sum_{k=1}^{K} \left(\sum_{i=1}^{n} w_i g_p(x_{ik}) - g_p(y_k) \right)^2 \right] \tag{2.17}$$

$$\text{s.t.} \sum_{i=1}^{n} w_i = 1, w_i \geq 0, i = 1, \ldots, n,$$

$$\text{plus conditions on } p.$$

The problem at the inner level is the same as (2.14) with a fixed g_p, which is a QP problem. At the outer level we have a global optimization problem with respect to a single parameter p. It is solved by using one of the standard methods. We recommend deterministic Pijavski-Shubert method.

Example 2.39 Determine the weights and the generating function of a family of weighted power means. We have $g_p(t) = t^p$, and hence solve bi-level optimization problem

$$\min_{p\in[-\infty,\infty]} \left[\min_{\mathbf{w}} \sum_{k=1}^{K} \left(\sum_{i=1}^{n} w_i x_{ik}^p - y_k^p \right)^2 \right] \tag{2.18}$$

$$\text{s.t. } \sum_{i=1}^{n} w_i = 1, w_i \geq 0, i = 1, \ldots, n.$$

Of course, for numerical purposes we need to limit the range for p to a finite interval, and treat all the limiting cases $p \to \pm\infty$, $p \to 0$ and $p \to -1$.

A different situation arises when the parametric form of g is not given. The approach proposed in [Bel03] is based on approximation of g with a monotone linear spline (see [Bel00]), as

$$g(t) = \sum_{j=1}^{J} c_j B_j(t), \tag{2.19}$$

where B_j are appropriately chosen basis functions, and c_j are spline coefficients. The monotonicity of g is ensured by imposing linear restrictions on spline coefficients, in particular non-negativity, as in [Bel00]. Further, since the generating function is defined up to an arbitrary linear transformation, one has to fix a particular g by specifying two interpolation conditions, like $g(a) = 0$, $g(b) = 1$, $a, b \in]0, 1[$, and if necessary, properly model asymptotic behavior if $g(0)$ or $g(1)$ are infinite.

After rearranging the terms of the sum, the problem of identification becomes (subject to linear conditions on $\mathbf{c}, \mathbf{w}$)

$$\min_{\mathbf{c},\mathbf{w}} \sum_{k=1}^{K} \left(\sum_{j=1}^{J} c_j \left[\sum_{i=1}^{n} w_i B_j(x_{ik}) - B_j(y_k) \right] \right)^2. \tag{2.20}$$

For a fixed $\mathbf{c}$ (i.e., fixed g) we have a quadratic programming problem to find $\mathbf{w}$, and for a fixed $\mathbf{w}$, we have a quadratic programming problem to find $\mathbf{c}$. However if we consider both $\mathbf{c}, \mathbf{w}$ as variables, we obtain a difficult global optimization problem. We convert it into a bi-level optimization problem

$$\min_{\mathbf{c}} \min_{\mathbf{w}} \sum_{k=1}^{K} \left(\sum_{j=1}^{J} c_j \left[\sum_{i=1}^{n} w_i B_j(x_{ik}) - B_j(y_k) \right] \right)^2, \tag{2.21}$$

where at the inner level we have a QP problem and at the outer level we have a non-linear problem with multiple local minima. When the number of spline coefficients J is not very large (<10), this problem can be efficiently solved by using deterministic global optimization methods from Sect. 1.7.4. If the number of variables is small and J is large, then reversing the order of minimization (i.e., using $\min_{\mathbf{w}} \min_{\mathbf{c}}$) is more efficient.

2.4 Other Means

Besides weighted quasi-arithmetic means, there exist very large families of other means, some of which we will mention in this section. A comprehensive reference to the topic of means is [Bul03]. However we must note that not all these means are monotone functions, so technically they are not aggregation functions. Still some members of these families are aggregation functions, and we will mention the sufficient conditions for monotonicity, if available.

In Chap. 7 we will formulate less restrictive monotonicity conditions and will examine some of the mentioned means in the context of weak monotonicity. We will also present generalisations and special cases of some of the means mentioned below in Chap. 6. Most of the mentioned means do not require $\mathbf{x} \in [0, 1]^n$, but we will assume $\mathbf{x} \geq \mathbf{0}$. The values at $\mathbf{x} = \mathbf{0}$ are defined by continuity.

2.4.1 Gini Means

Definition 2.40 (*Lehmer mean*) Let $q \in \mathbb{R}$ and $\mathbf{w} \in \mathbb{R}^n$, $\mathbf{w} \geq 0$. Weighted Lehmer mean is the function

$$L_{\mathbf{w}}^q(\mathbf{x}) = \frac{\sum_{i=1}^n w_i x_i^{q+1}}{\sum_{i=1}^n w_i x_i^q} \tag{2.22}$$

Definition 2.41 (*Gini mean*) Let $p, q \in \mathbb{R}$ and $\mathbf{w} \in \mathbb{R}^n$, $\mathbf{w} \geq 0$. Weighted Gini mean is the function

$$G_{\mathbf{w}}^{p,q}(\mathbf{x}) = \begin{cases} \left(\dfrac{\sum_{i=1}^n w_i x_i^p}{\sum_{i=1}^n w_i x_i^q} \right)^{1/p-q} & \text{if } p \neq q, \\[4ex] \left(\prod_{i=1}^n x_i^{w_i x_i^p} \right)^{1/\sum_{i=1}^n w_i x_i^p} & \text{if } p = q. \end{cases} \tag{2.23}$$

Properties

- $G_{\mathbf{w}}^{p,q} = G_{\mathbf{w}}^{q,p}$, so we assume $p \geq q$;
- $\lim\limits_{p \to q} G_{\mathbf{w}}^{p,q} = G_{\mathbf{w}}^{q,q}$;

- $\lim_{p \to \infty} G_{\mathbf{w}}^{p,q}(\mathbf{x}) = \max(\mathbf{x})$;
- $\lim_{q \to -\infty} G_{\mathbf{w}}^{p,q}(\mathbf{x}) = \min(\mathbf{x})$;
- If $p_1 \leq p_2, q_1 \leq q_2$, then $G_{\mathbf{w}}^{p_1,q_1} \leq G_{\mathbf{w}}^{p_2,q_2}$.

Special cases

- Setting $q = 0$ and $p \geq 0$ leads to weighted power means $G_{\mathbf{w}}^{p,0} = M_{\mathbf{w},[p]}$.
- Setting $p = 0$ and $q \leq 0$ also leads to weighted power means $G_{\mathbf{w}}^{0,q} = M_{\mathbf{w},[q]}$.
- Setting $q = p - 1$ leads to *counter-harmonic* means, also called *Lehmer means*.
  For example, when $n = 2$, $G_{(\frac{1}{2},\frac{1}{2})}^{q+1,q}(x_1, x_2) = \frac{x_1^{q+1}+x_2^{q+1}}{x_1^q+x_2^q}$, $q \in \mathbb{R}$.
- When $q = 1$ we obtain the *contraharmonic* mean $G_{(\frac{1}{2},\frac{1}{2})}^{2,1}(x_1, x_2) = \frac{x_1^2+x_2^2}{x_1+x_2}$.

Note 2.42 Counter-harmonic means (and hence Gini means in general) are not monotone, except in some special cases (e.g., power means) (Figs. 2.12, 2.13, 2.14 and 2.15).

2.4.2 Bonferroni Means

Definition 2.43 (*Bonferroni mean*) [Bon50] Let $p, q \geq 0$ and $\mathbf{x} \geq \mathbf{0}$. Bonferroni mean is the function

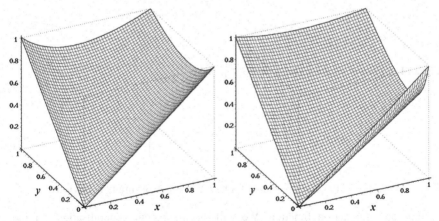

Fig. 2.12 3D plots of weighted Gini means $G_{(\frac{1}{2},\frac{1}{2})}^{2,1}$ and $G_{(\frac{1}{5},\frac{4}{5})}^{2,1}$ (both are weighted contraharmonic means). Note lack of monotonicity

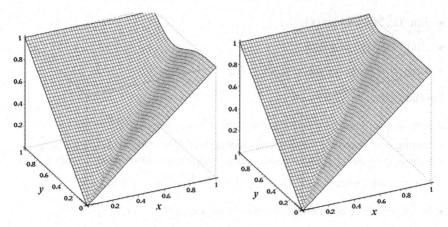

Fig. 2.13 3D plots of weighted Gini means $G^{5,4}_{(\frac{1}{5},\frac{4}{5})}$ and $G^{10,9}_{(\frac{1}{5},\frac{4}{5})}$ (both are weighted counter-harmonic means)

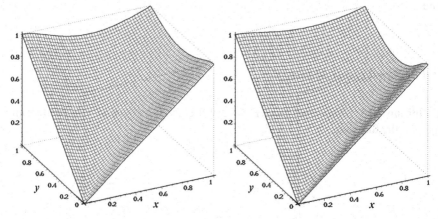

Fig. 2.14 3D plot of weighted Gini means $G^{2,2}_{(\frac{1}{2},\frac{1}{2})}$ and $G^{2,2}_{(\frac{1}{3},\frac{2}{3})}$

$$B^{p,q}(\mathbf{x}) = \left(\frac{1}{n(n-1)} \sum_{i,j=1,i\neq j}^{n} x_i^p x_j^q \right)^{1/(p+q)}. \tag{2.24}$$

Extension to $B^{p,q,r}(\mathbf{x})$ and products of a larger number of inputs is obvious.

It is an aggregation function. We will discuss recent generalisations of the Bonferroni mean in Sect. 6.4 (Figs. 2.16 and 2.17).

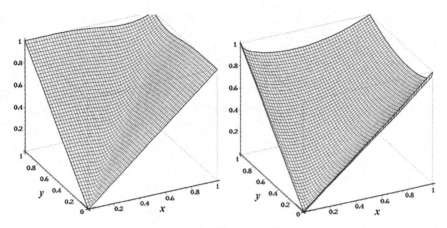

Fig. 2.15 3D plots of weighted Gini means $G^{5,5}_{(\frac{1}{2},\frac{1}{2})}$ and $G^{3,0.5}_{(\frac{1}{2},\frac{1}{2})}$

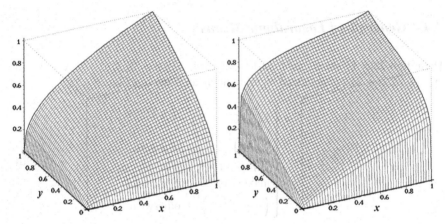

Fig. 2.16 3D plots of Bonferroni means $B^{3,2}$ and $B^{10,2}$

2.4.3 Heronian Mean

Definition 2.44 (*Heronian mean*) Heronian mean is the function

$$HR(\mathbf{x}) = \frac{2}{n(n+1)} \sum_{i=1}^{n} \sum_{j=i}^{n} \sqrt{x_i x_j}. \qquad (2.25)$$

It is an aggregation function. For $n = 2$ we have $HR = \frac{1}{3}(2M + G)$.

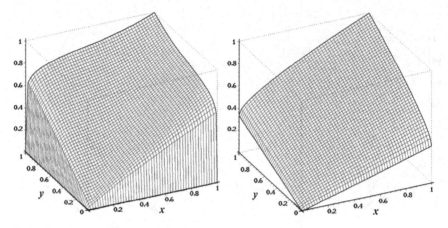

Fig. 2.17 3D plots of Bonferroni mean $B^{5,0.5}$ and the Heronian mean

2.4.4 Generalized Logarithmic Means

Definition 2.45 (*Generalized logarithmic mean*) Let $n = 2$, $x, y > 0$, $x \neq y$ and $p \in [-\infty, \infty]$. The generalized logarithmic mean is the function

$$
L^p(x, y) = \begin{cases}
\frac{y-x}{\log y - \log x}, & \text{if } p = -1, \\[2mm]
\frac{1}{e}\left(\frac{y^y}{x^x}\right)^{1/(y-x)}, & \text{if } p = 0, \\[2mm]
\min(x, y), & \text{if } p = -\infty, \\
\max(x, y), & \text{if } p = \infty, \\[2mm]
\left(\frac{y^{p+1} - x^{p+1}}{(p+1)(y-x)}\right)^{1/p} & \text{otherwise.}
\end{cases}
\tag{2.26}
$$

For $x = y$, $L^p(x, x) = x$.

Note 2.46 Generalized logarithmic means are also called Stolarsky means, sometimes L^p is called L^{p+1}.

Note 2.47 The generalized logarithmic mean is symmetric. The limiting cases $x = 0$ depend on p, although $L^p(0, 0) = 0$.

Special cases

- The function $L^0(x, y)$ is called *identric* mean;
- $L^{-2}(x, y) = G(x, y)$, the geometric mean;
- L^{-1} is called the logarithmic mean;
- $L^{-1/2}$ is the power mean with $p = -1/2$;
- L^1 is the arithmetic mean;
- Only $L^{-1/2}$, L^{-2} and L^1 are quasi-arithmetic means.

Note 2.48 For each value of p the generalized logarithmic mean is strictly increasing in x, y, hence they are aggregation functions (Figs. 2.18, 2.19 and 2.20).

Note 2.49 Generalized logarithmic means can be extended for n arguments, see Sect. 2.4.9.

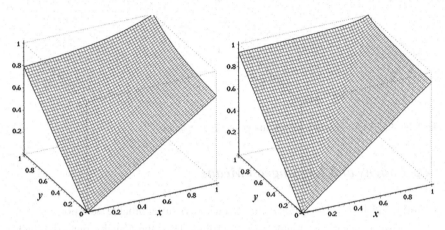

Fig. 2.18 3D plots of generalized logarithmic means L^{10} and L^{50}

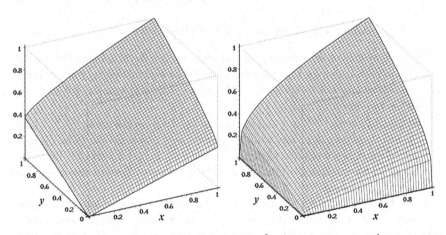

Fig. 2.19 3D plots of generalized logarithmic means L^{0}(identric mean) and L^{-1} (logarithmic mean)

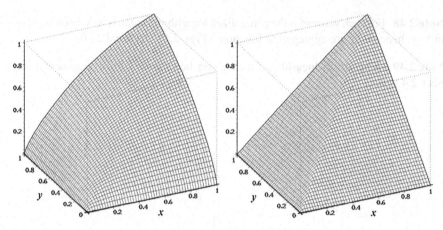

Fig. 2.20 3D plots of generalized logarithmic means L^{-5} and L^{-100}

2.4.5 Cauchy and Lagrangean Means

Definition 2.50 (*Cauchy mean*) Let us take two differentiable functions $g, h :$ $\mathbb{I} \to \mathbb{R}$ such that $g' \neq 0$ and $\frac{g'}{h'}$ is invertible. Then the Cauchy mean is given for $x \neq y$ by

$$C^{g,h}(x, y) = \left(\frac{g'}{h'}\right)^{-1} \left(\frac{g(x) - g(y)}{h(x) - h(y)}\right). \tag{2.27}$$

For $x = y$ the definition is augmented with $C^{g,h}(x, x) = x$.

The Cauchy means are continuous, symmetric and strictly increasing. The special case of $h = Id$ is called the Lagrangean mean L^g. The generalized logarithmic meas are Lagrangean means L^g with $g(t) = t^{p+1}$, $p \neq -1, 0$, $g(t) = \log(t)$ for $p = -1$, and $g(t) = t \cdot \log t$ for $p = 0$.

The Cauchy mean $C^{g,h}$ is a φ-transform of the Lagrangean mean $L^{g \circ h^{-1}}$ with $\varphi = h$. Two Lagrangean means with the generators g and h are the same mean if and only if the generators are related as $h' = ag' + b$, $a, b \in \mathbb{R}$, which is the same relation as for the generators of the quasi-arithmetic means. For this reason we can assume that g is convex.

Some Lagrangean (resp. Cauchy) means are quasi-arithmetic means (e.g., the arithmetic and geometric means), but some are not. For instance the harmonic mean is not Lagrangean, and the logarithmic mean is not quasi-arithmetic. The Cauchy mean $C^{g,h}$ is the quasi-arithmetic mean with the generator h if $h = g'/h'$. Homogeneous Lagrangean means are necessarily generalized logarithmic means. The Lagrangean mean generated by $g(t) = t^{p+1}$ is called Stolarsky mean. The Cauchy mean generated by two power functions $g(t) = t^p$, $h(t) = t^s$ is called the extended mean (sometimes also referred to as Stolarsky mean). For more details about these means refer to [Bul03, Mat11].

2.4.6 Mean of Bajraktarevic

Definition 2.51 (*Mean of Bajraktarevic*) Let $\mathbf{w}(t) = (w_1(t), \ldots, w_n(t))$ be a vector of weight functions $w_i : \mathbb{I} \to [0, \infty[$, and let $g : \mathbb{I} \to [-\infty, \infty]$ be a strictly monotone function. The mean of Bajraktarevic is the function

$$f(\mathbf{x}) = g^{-1} \left(\frac{\sum\limits_{i=1}^{n} w_i(x_i) g(x_i)}{\sum\limits_{i=1}^{n} w_i(x_i)} \right). \tag{2.28}$$

2.4.7 Mixture Functions

The Bajraktarevic mean is also called a *mixture* function [MPR03] when $g(t) = t$. The function g is called the generating function of this mean. If $w_i(t) = w_i$ are constants for all $i = 1, \ldots, n$, it reduces to the quasi-arithmetic mean. The special case of Gini mean $G^{p,q}$ is obtained by taking $w_i(t) = w_i t^q$ and $g(t) = t^{p-q}$ if $p > q$, or $g(t) = \log(t)$ if $p = q$.

The mean of Bajraktarevic is not generally an aggregation function because it fails the monotonicity condition. The following sufficient condition for monotonicity of mixture functions has been established in [MPR03].

Let weight functions $w_i(t) > 0$ be differentiable and monotone increasing, and $g(t) = t$. If $w_i'(t) \leq w_i(t)$ for all $t \in \mathbb{I}$ and all $i = 1, \ldots, n$, then f in (2.28) is monotone increasing (i.e., an aggregation function).

The other conditions were established in [MS06, MSV08]. Let $\mathbb{I} = [0, 1]$. Then sufficient conditions for monotonicity are: $w(x) \geq w'(x)(1 - x)$ for all $x \in [0, 1]$, or, if we fix the dimension n of the domain, $\frac{w^2(x)}{(n-1)w(1)} + w(x) \geq w'(x)(1 - x), x \in [0, 1], n > 1$.

Analogous results have been obtained for decreasing weighting functions using duality (with respect to the standard negation). Taking the dual weighting function $w^d(x) = w(1 - x)$, the resulting mixture function is the dual to f_w; that is, $f_{w^d} = 1 - f_w$. Additionally, f_w is invariant to scaling of the weight functions (i.e., $f_{\alpha w} = f_w \, \forall \alpha \in \mathbb{R} \setminus \{0\}$).

In this context we also state a result which ensures that by making the weights of an averaging function dependent on the arguments, as in mixture functions, we do not change the averaging behaviour of that function.

Proposition 2.52 *Let $f_\mathbf{w}$ be an averaging aggregation function with constant weighting vector $\mathbf{w}$. Then the function $f_{\mathbf{u}(\mathbf{x})}$ obtained from $f_\mathbf{w}$ by allowing the weights to depend on $\mathbf{x}$, and using normalisation of weights $w_i = \frac{u_i(\mathbf{x})}{\sum u_i(\mathbf{x})}$ (and assuming $u_i \geq 0$) is bounded by* $\min(\mathbf{x}) \leq f_{\mathbf{u}(\mathbf{x})}(\mathbf{x}) \leq \max(\mathbf{x})$.

Proof For every fixed $\mathbf{x}$ the weights $w_i(\mathbf{x}) = \frac{u_i(\mathbf{x})}{\sum u_i(\mathbf{x})}$ are fixed, non-negative and add to one. Then the value of $f_{\mathbf{w}}(\mathbf{x})$ with such weights is bounded by the mimimum and maximum, because $f_{\mathbf{w}}$ is averaging (for every choice of $\mathbf{w}$).   □

2.4.8 Compound Means

In this section we will be talking about bivariate means, although extensions to triviariate and multivatiate functions are possible.

Consider the arithmetic and geometric means M and G and take (strictly) positive arguments $x, y > 0$. Consider the sequence defined by $x_0 = x$, $y_0 = y$ and

$$x_n = G(x_{n-1}, y_{n-1}), \quad y_n = M(x_{n-1}, y_{n-1}), \quad n > 0. \tag{2.29}$$

Definition 2.53 (*Arithmetico-geometric means*) The mean $G \otimes M(x, y)$ is the common value of the limits $\lim_{n \to \infty} x_n = \lim_{n \to \infty} y_n$, and it is called the Arithmetico-geometric mean, or AGM.

The iterative process which leads to the sequences $x_n, x_n, n = 0, 1, \ldots$ is called Gaussian iterations. A detailed analysis of the AGM is given in [BB87], see also [Bul03]. Among the properties of the AGM we list the following.

- $G \otimes M$ is symmetric;
- $G \otimes M(x, y) = G \otimes M(x_n, y_n)$ for all $n \geq 0$;
- $G \otimes M$ is (positively) homogeneous;
- $G \leq G \otimes M \leq M$.

One can extend the AGM to other means A and B using the following definition.

Definition 2.54 (*Compound means*) Let A and B be bivariate means. Then define the sequences for $x, y > 0$ by $x_0 = x$, $y_0 = y$ and

$$x_n = A(x_{n-1}, y_{n-1}), \quad y_n = B(x_{n-1}, y_{n-1}), \quad n > 0.$$

If both the limits $\lim_{n \to \infty} x_n$ and $\lim_{n \to \infty} y_n$ are equal, then this common value is called the compound mean $A \otimes B$.

Of course, the AGM is the compound mean with $A = G$ and $B = M$. Another interesting case is $H \otimes M = G$. On the other hand $\max \otimes \min$ does not exist (unless $x = y$). If both A and B are symmetric, then $A \otimes B = B \otimes A$ if the compound mean exists. The compound mean is not associative, because $A \otimes (B \otimes B) = A \otimes B$.

The following result establishes the existence of the compound means (see [Bul03], p. 415).

Proposition 2.55 *If two means A and B are continuous and satisfy* min $< A \le$
$B <$ max *(i.e., the means are strictly internal and comparable), then their compound
means exist.*

In particular the compounds of the power means exist.

2.4.9 Extending Bivariate Means to More Than Two Arguments

Some of the means, like the logarithmic mean, are defined for two arguments only. An
interesting question is how to extend these definitions to more than two arguments.
There are a few methods here, and two are presented in the sequel.

Let us make a change of variables and set $x_i = e^{u_i}$. Recall that

$$e^u = 1 + u + \frac{u^2}{2!} + \frac{u^3}{3!} \cdots$$

We easily obtain for the arithmetic and geometric means respectively

$$M(x_1, x_2) = 1 + \frac{u_1 + u_2}{2} + \frac{u_1^2 + u_2^2}{2 \cdot 2!} + \frac{u_1^3 + u_2^3}{2 \cdot 3!} + \cdots ,$$

$$G(x_1, x_2) = \sqrt{e^{u_1} e^{u_2}} = 1 + \frac{u_1 + u_2}{2} + \frac{(u_1 + u_2)^2}{2^2 \cdot 2!} + \frac{(u_1 + u_2)^3}{2^3 \cdot 3!} + \cdots .$$

For the logaritmic mean L^{-1} we obtain

$$L^{-1}(x_1, x_2) = \frac{e^{u_1} - e^{u_2}}{u_1 - u_2} = 1 + \frac{u_1 + u_2}{2} + \frac{u_1^2 + u_1 u_2 + u_2^2}{3 \cdot 2!}$$
$$+ \frac{u_1^3 + u_1^2 u_2 + u_1 u_2^2 + u_2^3}{4 \cdot 3!} + \cdots .$$

The generic form of the polynomial term for $m > 1$ is

$$B_m u_1^m + B_{m-1} u_1^{m-1} u_2 + B_{m-2} u_1^{m-2} u_2^2 + \cdots + B_0 u_2^m$$

divided by the sum of its coefficients $B_m + B_{m+1} + \cdots + B_0$. These coefficients
characterize each mean completely. For the arithmetic mean we have $B_0 = B_m = 1$
and the rest are zeros. For the geometric mean $B_i = \binom{m}{i}$, the binomial coefficients,
and for the logarithmic mean all $B_i = 1$.

It is proposed to generalize the logarithmic mean by preserving the simple structure of the coefficients B_i [Mus02], so that

$$L^{-1}(x_1, \ldots, x_n) = 1 + \frac{u_1 + \cdots + u_n}{n} \tag{2.30}$$
$$+ \frac{u_1^2 + u_1 u_2 + \ldots u_1 u_n + u_2^2 + \cdots + u_n^2}{\binom{n+1}{2} \cdot 2!} + \cdots$$
$$+ \frac{u_1^m + u_1^{m-1} u_2 + \ldots u_1^{m-1} u_n + \cdots + u_n^m}{\binom{n+m-1}{m} \cdot m!} + \cdots$$

In this series the polynomial term has expression

$$P(n, m) = \sum_{i_1 + i_2 + \cdots + i_n = m} u_1^{i_1} u_2^{i_2} \ldots u_n^{i_n}$$

so that all coefficients B are equal to 1. The numbers $\binom{n+m-1}{m}$ correspond to the number of summands. Mustonen [Mus02] transformed (2.30) to a closed form

$$L^{-1}(x_1, \ldots, x_n) = (n-1)! \sum_{i=1}^{n} \frac{x_i}{\prod_{j \neq i} \log(x_i/x_j)} \tag{2.31}$$

when all x_i are mutually different positive numbers (see also [Mer04] for a proof). This extension is the same as in [Neu94]. Furthermore, by comparing the above expressions to the divided differences, Mustonen [Mus02] provides a recursive formula

$$L^{-1}(x_1, \ldots, x_n) = (n-1) \frac{L^{-1}(x_2, \ldots, x_n) - L^{-1}(x_1, \ldots, x_{n-1})}{\log x_n/x_1} \tag{2.32}$$

for $n = 2, 3, \ldots$

There are alternative formulas for extending the logarithmic mean, for example [Pit85]:

$$L^{-1}(x_1, \ldots, x_n) = \left((n-1) \sum_{i=1}^{n} \frac{x_i^{n-2} \log x_i}{\prod_{j \neq i} \log(x_i/x_j)} \right)^{-1}.$$

The extension based on the method of Mustonen was also applied to the generalized logarithmic mean L^p in (2.26) (and even to a more general case of Stolarsky-Tobey means) by using

$$L^p(x, y) = \left(\frac{L^{-1}(x^{p+1}, y^{p+1})}{L^{-1}(x, y)} \right)^{\frac{1}{p}},$$

and then writing the left-hand side as the ratio of n-variate logarithmic means.

A different approach, which also includes a way of assigning the weights to the inputs, was recently presented in [Duj15] and then extended in [BD15, DB15]. It is based on multiple invocations of the bivariate function in order to approximate the weighted n-variate idempotent function with any desired accuracy. This is similar to recursive functions (Definition 1.41) but invocations of the bivariate function happen in a different order which allows one to introduce a weighting vector.

Suppose we want to approximate (or construct for this matter) a weighted n-variate idempotent function with the weighting vector $\mathbf{w}$, by using only the unweighted bivariate idempotent function. For this let us fix a number of levels $L \geq 1$ and create an auxiliary vector of arguments $\mathbf{X}$ of size 2^L, in which the inputs from $\mathbf{x}$ are repeated according to the weights. That is, x_1 is taken k_1 times, x_2 is taken k_2 times and so on, and $\frac{k_1}{2^L} \approx w_1$, $\frac{k_2}{2^L} \approx w_2$, ..., and $\sum k_i = 2^L$. One way of doing so is to take $k_i = \lfloor w_i 2^L + \frac{1}{2} \rfloor$, $i = 1, \ldots, n-1$ and $k_n = 2^L - k_1 - k_2 - \cdots - k_{n-1}$.

This process of replicating each argument x_i a suitable number of times consistent with the weight w_i was presented earlier in [CMY04].

Next, let us build a binary tree with L levels presented in Fig. 2.21, where at each node a value is produced by aggregating the values of two children nodes with the given bivariate averaging function f (denoted by B in this figure). We start from the leaves of the tree which contain the elements of the vector $\mathbf{X}$. The value at the root node will be the desired output of the n-variate weighted function.

Of course, the output value will depend on the order in which the arguments x_i are processed, and to remove such dependency one can reorder $\mathbf{x}$ (say, in the decreasing order).

It is not difficult to check that the resulting function F is (a) idempotent, and (b) monotone increasing, as long as the bivariate function f is idempotent and monotone increasing. Furthermore, if f is homogeneous and/or shift-invariant, so is the resulting function F, and if f is bounded by two idempotent functions $A \leq f \leq B$, then so will be F. F will have the same absorbing element as f (if any). If f is a

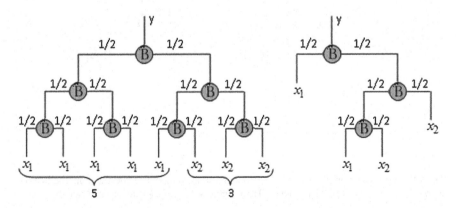

Fig. 2.21 Representation of a weighted arithmetic mean in a binary tree construction. The tree on the *right* is pruned by using idempotency

quasi-arithmetic mean, then F will be the corresponding n-variate quasi-arithmetic mean. Duality and φ-transforms are also preserved in the binary tree construction. If we select equal weights, so that the numbers k_i are approximately equal (they will actually be equal if n is a power of two), one gets the approximately unweighed extension of the given bivariate function f, and the accuracy of matching the weights is controlled by the number of levels L. For details of the proofs see [BD15, DB15].

A straightforward algorithm for doing so, which starts from the vector $\mathbf{X}$ computed as before, is very simple:

1. Compute $k_i = \lfloor w_i 2^L + \frac{1}{2} \rfloor, i = 1, \ldots, n-1$ and $k_n = 2^L - k_1 - k_2 - \cdots - k_{n-1}$. create the array $X := (x_1, \ldots, x_1, \ldots, x_n, \ldots, x_n)$ by taking k_1 copies of x_1, k_2 copies of x_2 and so on;
2. $N := 2^L$;
3. Repeat L times:

   (a) $N := N/2$;
   (b) For $i = 1 \ldots N$ do $X[i] := f(X[2i-1], X[2i])$;

4. Return $X[1]$.

One practical disadvantage of this algorithm is that its computational complexity is $O(2^L)$ in terms of the number of invocations of f. This makes it much slower than an explicit formula like (2.31). However it is possible to appropriately prune the binary tree by relying on idempotency of f. Indeed no invocation of f is necessary if both of its arguments are equal. Such a pruning was presented in [Duj15] in the special case of $n = 2$ where the aim was to construct a weighed bivariate mean from an unweighed one. Below we present a different (and more general) algorithm for the n-variate case whose complexity is $O(L(n-1))$, following [BD15]. This

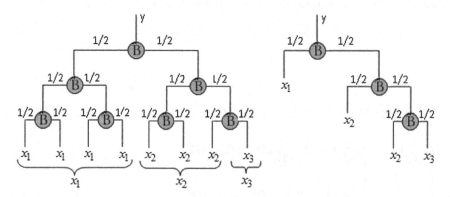

Fig. 2.22 Representation of a weighted 3-variate mean in a binary tree construction. The tree on the *right* is pruned by using idempotency. The weights $\mathbf{w} = (\frac{1}{2}, \frac{3}{8}, \frac{1}{8})$ are matched exactly

complexity is the upper bound, as at each level of the binary tree one can get at most $n - 1$ nodes with different values of the child nodes, so that pruning is impossible and f must be executed (Fig. 2.22).

The algorithm is a recursive depth-first traversal of the binary tree. A branch is pruned if it is clear that all its leaves have exactly the same value, and by idempotency this is the value of the root node of that branch.

Pruned tree aggregation (PTA) algorithm

function $node(m, N, k, x)$

1. If $N[k] \geq 2^m$ then do:

   (a) $N[k] := N[k] - 2^m$;
   (b) $y := x[k]$;
   (c) If $N[k] = 0$ then $k := k + 1$;
   (d) return y;

   else
2. return $f(node(m - 1, N, k, x), node(m - 1, N, k, x))$.

function $PTA(n, w, x, L)$

1. create the array $N = (k_1, k_2, \ldots, k_n)$ by using
   $k_i := \lfloor w_i 2^L + \frac{1}{2} \rfloor, i = 1, \ldots, n - 1$ and $k_n := 2^L - k_1 - k_2 - \cdots - k_{n-1}$;
2. $k := 1$;
3. return $node(L, N, k, x)$.

To see the complexity of this algorithm note that f is never executed if its arguments are the same, i.e., all the branches of the binary tree that can be pruned are pruned. Also note that both N and k are input-output parameters, so that the two arguments of f at step 2 are different as N and k change from one invocation of the function $node$ to another. The C++ code implementing this algorithm is listed in Fig. 2.23.

The PTA algorithm is numerically efficient, and its execution time is comparable to, or could be even smaller than that of analytical formulas like (2.31). In addition, it is a universal method applicable to any bivariate mean, and it allows a very intuitive introduction of weights as multiplicities of the inputs.

```
double node(double x[], long int N[], long int C, int & k,
            double(*F)(double,double))
{
/* recursive function in the binary tree processing
Parameters: x - input vector, N vector of multiplicities of  x
m current level of recursion counted from L (root node) to 0 (leaves)
k - input-output parameter, the current index of x being processed */
   if(N[k]>= C) {  /* we use idempotency here to prune the tree */
       N[k] -= C;
       if(N[k]<=0) return x[k++]; else return x[k];
   }
    C /= 2;
   /* tree not pruned, process the children nodes */
   return F( node(x,N,C,k,F), node(x,N,C,k,F) );
}


double nvariatef(double x[], double w[], int n,
                 double(*F)(double,double), int L)
/* Function F is the symmetric base aggregator.
 w[ ] = array of weights of inputs x[ ], n is the dimension of x and w.
 the weights must add to one and be non-negative.
 L = number of binary tree levels
 Run time = O[(n-1)L]   */
{
   long int t=0;
   int k=0;
   long int C=11<<L;    /* C=2^m */
   long int N[n]; /* multiplicities of x based on the weights */
   for(int i=0;i<n-1;i++)  {
      N[i]=w[i]*C+0.5;
      t+=N[i]; }
   N[n-1]=C-t;
   return node(x,N,C,k,F);
}
...
/* calling the function with f=HR*/
double binaryf(double x, double y) { return (x+y+sqrt(x*y))/3;}
double x[4]={0.2,0.2,0.4,0.8};
double w[4]={0.1,0.2,0.3,0.4};
int n=4, L=4;
double y=nvariatef(x,w,n,&binaryf,L);
```

Fig. 2.23 A C++ implementation of the pruned binary tree algorithm

References

[Acz48]   J. Aczél, On mean values. Bull. Am. Math. Soc. **54**, 392–400 (1948)

[Bel00]   G. Beliakov, Shape preserving approximation using least squares splines. Approx. Theory Appl. **16**, 80–98 (2000)

[Bel03]   G. Beliakov, How to build aggregation operators from data? Int. J. Intell. Syst. **18**, 903–923 (2003)

[Bel05]   G. Beliakov, Learning weights in the generalized OWA operators. Fuzzy Optim. Decis. Making **4**, 119–130 (2005)

[BCJ10]   G. Beliakov, T. Calvo, S. James, On Lipschitz properties of generated aggregation functions. Fuzzy Sets Syst. **161**, 1437–1447 (2010)

[BD15]   G. Beliakov, J.J. Dujmovic, Extension of bivariate means to weighted means of several arguments by using binary trees, submitted (2015)

[BJ13]   G. Beliakov, S. James, Stability of weighted penalty-based aggregation functions. Fuzzy Sets Syst. **226**, 1–18 (2013)

[BJGRM15]   G. Beliakov, S. James, D. Góomez, J.T. Rodríguez, J. Montero, Learning stable weights for data of varying dimension. In: *Proceedings of the AGOP Conference*, Katowice, Poland (2015)

[BMV04]   G. Beliakov, R. Mesiar, L. Valaskova, Fitting generated aggregation operators to empirical data. Int. J. Uncertainty, Fuzziness Knowl. Based Syst. **12**, 219–236 (2004)

[BPC07]   G. Beliakov, A. Pradera, T. Calvo, *Aggregation Functions: A Guide for Practitioners*, vol. 221, Studies in Fuzziness and Soft Computing (Springer, Berlin, 2007)

[Bon50]   C. Bonferroni, Sulle medie multiple di potenze. Bollettino Matematica Italiana **5**, 267–270 (1950)

[BB87]   J.M. Borwein, P.B. Borwein, *PI and the AGM: A Study in Analytic Number Theory and Computational Complexity* (Wiley, New York, 1987)

[Bul03]   P.S. Bullen, *Handbook of Means and Their Inequalities* (Kluwer, Dordrecht, 2003)

[CKKM02]   T. Calvo, A. Kolesárová, M. Komorníková, R. Mesiar, Aggregation operators: properties, classes and construction methods. In: *Aggregation Operators*. New Trends and Applications, ed. by T. Calvo, G. Mayor, R. Mesiar (Physica—Verlag, Heidelberg, 2002), pp. 3–104

[CM99]   T. Calvo, G. Mayor, Remarks on two types aggregation functions. Tatra Mount. Math. Publ. **16**, 235–254 (1999)

[CMY04]   T. Calvo, R. Mesiar, R.R. Yager, Quantitative weights and aggregation. IEEE Trans. Fuzzy Syst. **12**, 62–69 (2004)

[Cal+00]   T. Calvo et al., Generation on weighting triangles associated with aggregation functions. Int. J. Uncertainty Fuzziness Knowl. Based Syst. **8**, 417–451 (2000)

[CM09]   M. Cardin, M. Manzi, Supermodular and ultramodular aggregation evaluators. In: *Proceedings of the 5th International Summer School of Aggregation Operators AGOP'09*, July 6–11. Palma de Mallorca, Spain, 2009

[Cau21]   A.L. Cauchy, Cours d'analyse de l'Ecole Royale Polytechnique. Analyse algébrique **1** (1821)

[Chi29]   O. Chisini, Sul concetto di media. Periodico di Matematiche **4**, 106–116 (1929)

[DP85]   D. Dubois, H. Prade, A review of fuzzy set aggregation connectives. Inf. Sci. **36**, 85–121 (1985)

[DP04]   D. Dubois, H. Prade, On the use of aggregation operations in information fusion processes. Fuzzy Sets Syst. **142**, 143–161 (2004)

[Duj15]   J.J. Dujmovic, An efficient algorithm for general weighted aggregation. In: *Proceedings of the AGOP Conference, Katowice, Poland* (2015)

[Duj73]   J.J. Dujmovic, Two integrals related to means. Univ. Beograd. Publ. Elektrotechn. Fak. (1973)

[Duj74]   J.J. Dujmovic, Weighted conjunctive and disjunctive means and their application in system evaluation. Univ. Beograd. Publ. Elektrotechn. Fak. 147–158 (1974)

[Duj05]   J.J. Dujmovic, Seven flavors of andness/orness. In: *EUROFUSE* Belgrade, 2005

[Duj07]   J.J. Dujmovic, Continuous preference logic for system evaluation. IEEE Trans. Fuzzy
             Syst. **15**, 1082–1099 (2007)

[DB15]   J.J. Dujmovic, G. Beliakov, Idempotent weighted aggregation based on binary aggre-
             gation trees, submitted (2015)

[Eis71]   C. Eisenhart, The development of the concept of the best mean of a set of measure-
             ments from antiquity to the present day. In: *American Statistical Association Presi-
             dential Address*, http://www.york.ac.uk/depts/maths/histstat/eisenhart.pdf (1971)

[FSM03]   J.M. Fernández Salido, S. Murakami, Extending Yager's orness concept for the OWA
             aggregators to other mean operators. Fuzzy Sets Syst. **139**, 515–542 (2003)

[FM97]   J. Fodor, J.-L. Marichal, On nonstrict means. Aequationes Mathematicae **54**,
             308–327 (1997)

[FR94]   J. Fodor, M. Roubens, *Fuzzy Preference Modelling and Multicriteria Decision Sup-
             port* (Kluwer, Dordrecht, 1994)

[Gin58]   C. Gini, *Le Medie*. Milan (Russian translation, Srednie Velichiny, Statistica, Moscow,
             1970): Unione Tipografico-Editorial Torinese, 1958

[GMRR12]   D. Gómez, J. Montero, J.T. Rodríguez, K. Rojas, Stability in aggregation operators.
             In: *Proceedings of IPMU*, Catania, Italy, 2012

[GRMRB14]   D. Gómez, K. Rojas, J. Montero, J.T. Rodríguez, G. Beliakov, Consistency and
             stability in aggregation operators: an application to missing data problems. Int. J.
             Comput. Intell. Syst. **7**, 595–604 (2014)

[KSP14]   A. Kishor, A.K. Singh, N.R. Pal, Orness measure of OWA operators: a new approach.
             IEEE Trans. Fuzzy Syst. **22**, 1039–1045 (2014)

[Kom01]   M. Komorníková, Aggregation operators and additive generators. Int. J. Uncertainty
             Fuzziness Knowl Based Syst. **9**, 205–215 (2001)

[Liu06]   X. Liu, An orness measure for quasi-arithmetic means. IEEE Trans. Fuzzy Syst. **14**,
             837–848 (2006)

[MPR03]   R.A. Marques Pereira, R.A. Ribeiro, Aggregation with generalized mixture operators
             using weighting functions. Fuzzy Sets Syst. **137**, 43–58 (2003)

[Mat11]   J. Matkowski, A mean-value theorem and its applications. J. Math. Anal. Appl. **373**,
             227–234 (2011)

[MC97]   G. Mayor, T. Calvo, Extended aggregation functions. In: *IFSA'97*. vol. 1, Prague,
             1997, pp. 281–285

[Mer04]   J. Merikoski, Extending means of two variables to several variables. J. Inequal. Pure
             Appl. Math. **5**, Article 65 (2004)

[MS06]   R. Mesiar, J. Spirková, Weighted means and weighting functions. Kybernetik **42**,
             151–160 (2006)

[MSV08]   R. Mesiar, J. Spirková, L. Vavríková, Weighted aggregation operators based on min-
             imization. Inf. Sci. **178**, 1133–1140 (2008)

[Mus02]   S. Mustonen, *Logarithmic Mean for Several Arguments*, http://www.survo.fi/papers/
             logmean.pdf (2002)

[Neu94]   E. Neuman, The weighted logarithmic mean. J. Math. Anal. Appl. **188**, 885–900
             (1994)

[Pit85]   A.O. Pittenger, The logarithmic mean in n variables. Am. Math. Monthly **92**, 99–104
             (1985)

[PTC02]   A. Pradera, E. Trillas, T. Calvo, A general class of triangular norm-based aggregation
             operators: quasi-linear T-S operators. Int. J. Approx. Reasoning **30**, 57–72 (2002)

[RA14]   J.T. Rickard, J. Aisbett, New classes of threschold aggregation funcitons based upon
             the Tsallis q-exponential with applications to perceptual computing. IEEE Trans.
             Fuzzy Syst. **22**, 672–684 (2014)

[RGMR13]   K. Rojas, D. Gómez, J. Montero, J.T. Rodríguez, Strictly stable families of aggrega-
             tion operators. Fuzzy Sets Syst. **228**, 44–63 (2013)

[Rub71]   E. Rubin, Quantitative commentary on thucydides. Am. Stat. **25**, 52–54 (1971)

[Tor02]   V. Torra, Learning weights for the quasi-weighted means. IEEE Trans. Fuzzy Systems
            10, 653–666 (2002)
[TY05]   L. Troiano, R.R. Yager, A meaure of dispersion for OWA operators. In: *Proceedings of
            the 11th IFSA World Congress*, ed. by Y. Liu, G. Chen, M. Ying (Tsinghua University
            Press, Springer, Beijing China, 2005), pp. 82–87
[Yag88]   R.R. Yager, On ordered weighted averaging aggregation operators in multicriteria
            decision making. IEEE Trans. Syst. Man Cybern. **18**, 183–190 (1988)
[Yag91]   R.R. Yager, Connectives and quantifiers in fuzzy sets. Fuzzy Sets Syst. **40**, 39–76
            (1991)
[Yag95]   R.R. Yager, Measures of entropy and fuzziness related to aggregation operators. Inf.
            Sci. **82**, 147–166 (1995)

Chapter 3
Ordered Weighted Averaging

Abstract The focus of this chapter is on OWA functions. The formal definitions and the main properties of OWA are presented. Some extensions of the OWA functions are discussed in detail. Various methods of fitting OWA functions to empirical data are presented. This chapter ends with the discussion of the median functions and order statistics as the special cases of OWA functions.

3.1 Definitions

Ordered weighted averaging functions (OWA) belong to the class of averaging aggregation functions. They differ to the weighted arithmetic means in that the weights are associated not with the particular inputs, but with their magnitude. In some applications, all inputs are equivalent, and the importance of an input is determined by its value. For example, when a robot navigates obstacles using several sensors, the largest input (the closest obstacle) is the most important. OWA are symmetric aggregation functions that allocate weights according to the input value. Thus OWA can emphasize the largest, the smallest or mid-range inputs. They have been introduced by Yager [Yag88] and have become very popular in the fuzzy sets community [EM14, YK97, YKB11]

We recall the notation $\mathbf{x}_\searrow$, which denotes the vector obtained from $\mathbf{x}$ by arranging its components in *non-increasing* order $x_{(1)} \geq x_{(2)} \geq \cdots \geq x_{(n)}$.

Definition 3.1 *(OWA)* For a given weighting vector $\mathbf{w}$, $w_i \geq 0$, $\sum w_i = 1$, the OWA function is given by

$$OWA_{\mathbf{w}}(\mathbf{x}) = \sum_{i=1}^{n} w_i x_{(i)} = \langle \mathbf{w}, \mathbf{x}_\searrow \rangle. \tag{3.1}$$

Calculation of the value of an OWA function involves using a `sort()` operation.

Definition 3.2 *(Reverse OWA)* Given an OWA function $OWA_{\mathbf{w}}$, the reverse OWA is $OWA_{\mathbf{w}_d}$ with the weighting vector $\mathbf{w}_d = (w_n, w_{n-1}, \ldots, w_1)$.

© Springer International Publishing Switzerland 2016
G. Beliakov et al., *A Practical Guide to Averaging Functions*,
Studies in Fuzziness and Soft Computing 329,
DOI 10.1007/978-3-319-24753-3_3

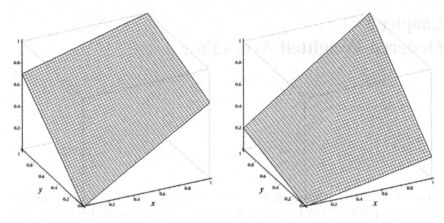

Fig. 3.1 3D plots of OWA functions $OWA_{(0.7,0.3)}$ and $OWA_{(0.2,0.8)}$

3.2 Main Properties

- As with all averaging aggregation functions, OWA are increasing (strictly increasing if all weights are positive) and idempotent (Fig. 3.1);
- The dual of an OWA function is the *reverse* OWA, with the vector of weights $\mathbf{w}_d = (w_n, w_{n-1}, \ldots, w_1)$.
- OWA functions are continuous, symmetric, homogeneous and shift-invariant;
- OWA functions do not have neutral or absorbing elements, except for the special cases min and max;
- The OWA functions are special cases of the Choquet integral (see Sect. 4.1) with respect to symmetric fuzzy measures.

Special cases

- If all weights are equal, $w_i = \frac{1}{n}$, OWA becomes the arithmetic mean $OWA_{\mathbf{w}}(\mathbf{x}) = M(\mathbf{x})$;
- If $\mathbf{w} = (1, 0, \ldots, 0)$, then $OWA_{\mathbf{w}}(\mathbf{x}) = \max(\mathbf{x})$;
- If $\mathbf{w} = (0, \ldots, 0, 1)$, then $OWA_{\mathbf{w}}(\mathbf{x}) = \min(\mathbf{x})$;
- If $\mathbf{w} = (\alpha, 0, \ldots, 0, 1 - \alpha)$, then OWA becomes the **Hurwizc aggregation function**, $OWA_{\mathbf{w}}(\mathbf{x}) = \alpha \max(\mathbf{x}) + (1 - \alpha) \min(\mathbf{x})$; For $\alpha = \frac{1}{2}$ it is called the half-range function;
- If $w_i = 0$ for all i except the k-th, and $w_k = 1$, then OWA becomes k-th order statistic, $OWA_{\mathbf{w}}(\mathbf{x}) = x_{(k)}$.
- If for an odd n $\mathbf{w} = (0, \ldots, 0, 1, 0, \ldots, 0)$ and for an even n $\mathbf{w} = (0, \ldots, 0, \frac{1}{2}, \frac{1}{2}, 0, \ldots, 0)$, then $OWA_{\mathbf{w}}(\mathbf{x}) = Med(\mathbf{x})$, the **median** (see Sect. 3.6.1);
- If we take weighting vectors with several non-zero values, for example $\mathbf{w} = (0, \ldots, 0, \frac{1}{6}, \frac{1}{3}, \frac{1}{3}, \frac{1}{6}, 0, \ldots, 0)$, we obtain the **central OWA** function proposed in [Yag07a] and later investigated in [ZSA08]. The central OWA takes into account several central inputs. In [Yag07a] the weights were also symmetric

$(w_j = w_{n+1-j})$, strongly decaying ($w_i < w_j$ if either $i < j \leq (n+1)/2$ or $i > j \geq (n+1)/2$), and inclusive ($w_j > 0$).

- One special case of centered OWA is the α-**trimmed mean** ($0 \leq \alpha \leq \frac{1}{2}$), which can be written as an OWA function with the weights

$$\mathbf{w} = \left(0, \ldots, 0, \frac{1}{n - 2[n\alpha]}, \ldots, \frac{1}{n - 2[n\alpha]}, 0, \ldots, 0\right),$$

where $[n\alpha]$ is the nearest integer no greater than $n\alpha$;

- The "Olympic OWA" has the weighting vector $\mathbf{w} = (0, \frac{1}{n-2}, \ldots, \frac{1}{n-2}, 0)$ and is a special case of trimmed means;

- Another interesting case is the **Winsorized mean**, in which extreme low and high values are replaced. An $\alpha\%$ Winsorized mean can be expressed as an OWA function with the weights

$$\mathbf{w} = (0, \ldots, 0, ([n\alpha] + 1)a, a, \ldots, a, ([n\alpha] + 1)a, 0, \ldots, 0),$$

with $a = \frac{1}{n}$.

As special cases of OWA, all the mentioned functions are symmetric, homogeneous and shift-invariant aggregation functions. Interestingly, the actual calculation of α-trimmed and Winsorized means can be achieved without sorting the arguments $\mathbf{x}$ (which has the complexity of $O(n \log n)$ and could be time consuming for repetitive calculations and large n). Indeed, the Winsorized mean can be written as

$$WM(\mathbf{x}) = \sum_{i=1}^{n} \min(\max(x_i, x_{([n\alpha]+1)}), x_{(n-[n\alpha])}),$$

where $x_{([n\alpha]+1)}, x_{(n-[n\alpha])}$ are order statistics that are computed in $O(n)$ operations (using quickselect algorithm). Then the α-trimmed mean is

$$TM(\mathbf{x}) = \frac{1}{n - 2[n\alpha]} \left(nWM(\mathbf{x}) - [n\alpha](x_{([n\alpha]+1)} + x_{(n-[n\alpha])})\right).$$

3.2.1 Orness Measure

The general expression for the measure of orness, given in (2.1), translates into the following simple formula

$$orness(OWA_{\mathbf{w}}) = \sum_{i=1}^{n} w_i \frac{n-i}{n-1} = OWA_{\mathbf{w}}\left(1, \frac{n-2}{n-1}, \ldots, \frac{1}{n-1}, 0\right). \quad (3.2)$$

Here is a list of additional properties involving the orness value.

- The orness of OWA and its dual are related by

$$orness(OWA_\mathbf{w}) = 1 - orness(OWA_{\mathbf{w}_d}).$$

- An OWA function is self-dual if and only if $orness(OWA_\mathbf{w}) = \frac{1}{2}$.
- In the special cases $orness(\max) = 1$, $orness(\min) = 0$, and $orness(M) = \frac{1}{2}$. Furthermore, the orness of OWA is 1 only if it is the max function and 0 only if it is the min function. However orness can be $\frac{1}{2}$ for an OWA different from the arithmetic mean, which is nevertheless self-dual.
- If the weighting vector is non-decreasing, i.e., $w_i \leq w_{i+1}, i = 1, \ldots, n-1$, then $orness(OWA_\mathbf{w}) \in [0, \frac{1}{2}]$. If the weighting vector is non-increasing, then $orness(OWA_\mathbf{w}) \in [\frac{1}{2}, 1]$.
- If two OWA functions with weighing vectors $\mathbf{w}_1$, $\mathbf{w}_2$ have their respective orness values O_1, O_2, and if $\mathbf{w}_3 = a\mathbf{w}_1 + (1-a)\mathbf{w}_2, a \in [0, 1]$, then OWA function with the weighting vector $\mathbf{w}_3$ has orness value [Ahn06]

$$orness(OWA_{\mathbf{w}_3}) = aO_1 + (1-a)O_2.$$

Note 3.3 Of course, to determine an OWA weighting vector with the desired orness value, one can use many different combinations of $\mathbf{w}_1, \mathbf{w}_2$, which all result in different $\mathbf{w}_3$ but with the same orness value.

Example 3.4 The measure of orness for some special weighting vectors has been precalculated in [Ahn06]

$$w_i = \frac{1}{n}\sum_{j=i}^{n}\frac{1}{j}, \quad orness(OWA_\mathbf{w}) = \frac{3}{4},$$

$$w_i = \frac{2(n+1-i)}{n(n+1)}, \quad orness(OWA_\mathbf{w}) = \frac{2}{3}.$$

Note 3.5 The classes of recursive and iterative OWA functions, which have the same orness value for any given n, was investigated in [TY05b].

Yager has defined andlike and orlike S-OWA functions respectively [Yag93] using parameter $-1 \leq \alpha \leq 1$. The weighting vectors of S-OWA functions are given by

$$w_1 = \frac{1-\alpha}{n} + \alpha, \; w_i = \frac{1-\alpha}{n}, i = 2, \ldots, n,$$

where orlike OWA will have a positive α and andlike OWA will have a negative α. The orness of these functions is $orness(OWA_\mathbf{w}) = \alpha + \frac{1}{2}(1-\alpha)$.

Recently the concept of the orness measure has been revisited in [KSP14]. Here the authors proposed an axiomatic framework for orness measure based on the following axioms.

O1 Fuzzy OR operator: $orness(OWA_{\mathbf{w}}) = 1$ if and only if $\mathbf{w} = \mathbf{w}^* = (1, 0, \ldots, 0)$.

O2 Fuzzy AND operator: $orness(OWA_{\mathbf{w}}) = 0$ if and only if $\mathbf{w} = \mathbf{w}_* = (0, \ldots, 0, 1)$.

O3 If $\mathbf{w} = (w_1, w_w, \ldots, w_n)$ and $\mathbf{w}' = (w_1, \ldots, w_j - \varepsilon, \ldots, w_k + \varepsilon, \ldots, w_n)$, where $\varepsilon > 0$ and $j < k$, then $orness(OWA_{\mathbf{w}}) > orness(OWA_{\mathbf{w}'})$.

O4 The $orness(OWA_{\mathbf{w}}) = \frac{1}{2}$ if and only if $\sum_{i=1}^{n'} c_i w_i = -\sum_{i=n'+1}^{n} c_i w_i$ for $c_i = n' + 1 - i - \lfloor \frac{2i-1}{n} \rfloor - \lfloor \frac{2n'-1}{n} \rfloor \lfloor \frac{i-1}{n'} \rfloor$ and $n' = \lfloor \frac{n+1}{2} \rfloor$.

The axiom O4 is a generalisation of the following simpler axiom:

O4a. The arithmetic mean has orness $\frac{1}{2}$: $orness(\frac{1}{n}, \ldots, \frac{1}{n}) = \frac{1}{2}$.

Definition 3.6 (*Modified OWA orness value*) Let $\mathbf{w}$ be a weighting vector. Then the modified orness value is

$$orness(OWA_{\mathbf{w}}) = \sum_{i=1}^{n} d_i w_i, \tag{3.3}$$

where $d_i = \frac{1}{2n'} \left(2n' + 1 - i - \lfloor \frac{2i-1}{n} \rfloor - \lfloor \frac{2n'-1}{n} \rfloor \lfloor \frac{i-1}{n'} \rfloor \right)$ and $n' = \lfloor \frac{n+1}{2} \rfloor$.

The vector $\mathbf{d}$ has the following properties.

- For an even n, $d_i = \frac{1}{n}(n + 1 - i - \lfloor \frac{2i-1}{n} \rfloor)$.
- For an odd n, $d_i = \frac{1}{n+1}(n + 2 - i - \lfloor \frac{2i-1}{n} \rfloor - \lfloor \frac{2i-2}{n+1} \rfloor)$.
- The values d_i form a monotonically decreasing sequence, with $d_1 = 1$ and $d_n = 0$.
- $d_i + d_{n-i+1} = 1$.
- $\sum d_i = \frac{n}{2}$.

The modified measure of orness in Definition 3.6 satisfies the four axioms O1–O4, as shown in [KSP14], where an interpretation of this quantity in terms of the normalised sum of moments is presented. Furthermore, the following results hold.

- $orness(OWA_{\mathbf{w}}) = 1 - orness(OWA_{\mathbf{w}_d})$
- If $\mathbf{w}$ is a buoyancy measure, i.e., $w_i \geq w_j$ for all $i < j$, then $orness(OWA_{\mathbf{w}}) \geq \frac{1}{2}$.

Finally OWA is a buoyancy measure if and only if it is a convex function, if and only if it is a subadditive function $OWA(\mathbf{x} + \mathbf{y}) \leq OWA(\mathbf{x}) + OWA(\mathbf{y})$, see Note 4.20.

3.2.2 Entropy

We recall the definition of weights dispersion (entropy), Definition 2.38, p. 76

$$Disp(\mathbf{w}) = -\sum_{i=1}^{n} w_i \log w_i.$$

It measures the degree to which all the information (i.e., all the inputs) is used in the aggregation process. The entropy is used to define the weights with the maximal entropy (functions called MEOWA), subject to a predefined orness value (details are in Sect. 3.4).

- If the orness is not specified, the maximum of $Disp$ is achieved at $w_i = \frac{1}{n}$, i.e., the arithmetic mean, and $Disp(\frac{1}{n}, \ldots, \frac{1}{n}) = \log n$.
- The minimum value of $Disp$, 0, is achieved if and only if $w_i = 0, i \neq k$, and $w_k = 1$, i.e., the order statistic, see Sect. 3.6.2.
- The entropy of an OWA and its dual (reverse OWA) coincide, $Disp(\mathbf{w}) = Disp(\mathbf{w}_d)$.

Similarly to the case of weighted quasi-arithmetic means, weighting triangles (see Sect. 2.3.5) should be used if one needs to work with families of OWAs (i.e., OWA extended aggregation functions in the sense of Definition 1.6). We also note that there are other types of entropy (e.g., Rényi entropy) used to quantify weights dispersion, see, e.g., [Yag95]. One such measure of dispersion was presented in [TY05a] and is calculated as

$$\rho(\mathbf{w}) = \frac{\sum\limits_{i=1}^{n} \frac{i-1}{n-1}(w_{(i)} - w_{(i+1)})}{\sum\limits_{i=1}^{n}(w_{(i)} - w_{(i+1)})} = \frac{1}{n-1} \frac{1 - w_{(1)}}{w_{(1)}},$$

where $w_{(i)}$ denotes the i-th largest weight. A related measure of weights dispersion is $\tilde{\rho}(\mathbf{w}) = 1 - w_{(1)}$ [Yag93]. Another useful measure is weights variance [FM03], see Eq. (3.16) on p. 118.

3.3 Other Types of OWA Functions

3.3.1 Neat OWA

OWA functions have been generalized to functions whose weights depend on the aggregated inputs.

Definition 3.7 (*Neat OWA*) An OWA function whose weights are defined by

$$w_i = \frac{x_{(i)}^p}{\sum\limits_{i=1}^{n} x_{(i)}^p},$$

with $p \in]-\infty, \infty[$ is called a neat OWA.

Note 3.8 Neat OWA functions are counter-harmonic means (see p. 83). We remind that they are not monotone (hence not aggregation functions).

3.3.2 Generalized OWA

Similarly to quasi-arithmetic means (Sect. 2.3), OWA functions have been generalized with the help of generating functions $g : [0, 1] \rightarrow [-\infty, \infty]$ as

Definition 3.9 (*Generalized OWA*) Let $g : [0, 1] \rightarrow [-\infty, \infty]$ be a continuous strictly monotone function and let **w** be a weighting vector. The function

$$GenOWA_{\mathbf{w},g}(\mathbf{x}) = g^{-1}\left(\sum_{i=1}^{n} w_i g(x_{(i)})\right) \tag{3.4}$$

is called a generalized OWA (also known as ordered weighted quasi-arithmetic mean [CKKM02]). As for OWA, $x_{(i)}$ denotes the i-th largest value of **x**.

Special cases

Ordered weighted geometric function was studied in [HHV03, XD02].

Definition 3.10 (*Ordered Weighted Geometric function (OWG)*) For a given weighting vector **w**, the OWG function is

$$OWG_{\mathbf{w}}(\mathbf{x}) = \prod_{i=1}^{n} x_{(i)}^{w_i}. \tag{3.5}$$

Note 3.11 Similarly to the weighted geometric mean, OWG is a special case of (3.4) with the generating function $g = \log$.

Definition 3.12 (*Ordered Weighted Harmonic function (OWH)*) For a given weighting vector **w**, the OWH function is

$$OWH_{\mathbf{w}}(\mathbf{x}) = \left(\sum_{i=1}^{n} \frac{w_i}{x_{(i)}}\right)^{-1}. \tag{3.6}$$

A large family of generalized OWA functions is based on power functions, similar to weighted power means [Yag04b]. Let g_r denote the family of power functions

$$g_r(t) = \begin{cases} t^r, & \text{if } r \neq 0, \\ \log(t), & \text{if } r = 0. \end{cases}$$

Definition 3.13 (*Power-based generalized OWA*) For a given weighting vector **w**, and a value $r \in \mathbb{R}$, the function

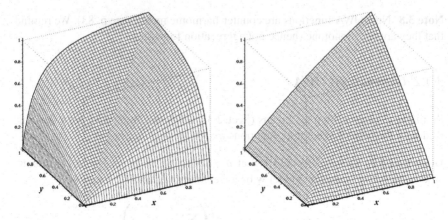

Fig. 3.2 3D plots of OWH functions $OWH_{(0.9,0.1)}$ and $OWH_{(0.2,0.8)}$

$$GenOWA_{\mathbf{w},[r]}(\mathbf{x}) = \left(\sum_{i=1}^{n} w_i x_{(i)}^r\right)^{1/r}, \tag{3.7}$$

if $r \neq 0$, and $GenOWA_{\mathbf{w},[r]}(\mathbf{x}) = OWG_{\mathbf{w}}(\mathbf{x})$ if $r = 0$, is called a power-based generalized OWA .

Of course, both OWG and OWH functions are special cases of power-based OWA with $r = 0$ and $r = -1$ respectively. The usual OWA corresponds to $r = 1$. Another special case is that of quadratic OWA, $r = 2$, given by

$$OWQ_{\mathbf{w}}(\mathbf{x}) = \sqrt{\sum_{i=1}^{n} w_i x_{(i)}^2}.$$

Other generating functions can also be used to define generalized OWA functions (Fig. 3.2).

Example 3.14 (*Trigonometric OWA*) Let $g_1(t) = \sin(\frac{\pi}{2}t)$, $g_2(t) = \cos(\frac{\pi}{2}t)$, and $g_3(t) = \tan(\frac{\pi}{2}t)$ be the generating functions. The trigonometric OWA functions are the functions

$$OWAS_{\mathbf{w}}(\mathbf{x}) = \frac{2}{\pi} \arcsin(\sum_{i=1}^{n} w_i \sin(\frac{\pi}{2}x_{(i)})),$$

$$OWAC_{\mathbf{w}}(\mathbf{x}) = \frac{2}{\pi} \arccos(\sum_{i=1}^{n} w_i \cos(\frac{\pi}{2}x_{(i)})), \text{ and}$$

$$OWAT_{\mathbf{w}}(\mathbf{x}) = \frac{2}{\pi} \arctan(\sum_{i=1}^{n} w_i \tan(\frac{\pi}{2}x_{(i)})).$$

Example 3.15 (Exponential OWA) Let the generating function be

$$g(t) = \begin{cases} \gamma^t, & \text{if } \gamma \neq 1, \\ t, & \text{if } \gamma = 1. \end{cases}$$

The exponential OWA is the function

$$OWAE_{\mathbf{w},\gamma}(\mathbf{x}) = \begin{cases} \log_\gamma \left(\sum_{i=1}^n w_i \gamma^{x_{(i)}} \right), & \text{if } \gamma \neq 1, \\ OWA_{\mathbf{w}}(\mathbf{x}), & \text{if } \gamma = 1. \end{cases}$$

Example 3.16 (Radical OWA) Let $\gamma > 0$, $\gamma \neq 1$, and let the generating function be

$$g(t) = \gamma^{1/t}.$$

The radical OWA is the function

$$OWAR_{\mathbf{w},\gamma}(\mathbf{x}) = \left(\log_\gamma \left(\sum_{i=1}^n w_i \gamma^{1/x_{(i)}} \right) \right)^{-1}.$$

3D plots of some generalized OWA functions are presented on Figs. 3.3, 3.4 and 3.5.

3.3.3 Weighted OWA

The weights in weighted means and in OWA functions represent different things. In weighted means w_i reflects the importance of the i-th input, whereas in OWA w_i reflects the importance of the i-th largest input. In [Tor97] Torra proposed a generalization of both weighted means and OWA, called WOWA. This aggregation function has two sets of weights $\mathbf{w}, \mathbf{p}$. Vector $\mathbf{p}$ plays the same role as the weighting vector in weighted means, and $\mathbf{w}$ plays the role of the weighting vector in OWA functions.

Consider the following motivation. A robot needs to combine information coming from n different sensors, which provide distances to the obstacles. The reliability of the sensors is known (i.e., we have weights $\mathbf{p}$). However, independent of their reliability, the distances to the nearest obstacles are more important, so irrespective of the reliability of each sensor, their inputs are also weighted according to their numerical value, hence we have another weighting vector $\mathbf{w}$. Thus both factors, the size of the inputs and the reliability of the inputs, need to be taken into account. WOWA provides exactly this type of aggregation function.

WOWA function becomes the weighted arithmetic mean if $w_i = \frac{1}{n}, i = 1, \ldots, n$, and becomes the usual OWA if $p_i = \frac{1}{n}, i = 1, \ldots, n$.

Definition 3.17 *(Weighted OWA)* Let $\mathbf{w}, \mathbf{p}$ be two weighting vectors, $w_i, p_i \geq 0$, $\sum w_i = \sum p_i = 1$. The following function is called Weighted OWA function

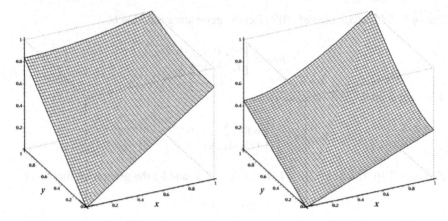

Fig. 3.3 3D plots of quadratic OWA functions $OWQ_{(0.7,0.3)}$ and $OWQ_{(0.2,0.8)}$

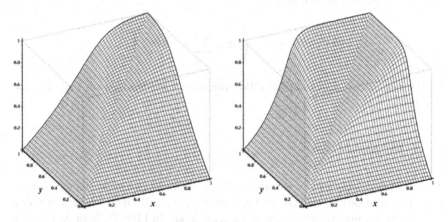

Fig. 3.4 3D plots of radical OWA functions $OWAR_{(0.9,0.1),100}$ and $OWAR_{(0.999,0.001),100}$

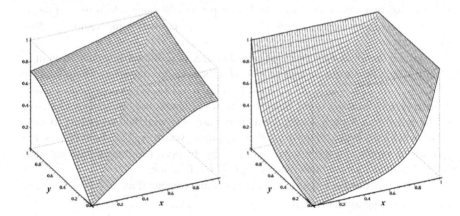

Fig. 3.5 3D plots of trigonometric OWA functions $OWAS_{(0.9,0.1)}$ and $OWAT_{(0.2,0.8)}$

$$WOWA_{\mathbf{w},\mathbf{p}}(\mathbf{x}) = \sum_{i=1}^{n} u_i x_{(i)},$$

where $x_{(i)}$ is the i-th largest component of $\mathbf{x}$, and the weights u_i are defined as

$$u_i = g\left(\sum_{j\in H_i} p_j\right) - g\left(\sum_{j\in H_{i-1}} p_j\right),$$

where the set $H_i = \{j | x_j \geq x_i\}$ is the set of indices of i largest elements of $\mathbf{x}$, and g is a monotone non-decreasing function with two properties:

1. $g(i/n) = \sum_{j\leq i} w_j, i = 0, \ldots, n$ (of course $g(0) = 0$);
2. g is linear if the points $(i/n, \sum_{j\leq i} w_j)$ lie on a straight line.

Note 3.18 The second condition is in the form formulated in [Tor97]. It can be reformulated as follows: 2a. g is linear if all w_i are equal.

Thus computation of WOWA involves a very similar procedure as that of OWA (i.e., sorting components of $\mathbf{x}$ and then computing their weighted sum), but the weights u_i are defined by using both vectors $\mathbf{w}, \mathbf{p}$, a special monotone function g, and depend on the components of $\mathbf{x}$ as well. One can see WOWA as an OWA function with the weights $\mathbf{u}$. Let us list some of the properties of WOWA.

- The weighting vector $\mathbf{u}$ satisfies $u_i \geq 0, \sum u_i = 1$.
- If $w_i = \frac{1}{n}$, then $WOWA_{\mathbf{w},\mathbf{p}}(\mathbf{x}) = M_{\mathbf{p}}(\mathbf{x})$, the weighted arithmetic mean.
- If $p_i = \frac{1}{n}$, $WOWA_{\mathbf{w},\mathbf{p}}(\mathbf{x}) = OWA_{\mathbf{w}}(\mathbf{x})$.
- WOWA is an idempotent aggregation function.

Of course, the weights $\mathbf{u}$ also depend on the generating function g. This function can be chosen as a linear spline (i.e., a broken line interpolant), interpolating the points $(i/n, \sum_{j\leq i} w_j)$ (in which case it automatically becomes a linear function if these points are on a straight line), or as a monotone quadratic spline, as was suggested in [Tor97, Tor00], see also [Bel01] where Schumaker's quadratic spline algorithm was used [Sch83], which automatically satisfies the straight line condition when needed.

It turns out that WOWA belongs to a more general class of Choquet integral based aggregation functions, discussed in Sect. 4.1, with respect to *distorted probabilities*, see Definition 4.42 [NT05, Tor98, TN07]. It is a piecewise linear function whose linear segments are defined on the simplicial partition of the unit cube $[0, 1]^n$: $S_i = \{\mathbf{x} \in [0, 1]^n | x_{p(j)} \geq x_{p(j+1)}\}$, where p is a permutation of the set $\{1, \ldots, n\}$. Note that there are exactly $n!$ possible permutations, the union of all S_i is $[0, 1]^n$, and the intersection of the interiors of $S_i \cap S_j = \emptyset, i \neq j$.

Recently there have been a number of alternative attempts to incorporate weights into OWA, some of which turned out to be mathematically equivalent. We refer the reader to the following publications [LJ14, Lla15, Mer11, YA14a]. Unfortunately

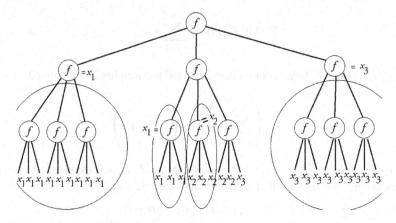

Fig. 3.6 Representation of a weighted tri-variate function f in a ternary tree construction. The weights are chosen as $\mathbf{p} = (\frac{12}{27}, \frac{5}{27}, \frac{10}{27})$ and $L = 3$. The circled branches are pruned by the algorithm

these generalizations of OWA do not produce idempotent functions except in a few special cases.

A rather different approach of introducing weights into OWA was reported in [Bel15]. It is based on the same strategy as in the extensions of bivariate means to weighted multivariate means presented in Sect. 2.4.9. The difference is that in OWA we start with an n-variate function, and therefore an n-ary tree needs to be constructed. We summarize the findings from [Bel15].

To incorporate a vector $\mathbf{p}$ of non-negative weights into a symmetric n-variate function we replicate the arguments a suitable number of times. We build an n-ary tree with L levels, as shown in Fig. 3.6. As the base symmetric aggregator f we take an OWA function $OWA_{\mathbf{w}}$ with the specified weights $\mathbf{w}$. Hence we create an auxiliary vector $\mathbf{X} = (x_1, \ldots, x_1, x_2, \ldots, x_2, \ldots, x_n, \ldots, x_n)$, so that x_1 is taken k_1 times, x_2 is taken k_2 times, and so on, and $\frac{k_1}{n^L} \approx p_1$, $\frac{k_2}{n^L} \approx p_2$, ..., and $\sum k_i = n^L$, where $L \geq 1$ is a specified number of levels of the tree. One way of doing so is to take $k_i = \lfloor p_i n^L + \frac{1}{n} \rfloor$, $i = 1, \ldots, n-1$ and $k_n = n^L - k_1 - k_2 - \cdots - k_{n-1}$.

Pruned n-Tree Aggregation (PnTA) Algorithm

function $node(n, m, N, K, x)$

1. If $N[K] \geq n^m$ then do:

   (a) $N[K] := N[K] - n^m$;
   (b) $y := x[K]$;
   (c) If $N[K] = 0$ then $K := K + 1$;
   (d) return y;

   else
2. for $i := 1, \ldots, n$ do
   $z[i] := node(n, m - 1, N, K, x)$

3. return $f(\mathbf{z})$.

   function $PnTA(n, w, p, L)$

1. create the array $N := (k_1, k_2, \ldots, k_n)$ by using
   $k_i := \lfloor p_i n^L + \frac{1}{n} \rfloor, i = 1, \ldots, n-1$, and $k_n := n^L - k_1 - \cdots - k_{n-1}$;
2. $K := 1$;
3. return $node(n, L, N, K, x)$.

The algorithm PnTA works in the same way as the PTA algorithm for binary trees (see p. 95). The vector of counters N helps determine whether there are more than n^m identical elements of the auxiliary array $\mathbf{X}$, in which case they are the leaves of a branch of the tree with m levels. This branch is pruned. The function f is executed only when some of its arguments are distinct, and since the elements of $\mathbf{X}$ are ordered, there are at most $n-1$ such possibilities at each level of the tree, hence the complexity of the algorithm is $O((n-1)L)$. The vector $\mathbf{X}$ needs to be sorted, which is equivalent to sorting the inputs $\mathbf{x}$ jointly with the multiplicities of the inputs N (i.e., using the components of $\mathbf{x}$ as the key).

Let us list some useful properties of the function f_p generated by the PnTA algorithm.

Theorem 3.19 *The weighted extension f_p of a function f by the PnTA algorithm preserves the intrinsic properties of the parent function f as follows:*

1. *f_p idempotent since f is idempotent;*
2. *if f is monotone increasing then f_p is monotone increasing;*
3. *if f is continuous then f_p is continuous;*
4. *if f is convex (resp. concave) then f_p is convex (resp. concave);*
5. *if f is homogeneous then f_p is homogeneous ;*
6. *if f is shift-invariant then f_p is shift-invariant ;*
7. *f_p has the same absorbing element as f (if any);*
8. *if f generates f_p then a φ-transform of f generates the corresponting φ-transform of f_p.*

Proof The proof easily follows from the properties of composition of the respective functions and idempotency of f. For the φ-transform notice that at each inner level of the tree the composition $\varphi^{-1} \circ \varphi = Id$, while φ is applied to the leaves of the tree and φ^{-1} is applied to the root. □

The other results are proven in [Bel15].

Theorem 3.20 *Let $f = OWA_w$. Then the algorithm PnTA generates the weighted function f_p which is the discrete Choquet integral (and is hence homogeneous and shift-invariant).*

Theorem 3.21 *Let $f = OWA_w$. Then the algorithm PnTA generates the weighted function f_p with the following properties:*

1. *for the weights* $w_i = \frac{1}{n}$ f_p *is the weighted arithmetic mean with the weights* **p***;*
2. *for the weights* $p_i = \frac{1}{n}$ f_p *is* OWA_w*;*
3. *when* $f = OWA_w = \min$ *(or* $= \max$*), and* $p_i > 0$ *for all i,* f_p *is also* min *(respectively,* max*);*
4. *when* $f = OWA_w = median$ *and n is odd,* f_p *is the weighted median (see Definition 3.46);*
5. *if* OWA_w *generates* f_p*, then the dual* OWA_w^d *generates the dual* f_p^d*, and in particular an OWA with the reverse weights generates the respective weighted OWA with the reverse weights.*

Theorem 3.22 *Let* $f = OWA_w$ *and let the weighting vector be decreasing (increasing). Then the algorithm PnTA generates a Choquet integral with respect to a submodular (supermodular) fuzzy measure.*

This last result is useful when constructing weighted norms from OWA with decreasing weights, see [BJL11, Yag10], see Sect. 4.1.3.

On the technical size we note that we do not need to sort the arguments in each OWA function in the n-ary tree as vector **x** is already sorted, hence only one sort operation for the inputs is required. Another note is that when the weights **p** are specified to m digits in base n, $L = m$ levels of the n-ary tree is sufficient to match these weighs exactly. For example if **p** are specified to 3 decimal places and $n = 10$, we only need to take $L = 3$. Therefore to match the weights to machine precision (e.g., 53 bits for data type double) n^L need not exceed the largest 64-bit integer, and hence the algorithm PnTA can be implemented with 64-bit data types. The source code in C++ is presented in Fig. 3.7.

By using Definition 3.9 we can introduce weights into generalized OWA functions in the same was as for OWA functions, by using the n-ary tree construction. This can be done in two ways: (a) by using $GenOWA_{\mathbf{w},g}$ function as f, or (b) by using a φ-transform of a weighted OWA with $\varphi = g$, that is, by applying g and g^{-1} only to the leaves and to the root node of the tree, relying on the preservation of φ-transforms. The second method is computationally more efficient as the functions g and g^{-1} need not be used in the middle of the tree, where they cancel each other.

This way we also obtain the special cases of weighted OWG and weighted power based generalized OWA functions.

3.4 How to Choose Weights in OWA

3.4.1 Methods Based on Data

The problem of identification of weights of OWA functions was studied by several authors [FY98, FM01, Xu05, YA14b, YK97]. A common feature of all methods is to eliminate nonlinearity due to reordering of the components of **x** by restricting the domain of this function to the simplex $S \subset \mathbb{I}^n$ defined by the inequalities

```
double OWA(int n, double x[],double w[])
{ /* no sorting is needed when used in the tree */
   double z=0;
   for(int i=0;i<n;i++) z+=x[i]*w[i];
   return z; }
double node(int n, double x[], long int N[], long int C, int & k,
      double w[], double(*F)(int, double [],double[]), double* z)
{  /* recursive function in the n-ary tree processing
   Parameters: x - input vector, N vector of multiplicities of  x
   m current level of recursion counted from L (root node)  to 0
   k - input-output parameter, the index of x being processed */
   if(N[k]==0) k++;
   if(N[k]>= C) {  /* we use idempotency to prune the tree */
       N[k] -= C;
       if(N[k]<=0) return x[k++]; else return x[k];    }
   C /= n;
   /* tree not pruned, process the children nodes */
   for(int i=0;i<n;i++) z[i]=node(n,x,N,C,k,w,F,z+n);
   return F(n,z,w);
}


double weightedf(double x[], double p[], double w[], int n,
         double(*F)(int, double[],double[]), int L)
/* Function F is the symmetric base aggregator.
 p[] = array of weights of inputs x[],
 w[] = array of weights for OWA, n = the dimension of x, p, w.
 the weights must add to one and be non-negative.
 L = number of binary tree levels. Run time = O[(n-1)L]  */
{  long int t=0, C=1;
   int k=0;
   for(int i=0;i<L;i++) C*=n;   /* C=n^L */
   sortpairs(x, x+n, p);
   long int N[n]; /* multiplicities of x based on the weights */
   for(int i=0;i<n-1;i++) { N[i]=p[i]*C+1./n;   t+=N[i]; }
   N[n-1]=C-t;
   double z[n*L]; /* working memory */
   return node(n,x,N,C,k,w,F,z);
}
/* example: calling the function */
int n=4, L=4;
double x[4]={0.2,0.2,0.4,0.8};
double w[4]={0.1,0.2,0.3,0.4};
double p[4]={0.3,0.2,0.1,0.4};
double y=weightedf(x,p,w,n,&OWA,L);
```

Fig. 3.7 A C++ implementation of the pruned n-ary tree algorithm PnTA. The function sortpairs (not shown) implements sorting of an array of pairs (x_i, p_i) in the order of decreasing x_i

$x_1 \geq x_2 \geq \cdots \geq x_n$. On that domain OWA function is a linear function (it coincides with the arithmetic mean). Once the coefficients of this function are found, OWA function can be computed on the whole $\mathbb{I}^n$ by using its symmetry. Algorithmically, it amounts to using an auxiliary data set $\{(\mathbf{z}_k, y_k)\}$, where vectors $\mathbf{z}_k = \mathbf{x}_{k\searrow}$. Thus identification of weights of OWA functions is a very similar problem to identification of weights of arithmetic means in Sect. 2.3.7. Depending on whether we use least squares or least absolute deviation criterion, we solve it by using either quadratic or linear programming techniques. In the first case we have the problem

$$\min \quad \sum_{k=1}^{K} \left(\sum_{i=1}^{n} w_i z_{ik} - y_k \right)^2 \tag{3.8}$$

$$\text{s.t.} \ \sum_{i=1}^{n} w_i = 1, w_i \geq 0, i = 1, \ldots, n.$$

In the second case we have

$$\min \quad \sum_{k=1}^{K} \left| \sum_{i=1}^{n} w_i z_{ik} - y_k \right| \tag{3.9}$$

$$\text{s.t.} \ \sum_{i=1}^{n} w_i = 1, w_i \geq 0, i = 1, \ldots, n,$$

which converts to a linear programming problem, see Sect. 1.7.1.

Filev and Yager [FY98] proposed a nonlinear change in variables to obtain an unrestricted minimization problem, which they propose to solve using nonlinear local optimization methods. Unfortunately the resulting nonlinear optimization problem is difficult due to a large number of local minimizers, and the traditional optimization methods are stuck in the local minima.

The approach relying on quadratic programming was used in [Bel03, Bel05, Tor02, Tor04, YF94], and it was shown to be numerically efficient and stable with respect to rank deficiency (e.g., when $K < n$, or the data are linearly dependent).

Often an additional requirement is imposed: the desired value of the measure of orness $orness(f) = \alpha \in [0, 1]$. This requirement is easily incorporated into a QP or LP problem as an additional linear equality constraint, namely

$$\sum_{i=1}^{n} w_i \frac{n-i}{n-1} = \alpha.$$

Preservation of ordering of the outputs

We may also require that the ordering of the outputs is preserved, i.e., if $y_j \leq y_k$ then we expect $f(\mathbf{x}_j) \leq f(\mathbf{x}_k)$ (see Sect. 1.6, p. 32). We arrange the data, so that the outputs are in non-decreasing order, $y_k \leq y_{k+1}, k = 1, \ldots, K - 1$. Then we define the additional linear constraints

$$\langle \mathbf{z}_{k+1} - \mathbf{z}_k, \mathbf{w} \rangle = \sum_{i=1}^{n} w_i (z_{i,k+1} - z_{ik}) \geq 0,$$

$k = 1, \ldots, K - 1$. We add the above constraints to problem (3.8) or (3.9) and solve it. The addition of an extra $K - 1$ constraints neither changes the structure of the optimization problem, nor does it drastically affect its complexity.

3.4.2 Methods Based on a Measure of Dispersion

Maximum entropy OWA

A different approach to choosing OWA weights was proposed in [O'H88] and followed in [FM01]. It does not use any empirical data, but various measures of weight entropy or dispersion. The measure of weights dispersion (see Definition 2.9 on p. 61, also see p. 106) is defined as

$$Disp(\mathbf{w}) = -\sum_{i=1}^{n} w_i \log w_i, \tag{3.10}$$

The idea is to choose for a given n such a vector of weights that maximizes the dispersion $Disp(\mathbf{w})$.

It is formulated as an optimization problem

$$\min \quad \sum_{i=1}^{n} w_i \log w_i \tag{3.11}$$

$$\text{s.t.} \quad \sum_{i=1}^{n} w_i = 1,$$

$$\sum_{i=1}^{n} w_i \frac{n-i}{n-1} = \alpha,$$

$$w_i \geq 0, i = 1, \ldots, n.$$

The solution is provided in [FM01] and is called Maximum Entropy OWA (MEOWA). Using the method of Lagrange multipliers, the authors obtain the following expressions for w_i:

$$w_i = (w_1^{n-i} w_n^{i-1})^{\frac{1}{n-1}}, i = 2, \ldots, n - 1, \tag{3.12}$$

$$w_n = \frac{((n-1)\alpha - n)w_1 + 1}{(n-1)\alpha + 1 - nw_1},$$

and w_1 being the unique solution to the equation

$$w_1[(n-1)\alpha + 1 - nw_1]^n = ((n-1)\alpha)^{n-1}[((n-1)\alpha - n)w_1 + 1] \qquad (3.13)$$

on the interval $(0, \frac{1}{n})$.

Note 3.23 For $n = 3$, we obtain $w_2 = \sqrt{w_1 w_3}$ independently of the value of α.

A different representation of the same solution was given in [CMM97]. Let t be the (unique) positive solution to the equation

$$dt^{n-1} + (d+1)t^{n-2} + \cdots + (d+n-2)t + (d+n-1) = 0, \qquad (3.14)$$

with $d = -\alpha(n-1)$. Then the MEOWA weights are identified from

$$w_i = \frac{t^i}{T}, \quad i = 1, \ldots, n, \quad \text{where } T = \sum_{j=1}^{n} t^j. \qquad (3.15)$$

Note 3.24 It is not difficult to check that both (3.12) and (3.15) represent the same set of weights, noting that $t = \sqrt[n-1]{\frac{w_n}{w_1}} = -\frac{1-d-nw_1}{d}$, or $w_1 = \frac{1+td-d}{n}$, and that substituting w_1 into (3.13) yields

$$1 - t^n = \frac{n(1-t)}{1 - d(1-t)},$$

which translates into

$$\frac{1 - t^n}{1 - t} - d(1 - t^n) - n = 0,$$

and then into

$$dt^n + t^{n-1} + t^{n-2} + \cdots + t + (1 - d - n) = 0.$$

After factoring out $(t - 1)$ we obtain (3.14).

Minimum variance OWA

Another popular characteristic of weighting vector is weights variance, defined as [FM03]

$$D^2(\mathbf{w}) = \frac{1}{n} \sum_{i=1}^{n} (w_i - M(\mathbf{w}))^2 = \frac{1}{n} \sum_{i=1}^{n} w_i^2 - \frac{1}{n^2}, \qquad (3.16)$$

where $M(\mathbf{w})$ is the arithmetic mean of $\mathbf{w}$.

Here one minimizes $D^2(\mathbf{w})$ subject to given orness measure. The resulting OWA function is called Minumum Variance OWA (MVOWA). Since adding a constant to the objective function does not change the minimizer, this is equivalent to the problem

$$\min \qquad \sum_{i=1}^{n} w_i^2 \qquad\qquad (3.17)$$

$$\text{s.t.} \qquad \sum_{i=1}^{n} w_i \frac{n-i}{n-1} = \alpha,$$

$$\sum_{i=1}^{n} w_i = 1, \, w_i \geq 0, i = 1, \ldots, n.$$

For $\alpha = \frac{1}{2}$ the optimal solution is always $w_j = \frac{1}{n}$, $j = 1, \ldots, n$. It is also worth noting that the optimal solution to (3.17) for $\alpha > \frac{1}{2}$, $\mathbf{w}^*$, is related to the optimal solution for $\alpha < \frac{1}{2}$, $\mathbf{w}$, by $w_i^* = w_{n-i+1}$, i.e., it gives the reverse OWA. Thus it is sufficient to establish the optimal solution in the case $\alpha < \frac{1}{2}$.

The optimal solution [FM03, Liu07] for $\alpha < \frac{1}{2}$ is given as the vector $\mathbf{w} = (0, 0, \ldots, 0, w_r, \ldots, w_n)$, i.e., $w_j = 0$ if $j < r$, and

$$w_r = \frac{6(n-1)\alpha - 2(n-r-1)}{(n-r+1)(n-r+2)},$$

$$w_n = \frac{2(2n - 2r + 1) - 6(n-1)\alpha}{(n-r+1)(n-r+2)},$$

and

$$w_j = w_r + \frac{j-r}{n-r}(w_n - w_r), \ r < j < n.$$

The index r depends on the value of α, and is found from the inequalities

$$n - 3(n-1)\alpha - 1 < r \leq n - 3(n-1)\alpha.$$

Recently it was established [Liu07] that the solution to the minimum variance OWA weights problem is equivalent to that of minimax disparity [WP05], i.e., the solution to

$$\min \qquad \left\{ \max_{i=1,\ldots,n-1} |w_i - w_{i-1}| \right\} \qquad\qquad (3.18)$$

$$\text{s.t.} \qquad \sum_{i=1}^{n} w_i \frac{n-i}{n-1} = \alpha,$$

$$\sum_{i=1}^{n} w_i = 1, \, w_i \geq 0, i = 1, \ldots, n.$$

We reiterate that the weights of OWA functions obtained as solutions to maximum entropy or minimum variance problems are fixed for any given n and orness measure, and can be precomputed. However, both criteria are also useful for data driven weights identification (in Sect. 3.4.1), if there are multiple optimal solutions. Then the solution maximizing $Disp(\mathbf{w})$ or minimizing $D(\mathbf{w})$ is chosen. Torra [Tor02] proposes to solve an auxiliary univariate optimization problem to maximize weights dispersion, subject to a given value of (3.10). On the other hand, one can fit the orness value α of MEOWA or MVOWA to empirical data, using a univariate nonlinear optimization method, in which at each iteration the vector $\mathbf{w}$ is computed using analytical solutions to problems (3.11) and (3.17).

Furthermore, it is possible to include both criteria directly into problem (3.8). It is especially convenient for the minimum variance criterion, as it yields a modified quadratic programming problem

$$\min \sum_{k=1}^{K} \left(\sum_{i=1}^{n} w_i z_{ik} - y_k \right)^2 + \lambda \sum_{i=1}^{n} w_i^2 \tag{3.19}$$

$$\text{s.t.} \qquad \sum_{i=1}^{n} w_i \frac{n-i}{n-1} = \alpha,$$

$$\sum_{i=1}^{n} w_i = 1, w_i \geq 0, i = 1, \ldots, n,$$

where $\lambda \geq 0$ is a user-specified parameter controlling the balance between the criterion of fitting the data and that of obtaining minimum variance weights.

3.4.3 Methods Based on Weight Generating Functions

Yager [Yag91, Yag96] has proposed to use monotone continuous functions Q : $[0, 1] \rightarrow [0, 1]$, $Q(0) = 0$, $Q(1) = 1$, called Basic Unit-interval Monotone (BUM) functions, or Regular Increasing Monotone (RIM) quantifiers [Yag91]. These functions generate OWA weights for any n using (see Sect. 2.3.5)

$$w_i = Q\left(\frac{i}{n}\right) - Q\left(\frac{i-1}{n}\right). \tag{3.20}$$

RIM quantifiers are fuzzy linguistic quantifiers[1] that express the concept of fuzzy majority. Yager defined such quantifiers for fuzzy sets "for all", "there exists", "identity", "most", "at least half", "as many as possible" as follows.

- "for all": $Q_{forall}(t) = 0$ for all $t \in [0, 1)$ and $Q_{forall}(1) = 1$.
- "there exists": $Q_{exists}(t) = 1$ for all $t \in (0, 1]$ and $Q_{exists}(0) = 0$.
- "identity": $Q_{Id}(t) = t$.

---

[1]I.e., Q is a monotone increasing function $[0, 1] \rightarrow [0, 1]$, $Q(0) = 0$, $Q(1) = 1$ whose value $Q(t)$ represents the degree to which t satisfies the fuzzy concept represented by the quantifier.

Other mentioned quantifiers are expressed by

$$Q_{a,b}(t) = \begin{cases} 0, & \text{if } t \le a, \\ \frac{t-a}{b-a} & \text{if } a < t < b, \\ 1, & \text{if } t \ge b. \end{cases} \tag{3.21}$$

Then we can choose pairs $(a, b) = (0.3, 0.8)$ for "most", $(a, b) = (0, 0.5)$ for "at least half" and $(a, b) = (0.5, 1)$ for "as many as possible".

Calculation of weights results in the following OWA:

- "for all": $\mathbf{w} = (0, 0, \dots, 0, 1)$, $OWA_{\mathbf{w}} = \min$.
- "there exists": $\mathbf{w} = (1, 0, 0, \dots, 0)$, $OWA_{\mathbf{w}} = \max$.
- "identity": $\mathbf{w} = (\frac{1}{n}, \dots, \frac{1}{n})$, $OWA_{\mathbf{w}} = M$.

Example 3.25 Consider linguistic quantifier "most", given by (3.21) with $(a, b) = (0.3, 0.8)$ and $n = 5$. The weighting vector is then $(0, 0.2, 0.4, 0.4, 0)$.

Weight generating functions are applied to generate weights of both quasi-arithmetic means and OWA functions. They allow one to compute the degree of orness of an OWA function in the limiting case

$$\lim_{n \to \infty} orness(f_n) = orness(Q) = \int_0^1 Q(t)dt.$$

Entropy and other characteristics can also be computed based on Q, see [TY06].

Yager [Yag07b] has proposed using generating, or stress functions (see also [Liu05]), defined by

Definition 3.26 (*Generating function of RIM quantifiers*) Let $q : [0, 1] \to [0, \infty]$ be an (integrable) function. It is a generating function of RIM quantifier Q, if

$$Q(t) = \frac{1}{K} \int_0^t q(u)du,$$

where $K = \int_0^1 q(u)du$ is the normalization constant. The normalized generating function will be referred to as $\tilde{q}(t) = \frac{q(t)}{K}$.

Note 3.27 The generating function has the properties of a density function (e.g., a probability distribution density, although Q is not necessarily interpreted as a probability). If Q is differentiable, we may put $q(t) = Q'(t)$. Of course, for a given Q, if a generating function exists, it is not unique.

Note 3.28 In general, Q needs not be continuous to have a generating function. For example, it may be generated by Dirac's delta function[2]

---

[2]This is an informal definition. The proper definition involves the concepts of distributions and measures, see, e.g., [Rud91].

$$\delta(t) = \begin{cases} \infty, \text{ if } t = 0, \\ 0 \quad \text{otherwise,} \end{cases}$$

constrained by $\int_{-\infty}^{\infty} \delta(t)dt = 1$.

By using the generating function we generate the weights as

$$\tilde{w}_i = \frac{1}{K} q\left(\frac{i}{n}\right) \frac{1}{n}.$$

Note that these weights provide an approximation to the weights generated by (3.20), and that they do not necessarily sum to one. To ensure the latter, we shall use the weights

$$w_i = \frac{\frac{1}{K} q\left(\frac{i}{n}\right) \frac{1}{n}}{\sum_{j=1}^{n} \frac{1}{K} q\left(\frac{i}{n}\right) \frac{1}{n}} = \frac{q\left(\frac{i}{n}\right)}{\sum_{j=1}^{n} q\left(\frac{i}{n}\right)}. \tag{3.22}$$

Equation (3.22) provides an alternative method for OWA weight generation, independent of Q, while at the same time it gives an approximation to the weights provided by (3.20). Various interpretations of generating functions are provided in [Yag07b], from which we quote just a few examples.

Example 3.29

- A constant generating function $q(t) = 1$ generates weights $w_i = \frac{1}{n}$, i.e., the arithmetic mean.
- Constant in range function $q(t) = 1$ for $t \le \beta$ and 0 otherwise, emphasizes the larger arguments, and generates the weights $w_i = \frac{1}{r}, i = 1, \ldots, r$ and $w_i = 0, i = r + 1, \ldots, n$, where r is the largest integer less or equal βn.
- Generating function $q(t) = 1$, for $\alpha \le t \le \beta$ and 0 otherwise, emphasizes the "middle" arguments, and generates the weights $w_i = \frac{1}{p-r}, i = r + 1, \ldots, p$ and 0 otherwise, with (for simplicity) $\alpha n = r$ and $\beta n = p$.
- Generating function with two tails $q(t) = 1$ if $t \in [0, \alpha]$ or $t \in [\beta, 1]$ and 0 otherwise, emphasizes both large and small arguments and yields $w_i = \frac{1}{r_1 + r_2}$ for $i = 1, \ldots, r_1$ and $i = n + 1 - r_2, \ldots, n$, and $w_i = 0, i = r_1 + 1, \ldots, n - r_2$, with $r_1 = \alpha n, r_2 = \beta n$ integers.
- Linear stress function $q(t) = t$ generates weights $w_i = \frac{i}{\sum_{j=1}^{n} i} = \frac{2i}{n(n+1)}$, which gives orness value $\frac{1}{3}$, compare to Example 3.4. It emphasizes smaller arguments.

Of course, by using the same approach (i.e., $Q(t)$ or $q(t)$) one can generate the weights of generalized OWA and weighted quasi-arithmetic means. However the interpretation and the limiting cases for the means will be different. For example the weighting vector $\mathbf{w} = (1, 0, \ldots, 0)$ results not in the *max* function, but in the projection to the first coordinate $f(\mathbf{x}) = x_1$.

Table 3.1 A data set with inputs of varying dimension

k	n_k	x_1	x_2	x_3	x_4	x_5	y
1	3	x_{11}	x_{21}	x_{31}			y_1
2	2	x_{12}	x_{22}				y_2
3	3	x_{13}	x_{23}	x_{33}			y_3
4	5	x_{14}	x_{24}	x_{34}	x_{44}	x_{54}	y_4
5	4	x_{15}	x_{25}	x_{35}	x_{45}		y_5
$\vdots$							

3.4.4 Fitting Weight Generating Functions

Weight generating functions allow one to compute weighting vectors of OWA and weighted means for any number of arguments, i.e., to obtain extended aggregation functions in the sense of Definition 1.6. This is very convenient when the number of arguments is not known a priori. Next we pose the question as to whether it is possible to learn weight generating functions from empirical data, similarly to determining weighting vectors of aggregation functions of a fixed dimension.

A positive answer was provided in [Bel05, BMV04]. The method consists in representing a weight generating function with a spline or polynomial, and fitting its coefficients by solving a least squares or least absolute deviation problem subject to a number of linear constraints. Consider a data set $(\mathbf{x}_k, y_k)$, $k = 1, \ldots, K$, where vectors $\mathbf{x}_k \in [0, 1]^{n_k}$ need not have the same dimension (see Table 3.1). This is because we are dealing with an extended aggregation function—a family of n-ary aggregation functions.

First, let us use the method of monotone splines, discussed in [Bel00]. We write

$$Q(t) = \sum_{j=1}^{J} c_j B_j(t), t \in (0, 1) \text{ and } Q(0) = 0, Q(1) = 1,$$

where functions $B_j(t)$ constitute a convenient basis for polynomial splines, [Bel00], in which the condition of monotonicity of Q is expressed as $c_j \geq 0$, $j = 1, \ldots, J$. We do not require Q to be continuous on $[0, 1]$ but only on $]0, 1[$. We also have two linear constraints

$$\sum_{j=1}^{J} c_j B_j(0) \geq 0, \quad \sum_{j=1}^{J} c_j B_j(1) \leq 1,$$

which convert to equalities if we want Q to be continuous on $[0, 1]$. Next put this expression in (3.20) to get

$$f(\mathbf{x}_k) = \sum_{i=1}^{n_k} z_{ik} \left(Q\left(\frac{i}{n_k}\right) - Q\left(\frac{i-1}{n_k}\right) \right)$$

$$= \sum_{i=2}^{n_k-1} z_{ik} \left(\sum_{j=1}^{J} c_j \left[B_j\left(\frac{i}{n_k}\right) - B_j\left(\frac{i-1}{n_k}\right) \right] \right)$$

$$+ z_{1k} \left[\sum_{j=1}^{J} c_j B_j\left(\frac{1}{n_k}\right) - 0 \right] + z_{n_k k} \left[1 - \sum_{j=1}^{J} c_j B_j\left(\frac{n_k-1}{n_k}\right) \right]$$

$$= \sum_{j=1}^{J} c_j \left(\sum_{i=2}^{n_k-1} z_{ik} \left[B_j\left(\frac{i}{n_k}\right) - B_j\left(\frac{i-1}{n_k}\right) \right] \right.$$

$$\left. + z_{1k} B_j\left(\frac{1}{n_k}\right) - z_{n_k k} B_j\left(\frac{n_k-1}{n_k}\right) \right) + z_{n_k k}$$

$$= \sum_{j=1}^{J} c_j A_j(\mathbf{x}_k) + z_{n_k k}.$$

The vectors $\mathbf{z}_k$ stand for $\mathbf{x}_k$ when we treat weighted arithmetic means, or $\mathbf{x}_{k\searrow}$ when we deal with OWA functions. The entries $A_j(\mathbf{x}_k)$ are computed from z_{ik} using expression in the brackets. Note that if Q is continuous on $[0, 1]$ the expression simplifies to

$$f(\mathbf{x}_k) = \sum_{j=1}^{J} c_j \left(\sum_{i=1}^{n_k} z_{ik} \left[B_j\left(\frac{i}{n_k}\right) - B_j\left(\frac{i-1}{n_k}\right) \right] \right).$$

Consider now the least squares approximation of empirical data. We obtain a quadratic programming problem

$$\text{minimize} \quad \sum_{k=1}^{K} \left(\sum_{j=1}^{J} c_j A_j(\mathbf{x}_k) + z_{n_k k} - y_k \right)^2 \tag{3.23}$$

$$\text{s.t.} \quad \sum_{j=1}^{J} c_j B_j(0) \geq 0, \quad \sum_{j=1}^{J} c_j B_j(1) \leq 1,$$

$$c_j \geq 0.$$

The solution is performed by QP programming methods (Sect. 1.7).

OWA aggregation functions and weighted arithmetic means are special cases of Choquet integral based aggregation functions, described in the next chapter. Choquet integrals are defined with respect to a fuzzy measure (see Definition 4.2). When the fuzzy measure is additive, Choquet integrals become weighted arithmetic means, and when the fuzzy measure is symmetric, they become OWA functions. There are special classes of fuzzy measures called k-additive measures (see Definition 4.50). We will discuss them in detail in Sect. 4.1.3, and in the remainder of this section we

will present a method for identifying weight generating functions that correspond
to symmetric 2- and 3-additive fuzzy measures. These fuzzy measures lead to OWA
functions with special weights distributions.

Proposition 3.30 [BMV04] *A Choquet integral based aggregation function with
respect to a symmetric 2-additive fuzzy measure is an OWA function whose weight
generating function is given by*

$$Q(t) = at^2 + (1 - a)t \text{ for some } a \in [-1, 1].$$

*Furthermore, such an OWA weighting vector is equidistant (i.e., $w_{i+1} - w_i = const$
for all $i = 1, \ldots, n - 1$).*

*A Choquet integral based aggregation function with respect to a symmetric
3-additive fuzzy measure is an OWA function whose weight generating function is
given by*

$$Q(t) = at^3 + bt^2 + (1 - a - b)t \text{ for some } a \in [-2, 4],$$

such that

- *if $a \in [-2, 1]$ then $b \in [-2a - 1, 1 - a]$;*
- *if $a \in]1, 4]$ then $b \in [-3a/2 - \sqrt{3a(4 - a)/4}, -3a/2 + \sqrt{3a(4 - a)/4}]$.*

Proposition 3.30 provides two parametric classes of OWA functions that correspond to 2- and 3-additive symmetric fuzzy measures. In these cases, rather than
fitting a general monotone non-decreasing function, we fit a quadratic or cubic function, identified by parameters a and b.

Interestingly, in the case of 2-additive symmetric fuzzy measure, we obtain the
following formula, a linear combination of OWA and the arithmetic mean

$$f(\mathbf{x}) = aOWA_{\mathbf{w}}(\mathbf{x}) + (1 - a)M(\mathbf{x}),$$

with $\mathbf{w} = (\frac{1}{n^2}, \frac{3}{n^2}, \ldots, \frac{2n-1}{n^2})$. In this case the solution is explicit, the optimal a is
given by [BMV04]

$$a = \max \left\{ -1, \min \left\{ 1, \frac{\sum\limits_{k=1}^{K} (y_k - U_k)V_k}{\sum\limits_{k=1}^{K} V_k^2} \right\} \right\},$$

where

$$U_k = \sum_{i=1}^{n_k} \frac{z_{ik}}{n_k}, \text{ and}$$

$$V_k = \frac{1}{n_k} \sum_{i=1}^{n_k} \left(\frac{2i - 1}{n_k} - 1 \right) z_{ik}.$$

For 3-additive symmetric fuzzy measures the solution is found numerically by solving a convex optimization problem in the feasible domain D in Proposition 3.30, which is the intersection of a polytope and an ellipse. Details are provided in [BMV04].

3.4.5 Choosing Parameters of Generalized OWA

Choosing weight generating functions

Consider the case of generalized OWA functions, where a given generating function g is given. As earlier, we have a data set $(\mathbf{x}_k, y_k)$, $k = 1, \ldots, K$, and we are interested in finding the weighting vector $\mathbf{w}$ that fits the data best. When we use the least squares, as discussed on p. 34, we have the following optimization problem

$$\min \sum_{k=1}^{K} \left(g^{-1} \left(\sum_{i=1}^{n} w_i g(z_{ik}) \right) - y_k \right)^2 \tag{3.24}$$

$$\text{s.t. } \sum_{i=1}^{n} w_i = 1, w_i \geq 0, i = 1, \ldots, n,$$

where $\mathbf{z}_k = \mathbf{x}_{k\searrow}$ (see Sect. 3.4). This problem is converted to a QP problem similarly to the case of weighted quasi-arithmetic means:

$$\min \sum_{k=1}^{K} \left(\sum_{i=1}^{n} w_i g(z_{ik}) - g(y_k) \right)^2 \tag{3.25}$$

$$\text{s.t. } \sum_{i=1}^{n} w_i = 1, w_i \geq 0, i = 1, \ldots, n.$$

This is a standard convex QP problem. This approach is presented in [Bel05].

If one uses the least absolute deviation (LAD) criterion (p. 34) we obtain a different optimization problem

$$\min \sum_{k=1}^{K} \left| \sum_{i=1}^{n} w_i g(z_{ik}) - g(y_k) \right| \tag{3.26}$$

$$\text{s.t. } \sum_{i=1}^{n} w_i = 1, w_i \geq 0, i = 1, \ldots, n.$$

This problem is subsequently converted into a linear programming problem as discussed in the Sect. 1.7.1.

As in the case of weighted quasi-arithmetic means, in the presence of multiple optimal solutions, one can use an additional criterion of the dispersion of weights [Tor02].

Preservation of ordering of the outputs

If we require that the ordering of the outputs be preserved, i.e., if $y_j \le y_k$ then we expect $f(\mathbf{x}_j) \le f(\mathbf{x}_k)$ (see Sect. 1.6, p. 32), then we arrange the data, so that the outputs are in a non-decreasing order, $y_k \le y_{k+1}, k = 1, \ldots, K - 1$. Then we define the additional linear constraints

$$\langle \mathbf{g}(z_{k+1}) - g(\mathbf{z}_k), \mathbf{w} \rangle = \sum_{i=1}^{n} w_i (g(z_{i,k+1}) - g(z_{ik})) \ge 0,$$

$k = 1, \ldots, K - 1$. We add the above constraints to problem (3.25) or (3.26) and solve the modified problem.

Choosing generating functions

Consider now the case where the generating function g is also unknown, and hence it has to be found based on the data. We study two cases: (a) when g is given algebraically, with one or more unknown parameters to estimate (e.g., $g_r(t) = t^r$, r unknown), and (b) when no specific algebraic form of g is given.

In the first case we solve the problem

$$\min_{r, \mathbf{w}} \sum_{k=1}^{K} \left(\sum_{i=1}^{n} w_i g_r(z_{ik}) - g_r(y_k) \right)^2 \tag{3.27}$$

$$\text{s.t. } \sum_{i=1}^{n} w_i = 1, w_i \ge 0, i = 1, \ldots, n,$$

plus conditions on r.

While this general optimization problem is non-convex and nonlinear (i.e., difficult to solve), we can convert it to a bi-level optimization problem (see Sect. 1.7.4)

$$\min_{r} \left[\min_{\mathbf{w}} \sum_{k=1}^{K} \left(\sum_{i=1}^{n} w_i g_r(z_{ik}) - g_r(y_k) \right)^2 \right] \tag{3.28}$$

$$\text{s.t. } \sum_{i=1}^{n} w_i = 1, w_i \ge 0, i = 1, \ldots, n,$$

plus conditions on r.

The problem at the inner level is the same as (3.25) with a fixed g_r, which is a QP problem. At the outer level we have a global optimization problem with respect to a single parameter r. It is solved by using one of the methods discussed in Sect. 1.7.4. We recommend deterministic Pijavski-Shubert method.

Example 3.31 Determine the weights and the generating function of a family of generalized OWA based on the power function, subject to a given measure of orness α. We have $g_r(t) = t^r$, and hence solve bi-level optimization problem

$$\min_{r\in[-\infty,\infty]} \left[\min_{\mathbf{w}} \sum_{k=1}^{K} \left(\sum_{i=1}^{n} w_i z_{ik}^r - y_k^r \right)^2 \right] \tag{3.29}$$

$$\text{s.t.} \sum_{i=1}^{n} w_i = 1, w_i \geq 0, i = 1, \ldots, n,$$

$$\sum_{i=1}^{n} w_i \left(\frac{n-i}{n-1} \right)^r = \alpha.$$

Of course, for numerical purposes we need to limit the range for r to a finite interval, and treat all the limiting cases $r \to \pm\infty, r \to 0$ and $r \to -1$.

A different situation arises when the parametric form of g is not given. The approach proposed in [Bel03, Bel05] is based on approximation of g with a monotone linear spline, as

$$g(t) = \sum_{j=1}^{J} c_j B_j(t), \tag{3.30}$$

where B_j are appropriately chosen basis functions, and c_j are spline coefficients. The monotonicity of g is ensured by imposing linear restrictions on spline coefficients, in particular non-negativity, as in [Bel00]. Further, since the generating function is defined up to an arbitrary linear transformation, one has to fix a particular g by specifying two interpolation conditions, like $g(a) = 0, g(b) = 1, a, b \in]0, 1[$, and if necessary, properly model asymptotic behavior if $g(0)$ or $g(1)$ are infinite.

After rearranging the terms of the sum, the problem of identification becomes (subject to linear conditions on $\mathbf{c}, \mathbf{w}$)

$$\min_{\mathbf{c},\mathbf{w}} \sum_{k=1}^{K} \left(\sum_{j=1}^{J} c_j \left[\sum_{i=1}^{n} w_i B_j(z_{ik}) - B_j(y_k) \right] \right)^2. \tag{3.31}$$

For a fixed $\mathbf{c}$ (i.e., fixed g) we have a quadratic programming problem to find $\mathbf{w}$, and for a fixed $\mathbf{w}$, we have a quadratic programming problem to find $\mathbf{c}$. However if we consider both $\mathbf{c}, \mathbf{w}$ as variables, we obtain a difficult global optimization problem. We convert it to a bi-level optimization problem

$$\min_{\mathbf{c}} \min_{\mathbf{w}} \sum_{k=1}^{K} \left(\sum_{j=1}^{J} c_j \left[\sum_{i=1}^{n} w_i B_j(z_{ik}) - B_j(y_k) \right] \right)^2,$$

where at the inner level we have a QP problem and at the outer level we have a non-linear problem with multiple local minima. When the number of spline coefficients J is not very large (<10), this problem can be efficiently solved by using deterministic global optimization methods from Sect. 1.7.4. If the number of variables is small and J is large, then reversing the order of minimization (i.e., using $\min_{\mathbf{w}} \min_{\mathbf{c}}$) is more efficient.

Choosing generating functions and weight generating functions

We remind the definition of Generalized OWA Definition 3.9. Consider the case of generating function $g(t) = t^r$, in which case

$$GenOWA_{\mathbf{w},[r]}(\mathbf{x}) = \left(\sum_{i=1}^{n} w_i z_i^r \right)^{1/r}.$$

Consider first a fixed r. To find a weight generating function $Q(t)$, we first linearize the least squares problem to get

$$\min_{\mathbf{c}} \sum_{k=1}^{K} \left(\sum_{j=1}^{J} c_j \sum_{i=1}^{n_k} \left[B_j \left(\tfrac{i}{n_k} \right) - B_j \left(\tfrac{i-1}{n_k} \right) \right] (z_{ik})^r - y_k^r \right)^2$$

$$\text{s.t.} \qquad \sum_{j=1}^{J} c_j B_j(0) = 0, \quad \sum_{j=1}^{J} c_j B_j(1) = 1,$$

$$c_j \geq 0.$$

This is a standard QP, which differs from (3.23) because z_{ik} and y_k are raised to power r (we considered a simpler case of $Q(t)$ continuous on $[0, 1]$).

Now, if the parameter r is also unknown, we determine it from data by setting a bi-level optimization problem

$$\min_{r} \min_{\mathbf{c}} \sum_{k=1}^{K} \left(\sum_{j=1}^{J} c_j \sum_{i=1}^{n_k} \left[B_j \left(\tfrac{i}{n_k} \right) - B_j \left(\tfrac{i-1}{n_k} \right) \right] (z_{ik})^r - y_k^r \right)^2$$

$$\text{s.t.} \qquad \sum_{j=1}^{J} c_j B_j(0) = 0, \quad \sum_{j=1}^{J} c_j B_j(1) = 1,$$

$$c_j \geq 0,$$

in which at the inner level we solve a QP problem with a fixed r, and at the outer level we optimize with respect to a single nonlinear parameter r, in the same way we did in Example 3.31.

For more complicated case of both generating functions g and Q given non-parametrically (as splines) we refer to [Bel05].

3.5 Induced OWA

The idea of using an auxiliary variable to re-order the inputs had its first inception in the image compression work of Mitchell and Estrakh [ME97] and a follow-up application which sorted the inputs by *fuzzy ranks* [ME98]. In these applications, the arguments were sorted by a function of their values rather than the values themselves. Yager and Filev then formally defined the Induced OWA (IOWA) in [YF99], denoting the auxiliary variable associated with each input as an inducing variable.

3.5.1 Definition

Definition 3.32 (*Induced OWA*) Given a weighting vector **w** and an inducing variable **z** the Induced Ordered Weighted Averaging (IOWA) function is

$$IOWA_{\mathbf{w}}(\langle x_1, z_1 \rangle, \ldots, \langle x_n, z_n \rangle) = \sum_{i=1}^{n} w_i x_{(i)}, \qquad (3.32)$$

where the (.) notation denotes the inputs $\langle x_1, z_1 \rangle$ reordered such that $z_{(1)} \geq z_{(2)} \geq \cdots \geq z_{(n)}$ and the convention that if q of the $z_{(i)}$ are tied, i.e. $z_{(i)} = z_{(i+1)} = \cdots = z_{(i+q-1)}$,

$$x_{(i)} = \frac{1}{q+1} \sum_{j=(i)}^{(i+q-1)} x_j.$$

The input pairs $\langle x_i, z_i \rangle$ may be two independent features of the same input, or can be related by some function, i.e. $z_i = f_i(x_i)$.

3.5.2 Properties

One immediately notices the similarity between the IOWA in Eq. (3.32) and the OWA defined in Eq. (3.1). With the exception of ties, the value obtained from an IOWA will be the same as that obtained from an OWA whose weights are permuted in a different order.

Proposition 3.33 *Given an inducing variable* **z** *where* $z_i = z_j \Leftrightarrow i = j$ *and a fixed input vector* **x**,
$$IOWA_{\mathbf{w}}(\langle \mathbf{x}, \mathbf{z} \rangle) = OWA_{\mathbf{u}}(\mathbf{x}).$$

where $u_{\sigma(i)} = u_{\sigma(1)}, \ldots, u_{\sigma(n)} = \mathbf{w}$, *and* $\sigma(i)$ *is some permutation of the values in the weighting vector* **u**.

Although the properties and behavior of the IOWA will be largely dependent on the inducing variable $\mathbf{z}$, in most cases it will exhibit the properties which have shown to hold for the standard OWA, namely monotonicity, idempotency and symmetry. There are, however, certain choices for $\mathbf{z}$ such that symmetry and even monotonicity may be violated. We will discuss these properties with respect to the input vector $\mathbf{x}$.

- **Monotonicity**: For a fixed vector $\mathbf{z}$, it will hold that

$$\mathbf{x} \le \mathbf{y} \Rightarrow IOWA_{\mathbf{w}}(\langle \mathbf{x}, \mathbf{z} \rangle) \le IOWA_{\mathbf{w}}(\langle \mathbf{y}, \mathbf{z} \rangle).$$

All values of the weighting vector $\mathbf{w}$ satisfy $w_i \ge 0$, hence an increase to any of the x_i cannot decrease the overall output. If we have $w_i > 0, \forall i$, the IOWA will be strictly monotone increasing.

Note 3.34 In some situations, however, the inducing variable $\mathbf{z}$ may change when $\mathbf{x}$ changes. As an example, suppose we have input pairs $\langle x_i, z_i \rangle_{i=1,...,3}$ obtained from 3 observation stations. The z_i is the reliability of the reading x_i, based partially on the value of x as well as external information. We want to aggregate the x_i, giving preference to the most reliable readings. We hence use an IOWA which orders the observations from most to least reliable in accordance with the values z_i, and define a weighting vector with decreasing weights. For a given reading, the input vector has an induced order of $z_2 \ge z_1 \ge z_3$, so the value x_2 is allocated the largest weight w_1. Now suppose the value of x_2 increases to a value that is unusual at Station 2 and is hence less reliable. The value of z_2 decreases and the induced order is now $z_1 \ge z_2 \ge z_3$. This results in a smaller weight being given to the input x_2 and monotonicity is violated.

- **Averaging/Idempotency**: The positive and normalized (i.e. $\sum_{i=1}^{n} w_i = 1$) values of the weighting vector $\mathbf{w}$ ensure that the IOWA function will be both averaging and idempotent. I.e.,

$$\min(\mathbf{x}) \le \text{IOWA}_{\mathbf{w}}(\langle \mathbf{x}, \mathbf{z} \rangle) \le \max(\mathbf{x}),$$

$$\text{IOWA}_{\mathbf{w}}(\langle t, z_1 \rangle, \langle t, z_2 \rangle, ..., \langle t, z_n \rangle) = t.$$

Note 3.35 In the study of aggregation functions, averaging behavior and idempotency are usually considered to be equivalent. Averaging behavior necessarily implies idempotency, however the property of idempotency is not sufficient for averaging behavior without monotonicity. As discussed previously in Note 3.34, an IOWA-type function may not necessarily be monotone, however this does not cause either of these properties to be lost.

- **Symmetry** : With respect to the input pairs $\langle \mathbf{x}, \mathbf{z} \rangle$, the initial indexing is unimportant, e.g. for $n = 2$,

$$\text{IOWA}_{\mathbf{w}}(\langle x_1, z_1 \rangle, \langle x_2, z_2 \rangle) = \text{IOWA}_{\mathbf{w}}(\langle x_2, z_2 \rangle, \langle x_1, z_1 \rangle).$$

With respect to the input vector **x**, however, certain choices of **z** may result in loss of the symmetry property. For instance, the z_i may be constant or a function that is somewhat dependant on the initial indexing, e.g. $z_i = f(i)$. Such choices can induce the same order. As will be shown below, inducing variables that give a fixed calculation order result in weighted arithmetic means. If such instances arise in practice, it may make more sense to perceive the problem in terms of arithmetic means with a particular weight associated to each input, rather than defining a so-called inducing variable.

- **Homogeneity and Shift-Invariance**: Standard OWA functions are both homogeneous and shift-invariant. For the induced OWA, this will once again be somewhat dependent on the choice of inducing variable. The z_i may be specifically associated with each x_i, but needn't be a function of the actual values, i.e. $x_i = x_j$ does not necessarily imply $z_i = z_j$. Given any input pair $\langle x_i, z_i \rangle$, let $z_{i+\lambda}$ be the value associated with the input $x_i + \lambda$ and $z_{\lambda i}$ be the value associated with the input $\lambda \cdot x_i$.

    The induced OWA will be shift-invariant if:

    $$z_i < z_j \Leftrightarrow z_{i+\lambda} < z_{j+\lambda}, \forall i, j.$$

    The induced OWA will be homogeneous if:

    $$z_i < z_j \Leftrightarrow z_{\lambda i} < z_{\lambda j}, \forall i, j (\lambda > 0).$$

In other words, provided λ does not change the relative ordering, IOWA will be stable for translations and homogenous. It should be noted that equalities should also be preserved, which follows from the above equations.

- **Dual**: As with the standard OWA, the dual of an IOWA can be defined as the IOWA with respect to the reverse weighting vector, $\mathbf{w}_d = (w_n, w_{n-1}, ..., w_1)$. It is also possible to define the dual function of an IOWA by the inducing variable $\mathbf{z}_d$ which induces the reverse ordering to **z**.

Special cases

Standard OWA   The Induced OWA includes the standard OWA as a special case whenever the ordering of the inducing variable corresponds exactly with the order of the input variable, e.g. $z_i = x_i$. Clearly, the special cases of the OWA: the minimum, maximum and median can be obtained through this choice and the appropriate selection of the weighting vector **w**;

   Reverse OWA   The reverse OWA is a function which orders the inputs in nondecreasing order $x_{(1)} \leq \cdots \leq x_{(n)}$. The inducing variable **z** can hence be chosen such that this order is achieved, e.g. $z_i = 1 - x_i$;

Weighted Arithmetic Mean Any selection of $\mathbf{z}$ that maintains the order of the input components z_i when arranged in non-increasing order, e.g. $z_i = n + 1 - i$ will result in the weighted arithmetic mean $\sum_{i=1}^{n} w_i x_i$.

3.5.3 Induced Generalized OWA

Following from the definition of the generalized OWA (Definition 3.3.2), Chiclana et al. introduced the natural extension of the IOWA to the IOWG [CHVHA04]. Of course, this corresponds to a special case of the Induced generalized OWA.

Definition 3.36 (*Induced Generalized OWA*) Given a weighting vector $\mathbf{w}$, an inducing variable $\mathbf{z}$ and a continuous strictly monotone function $g : \mathbb{I} \rightarrow [-\infty, \infty]$, the Induced Generalized OWA (I-GenOWA) function is

$$\textit{I-GenOWA}_{\mathbf{w},g}(\mathbf{x}) = g^{-1} \left(\sum_{i=1}^{n} w_i g(x_{(i)}) \right). \tag{3.33}$$

As for IOWA, $(.)$ notation denotes the inputs $\langle x_1, z_1 \rangle$ reordered such that $z_{(1)} \geq z_{(2)} \geq \cdots \geq z_{(n)}$ and the convention that if q of the $z_{(i)}$ are tied, i.e. $z_{(i)} = z_{(i+1)} = \cdots = z_{(i+q-1)}$,

$$x_{(i)} = \frac{1}{q+1} \sum_{j=(i)}^{(i+q-1)} x_j.$$

The induced generalized OWA was studied in [BJ11, MGL09]. With the I-GenOWA function, we can essentially use quasi-arithmetic mean with respect to an order-inducing variable. Special cases of quasi-arithmetic mean include power means and harmonic means, however we will provide only the geometric mean, since many studies have investigated its use and properties.

Definition 3.37 (*Induced Ordered Weighted Geometric function*) For a given weighting vector $\mathbf{w}$ and an inducing variable $\mathbf{z}$, the IOWG function is

$$\text{IOWG}_{\mathbf{w}}(\langle \mathbf{x}, \mathbf{z} \rangle) = \prod_{i=1}^{n} x_{(i)}^{w_i}. \tag{3.34}$$

where $(.)$ notation denotes the inputs $\langle x_1, z_1 \rangle$ reordered such that $z_{(1)} \geq z_{(2)} \geq \cdots \geq z_{(n)}$. The same convention is employed for ties as with IOWA and I-GenOWA.

In [CHVHA04], IOWG was used to aggregate the preferences of experts, where the inducing variable was given by either the importance, consistency or preference of a given expert.

3.5.4 Choices for the Inducing Variable

The process by which the arguments are reordered is of fundamental concern when considering induced aggregation functions. The choice for the inducing variable z will clearly depend on the situation to be modeled. In many cases, we may wish to give weight to observations that are more similar or "closer" to an object of interest $\mathbf{x}$, so we will let z_i represent some function of distance, i.e. $z_i = f(|x_i - \bar{x}_i|)$. In other cases, the z_i may be some representation of the reliability of the value x_i. If the z_i are constant, this essentially models a weighted arithmetic mean, however in many cases the accuracy of an observation may fluctuate. We consider some typical examples where induced aggregation functions can be useful, with particular focus on the choice of $\mathbf{z}$.

Standard auxiliary ordering

The inducing variable may simply be an attribute associated with the input $\mathbf{x}$ that is not considered in the actual aggregation process, but is informative about the object itself. For instance, consider the peer-review process for some journal and conference papers. Each reviewer allocates a score to each criterion, e.g. originality, relevance etc. Sometimes the reviewer also provides his/her evaluation of their own familiarity with the topic, e.g. very familiar, only marginally familiar etc. This last input, of course is not taken into account when aggregating the scores for submission, however could be taken into account in the weight allocation process. To give an overall score for each criterion, we can then use an IOWA where x_i is the score allocated by the i-th reviewer, z_i is the familiarity of the reviewer with the topic and $\mathbf{w}$ is a weighting vector with non-increasing weights such that the heavier weight is given to experts with more expertise on the given paper.

This allocation of weighting is different to providing the expert herself a weight based on her expertise, as the variability of $\mathbf{z}$ suggests that this may fluctuate depending on the paper she is marking.

Example 3.38 An editor for a journal considers two papers which have been evaluated by the same three reviewers.

	Paper 1		Paper 2	
	Score (x)	Expertise (z)	Score (x)	Expertise (z)
Reviewer 1	70	8	92	4
Reviewer 2	85	7	62	8
Reviewer 3	76	4	86	5

Using the weighting vector $\mathbf{w} = (0.5, 0.3, 0.2)$, the score for the first paper is

$$IOWA_{\mathbf{w}}(\langle \mathbf{x}, \mathbf{z} \rangle) = 0.5(70) + 0.3(85) + 0.2(76) = 75.7.$$

The score for the second paper is

$$IOWA_{\mathbf{w}}(\langle \mathbf{x}, \mathbf{z} \rangle) = 0.5(62) + 0.3(86) + 0.2(92) = 75.2.$$

Nearest-neighbor rules

Nearest-neighbor methods and their variants have been popularly applied to classification and function approximation problems. The underlying assumption is that objects described similarly by their features will belong to the same class or have the same function output. For instance, consider a classification problem that requires an object $\mathbf{x} = (x_1, x_2, ..., x_p)$ to be assigned to a class Y_1 or Y_2. Given a number of training data $\mathcal{D} = \{(\mathbf{x}_1, y_1), \ldots, (\mathbf{x}_K, y_K)\}$, we can identify the object $\mathbf{x}_j$ most similar to $\mathbf{x}$ and allocate the same label. Extensions can be made such that the class labels of a number of the objects in $\mathcal{D}$ are aggregated to determine the class of $\mathbf{x}$. Induced aggregation functions can be used to model this situation, with the weights and inducing variable often reflecting the similarity of each of the training data. Of course, the way similarity is calculated becomes very important.

The nearest-neighbor approach can also be used for function approximation. Suppose $\mathbf{x}_i \in \mathbb{I}^p$ and $y_i \in \mathbb{I}$. We assume the training data are generated by some function $f(\mathbf{x}_i) + \epsilon_i = y_i$ where ϵ_i are random errors, and then approximate $y_{\bar{\mathbf{x}}}$ by aggregating the y_i of the most similar $\mathbf{x}_i$. The induced OWA and induced aggregation functions in general can be used in this context, where the input vector comprises the y_i values taken from $\mathcal{D}$ and $\mathbf{z}$ represents a similarity or distance function. For example, a standard representation of the nearest-neighbor model is,

$$y_{\mathbf{x}} = IOWA_{\mathbf{w}}(\langle \mathbf{y}, \mathbf{z} \rangle) = \sum_{i=1}^{K} w_i y_{\eta(i)},$$

with $\mathbf{w} = (1, 0, ..., 0)$, and $z_i = \frac{1}{||\mathbf{x}_i - \bar{\mathbf{x}}||}$, the reciprocal of Euclidean distance between $\mathbf{x}$ and $\mathbf{x}_i$ associated with y_i.

A function which approximates $y_{\mathbf{x}}$ based on the k nearest-neighbors (known as the kNN method) could be calculated using the same function, with $w_i = \frac{1}{k}, i = 1, ..., k, w_i = 0$ otherwise. One can also consider weighted versions where the weights gradually decay, i.e. $w_1 \geq w_2 \geq \cdots \geq w_K$, e.g. $w_i = \frac{2(k+1-i)}{k(k+1)}$.

There exist alternative choices for the distance function, in particular, there are many suitable metrics depending on the spatial distribution of the data. Yager has studied the use of IOWA for modeling these situations in [Yag02a, Yag02b, Yag04a, YK03] as well as in [YF99] with Filev.

Time-series smoothing can also be handled within the framework of induced aggregation operators. It is similar to a nearest-neighbor problem with a single-dimension variable, t. In time-series smoothing, we want to estimate or smooth the value y_t at a point in time t based on the values y_i obtained at previous times t_i. Induced OWA operators allow a simple framework for modeling this smoothing process. For instance, the 3-day simple moving average on a data set $\mathcal{D} = \{\langle t_1, y_1 \rangle, \ldots, \langle t_n, y_n \rangle\}$ can be modeled by $IOWA_\mathbf{w}(t)$, with $w = (\frac{1}{3}, \frac{1}{3}, \frac{1}{3}, 0,$ $0, \ldots, 0)$, $z_i = \frac{1}{t - t_i}$.

Best-yesterday models

Extrapolation problems involve predicting future values based on previous observations. Weather prediction is an important example. In [YF99], Yager and Filev present the best-yesterday model for predicting stock market prices based on the opinions of multiple experts. We could aggregate their scores using a weighted mean, allocating a weight to the experts who seem more reliable, however an alternative is to use an IOWA operator, inducing the input vector based on the most accurate predictions from previous days. We consider an adapted example from [YF99] here.

Example 3.39 We have four experts who, daily, predict the next day's opening share price of the FUZ Company. Our data then consists of the predictions each day for each expert $i = 1, \ldots, 4$, $x_i(t)$ and the actual stock price of the FUZ Company each day $y(t)$. Our aggregated prediction could be the value obtained from the induced OWA

$$IOWA_\mathbf{w}(t) = \sum_{i=1}^{4} w_i x_{\eta(i)}(t),$$

where $z_i = -|x_i(t-1) - y(t-1)|$ and the weights w_i are non-increasing. This allocates more importance to the expert whose predictions were closest to the actual price yesterday.

Of course, we could order our experts by their accuracy for the past 2, 3, etc., days, or determine the weights using optimization techniques. In the Yager and Filev example, for instance, the fitted weighting vector was $\mathbf{w} = (0.2, 0.12, 0.08, 0.6)$ i.e., the best fitting weighted model gave more influence to the expert who was furthest from the mark the previous day.

Group decision making

IOWA operators can be useful in group decision making (GDM) problems for modeling concepts such as consensus. In [CHVHA07], the use of different inducing variables was considered for group decisions based on pair-wise preference matrices modeling multiple alternatives. The usual approach to varying weights is to have these reflect the importance of each expert. In the context of preference matrices, however it might also make sense to allocate importance to each input based on the consistency of each expert, inferred from how well their preferences satisfy transitivity etc. Another of the inducing variables presented took into account the overall preference for a particular alternative expressed by each expert.

The standard OWA is able to model majority concepts such as "most" or "80%" using weighting vectors based on linguistic quantifiers. In [PY06], it was proposed that consensus might be better achieved with inducing variables reflecting the *support* for each individual score. Consider the evaluations of 5 experts, $\mathbf{x} = (0.3, 0.1, 0.7, 0.9, 0.8)$. It makes sense that the score given by expert 5 is more representative of the group than say, expert 2. The support for evaluation x_i from x_j can be modeled simply using:

$$\text{Sup}(x_i, x_j) = \begin{cases} 1, & \text{if } |x_i - x_j| < \alpha; \\ 0 & \text{otherwise,} \end{cases}$$

where α is a desired threshold. The inducing variable based on support can then be given by,

$$z_i = \sum_{j=1, j \neq i}^{n} \text{Sup}(x_i, x_j).$$

In turn, weighting vectors with non-increasing weights can be specified such that experts with more support are allocated higher importance.

Multiple inducing variables

In [MS00], a generalization to multiple inducing variables was considered. Suppose we have N priorities and each of the inputs x_i are associated with N ratings or degrees of satisfaction with respect to these priorities. In this context, the order can be induced by some aggregation of these N scores, and a single inducing variable $\mathbf{z}^*$ can be considered as the vector of aggregated results.

Example 3.40 We want to measure the pollution levels at a beach, and we have multiple mobile sensors that report to a central computer for analysis. The reliability of the sensor readings depends somewhat on the time since they were transmitted, as well as the distance traveled in transmission and the local conditions as they were sent—for instance, varying water pressure, presence of animals etc. We hence decide to aggregate the pollution levels with an $IOWA_{\mathbf{w}}(\langle \mathbf{x}, \mathbf{z}^* \rangle)$ where $\mathbf{z}^*_i = f(z_{i1}, z_{i2}, z_{i3})$ is the aggregated inducing input associated with each pollution level input x_i.

3.6 Medians and Order Statistics

3.6.1 Median

In statistics, the median of a sample is a number dividing the higher half of a sample, from the lower half. The median of a finite list of numbers can be found by arranging all the numbers in increasing or decreasing order and picking the middle one. If the number of inputs is even, one takes the mean of the two middle values.

The median is a type of average which is more representative of a "typical" value than the mean. It essentially discards very high and very low values (outliers). For example, the median price of houses is often reported in the real estate market, because the mean can be influenced by just one or a few very expensive houses, and will not represent the cost of a "typical" house in the area.

Definition 3.41 (*Median*) The median is the function

$$Med(\mathbf{x}) = \begin{cases} \frac{1}{2}(x_{(k)} + x_{(k+1)}), & \text{if } n = 2k \text{ is even} \\ x_{(k)}, & \text{if } n = 2k - 1 \text{ is odd,} \end{cases}$$

where $x_{(k)}$ is the k-th largest (or smallest) component of $\mathbf{x}$.

Note 3.42 The median can be conveniently expressed as an OWA function with a special weighting vector. For an odd n let $w_{\frac{n+1}{2}} = 1$ and all other $w_i = 0$, and for an even n let $w_{\frac{n}{2}} = w_{\frac{n}{2}+1} = \frac{1}{2}$, and all other $w_i = 0$. Then $Med(\mathbf{x}) = OWA_{\mathbf{w}}(\mathbf{x})$.

One interesting median-based averaging function is the Hodges-Lehmann estimator, which is the median of all pairwise means of the components of $\mathbf{x}$,

$$Med(\mathbf{z}) \text{ with } z_k = \frac{x_i + x_j}{2}, 1 \le i < j \le n, k = 1, \ldots, \frac{n(n-1)}{2}.$$

Definition 3.43 (*a-Median*) Given a value $a \in [0, 1]$, the a-median is the function

$$Med_a(\mathbf{x}) = Med(x_1, \ldots, x_n, \overbrace{a, \ldots, a}^{n-1 \text{ times}}).$$

Note 3.44 a-medians are also the limiting cases of idempotent nullnorms. They have absorbing element a and are continuous, symmetric and associative (and, hence, bisymmetric). They can be expressed as

$$Med_a(\mathbf{x}) = \begin{cases} \max(\mathbf{x}), & \text{if } \mathbf{x} \in [0, a]^n, \\ \min(\mathbf{x}), & \text{if } \mathbf{x} \in [a, 1]^n, \\ a & \text{otherwise.} \end{cases}$$

The following construction involves a generating function, similar to quasi-arithmetic means (Sect. 2.3).

Definition 3.45 (*Quasi-median*) Given a continuous strictly monotone function $g : \mathbb{I} \to [-\infty, \infty]$, the quasi-median is the function

$$QMed_g(\mathbf{x}) = g^{-1}\left(Med(g(x_1), \ldots, g(x_n))\right).$$

Note that for an odd n the values of the quasi-median and the median coincide.

An attractive property of the medians is that they are applicable to inputs given on the ordinal scale, i.e., when only the ordering, rather than the numerical values

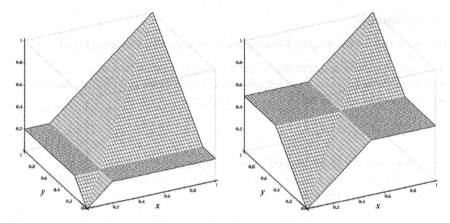

Fig. 3.8 3D plots of a-medians $Med_{0.2}$ and $Med_{0.5}$

matter. For example, one can use medians for aggregation of inputs like labels of fuzzy sets, such as *very high*, *high*, *medium*, *low* and *very low* (Fig. 3.8).

The concept of the weighted median was treated in detail in [Yag98].

Definition 3.46 (*Weighted median*) Let **w** be a weighting vector, and let **u** denote the vector obtained from **w** by arranging its components in the order induced by the components of the input vector **x**, such that $u_k = w_i$ if $x_i = x_{(k)}$ is the k-th largest component of **x**. The lower weighted median is the function

$$Med_\mathbf{w}(\mathbf{x}) = x_{(k)}, \tag{3.35}$$

where k is the index obtained from the condition

$$\sum_{j=1}^{k-1} u_j < \frac{1}{2} \text{ and } \sum_{j=1}^{k} u_j \geq \frac{1}{2}. \tag{3.36}$$

The upper weighted median is the function (3.35) where k is the index obtained from the condition

$$\sum_{j=1}^{k-1} u_j \leq \frac{1}{2} \text{ and } \sum_{j=1}^{k} u_j > \frac{1}{2}.$$

Note 3.47 It is convenient to describe calculation of $Med_\mathbf{w}(\mathbf{x})$ using the following procedure. Take the vector of pairs $((x_1, w_1), (x_2, w_2), \ldots, (x_n, w_n))$ and sort them in the order of decreasing x. We obtain $((x_{(1)}, u_1), (x_{(2)}, u_2), \ldots, (x_{(n)}, u_n))$. Calculate the index k from the condition (3.36). Return $x_{(k)}$.

Note 3.48 The weighted median can be obtained by using penalty-based construction outlined on p. 183.

Main properties

The properties of the weighted median are consistent with averaging functions:

- Weighted median is a continuous idempotent aggregation function;
- If all the weights $w_i = \frac{1}{n}$, weighted median becomes the ordinary median Med;
- If any weight $w_i = 0$, then

$$Med_{\mathbf{w}}(\mathbf{x}) = Med_{(w_1,\ldots,w_{i-1},w_{i+1},\ldots,w_n)}(x_1,\ldots,x_{i-1},x_{i+1},\ldots,x_n),$$

i.e., the input x_i can be dropped from the aggregation procedure;
- If any input value is repeated, one can use just a copy of this value and add the corresponding weights, namely if $x_i = x_j$ for some $i < j$, then

$$Med_{\mathbf{w}}(\mathbf{x}) = Med_{\tilde{\mathbf{w}}}(x_1,\ldots,x_{j-1},x_{j+1},\ldots,x_n),$$

where $\tilde{\mathbf{w}} = (w_1,\ldots,w_{i-1},w_i + w_j,w_{i+1},\ldots,w_{j-1},w_{j+1},\ldots,w_n)$.

As far as learning the weights of weighted medians from empirical data, Yager [Yag98] presented a gradient based local optimization algorithm. Given that such a method does not guarantee the globally optimal solution, it is advisable to combine it with a generic global optimization scheme, such as multistart local search or simulated annealing.

Based on the weighted median, Yager [Yag98] also defined an ordinal OWA function, using the following construction. We recall (see Sect. 3.1) that $OWA_{\mathbf{w}}(\mathbf{x}) = \langle \mathbf{w}, \mathbf{x}_{\searrow} \rangle$, i.e., the weighted mean of the vector $\mathbf{x}_{\searrow}$. By replacing the weighted mean with weighted median we obtain

Definition 3.49 (*Ordinal OWA*) The ordinal OWA function is

$$OOWA_{\mathbf{w}}(\mathbf{x}) = Med_{\mathbf{w}}(\mathbf{x}_{\searrow}).$$

Note 3.50 Since the components of the argument of the weighted median in Definition 3.49 are already ordered, calculation of the ordinal OWA is reduced to the formula
$$OOWA_{\mathbf{w}}(\mathbf{x}) = x_{(k)},$$

where k is the index obtained from the condition

$$\sum_{j=1}^{k-1} w_j < \frac{1}{2} \text{ and } \sum_{j=1}^{k} w_j \geq \frac{1}{2}.$$

A more general class of aggregation functions on an ordinal scale is that of weighted ordinal means, presented in [KMM07].

3.6.2 Order Statistics

Definition 3.51 (*Order statistic*) The k-th order statistic is the function

$$kOS(\mathbf{x}) = x_{(k)},$$

i.e., its value the k-th **smallest**[3] component of $\mathbf{x}$.

Note 3.52 The order statistics can be conveniently expressed as OWA functions with special weighting vectors. Let $\mathbf{w} = (0, 0, \ldots, 0, 1, 0 \ldots, 0)$, i.e., $w_i = 0$ for $i \neq n - k + 1$ and $w_{n-k+1} = 1$. Then $kOS(\mathbf{x}) = OWA_{\mathbf{w}}(\mathbf{x})$.

Note 3.53 To compute the order statistics, sorting is not required. It can be achieved by using a selection algorithm (e.g., Floyd and Rivest's algorithm) which has the worst-case complexity $O(n)$. This is significantly less expensive when n is large.

References

[Ahn06]   B.S. Ahn, On the properties of OWA operator weights functions with constant level of orness. IEEE Trans. Fuzzy Syst. **14**, 511–515 (2006)

[Bel00]   G. Beliakov, Shape preserving approximation using least squares splines. Approximation Theor. Appl. **16**, 80–98 (2000)

[Bel01]   G. Beliakov, "Shape preserving splines in constructing WOWA operators: Comment on paper by V. Torra in Fuzzy Sets and Systems 113, 389–396". *Fuzzy Sets and Systems* 121(2001), 549–550 (2000)

[Bel03]   G. Beliakov, How to build aggregation operators from data? Int. J. Intell. Syst. **18**, 903–923 (2003)

[Bel05]   G. Beliakov, Learning weights in the generalized OWA operators. Fuzzy Optim. Decis. Making **4**, 119–130 (2005)

[Bel15]   G. Beliakov, A method of introducing weights into OWA operators and other symmetric functions, in *Uncertainty Modeling. Dedicated to B. Kovalerchuk*, ed. by V. Kreinovich (Springer, Berlin, 2016)

[BJ11]   G. Beliakov, S. James, Induced ordered weighted averaging operators, in *Recent Developments in the Ordered Weighted Averaging Operators: Theory and Practice*, ed. by R.R. Yager, J. Kacprzyk, G. Beliakov (Springer, Heidelberg, 2011), pp. 29–47

[BJL11]   G. Beliakov, S. James, G. Li, Learning Choquet-integralbased metrics for semisupervised clustering. IEEE Trans. Fuzzy Syst. **19**, 562–574 (2011)

[BMV04]   G. Beliakov, R. Mesiar, L. Valaskova, Fitting generated aggregation operators to empirical data. Int. J. Uncertainty Fuzziness Knowl. Based Syst. **12**, 219–236 (2004)

[CKKM02] T. Calvo, A. Kolesárová, M. Komorníková, R. Mesiar, Aggregation operators: properties, classes and construction methods, in *Aggregation Operators. New Trends and Applications*, ed. by T. Calvo, G. Mayor, R. Mesiar (Physica - Verlag, Heidelberg, 2002), pp. 3–104

[CMM97]   M. Carbonell, M. Mas, G. Mayor, On a class of monotonic extended OWA operators, in *6th IEEE International Conference on Fuzzy Systems*, vol. III (Barcelona, Spain, 1997), pp. 1695–1700

---

[3]Note that in OWA, weighted median and ordinal OWA, $x_{(k)}$ denotes the k-th **largest** element of $\mathbf{x}$.

[CHVHA04]   F. Chiclana, E. Herrera-Viedma, F. Herrera, S. Alonso, Induced ordered weighted geometric operators and their usein the aggregation of multiplicative preference relations. Int. J. Intell. Syst. **19**, 233–255 (2004)

[CHVHA07]   F. Chiclana, E. Herrera-Viedma, F. Herrera, S. Alonso, Some induced ordered weighted averaging operators and their use for solving group decision-making problems based on fuzzy preference relations". Eur. J. Oper. Res. **182**, 383–399 (2007)

[EM14]   A. Emrouznejad, M. Marra, Ordered weighted averaging operators 1988–2014: a citation-based literature survey. Int. J. Intell. Syst. **29**, 994–1014 (2014)

[FY98]   D. Filev, R.R. Yager, On the issue of obtaining OWA operator weights. Fuzzy Sets Syst. **94**, 157–169 (1998)

[FM01]   R. Fuller, P. Majlender, An analytic approach for obtaining maximal entropy OWA operator weights. Fuzzy Sets Syst. **124**, 53–57 (2001)

[FM03]   R. Fuller, P. Majlender, On obtaining minimal variability OWA operator weights. Fuzzy Sets Syst. **136**, 203–215 (2003)

[HHV03]   F. Herrera, E. Herrera-Viedma, A study of the origin and uses of the ordered weighted geometric operator in multicriteria decision making. Int. J. Intell. Syst. **18**, 689–707 (2003)

[KSP14]   A. Kishor, A.K. Singh, N.R. Pal, Orness measure of OWA operators: a new approach. IEEE Trans. Fuzzy Syst. **22**, 1039–1045 (2014)

[KMM07]   A. Kolesárová, R. Mesiar, G. Mayor, Weighted ordinal means. Inf. Sci. **177**, 3822–3830 (2007)

[LJ14]   J. Lin, Y. Jiang, Some hybrid weighted averaging operators and their application to decision making. Inf. Fusion **16**, 18–28 (2014)

[Liu05]   X. Liu, On the properties of equidifferent RIM quantifier with generating function. Int. J. Gen. Syst. **34**, 579–594 (2005)

[Liu07]   X. Liu, The solution equivalence of minimax disparity and minimum variance problems for OWA operators. Int. J. Approx. Reasoning **45**, 68–81 (2007)

[Lla15]   B. Llamazares, Constructing Choquet integral-based operators that generalize weighted means and OWA operators. Inf. Fusion **23**, 131–138 (2015)

[MGL09]   J.M. Merigó, A.M. Gil-Lafuente, The induced generalized OWA operator. Inf. Sci. **179**, 729–741 (2009)

[Mer11]   J.M. Merigó, A unified model between the weighted average and the induced OWA operator. Expert Syst. Appl. **38**, 11560–11572 (2011)

[ME97]   H.B. Mitchell D.D. Estrakh, A modified OWA operator and its use in lossless DPCM image compression. Int. J. Uncertainty Fuzziness Knowl. Based Syst. **5**, 429–436 (1997)

[ME98]   H.B. Mitchell, D.D. Estrakh, An OWA operator with fuzzy ranks. Int. J. Intell. Syst. **13**, 59–81 (1998)

[MS00]   H.B. Mitchell, P.A. Schaefer, Multiple priorities in an induced ordered weighted averaging operator. Int. J. Intell. Syst. **15**, 317–327 (2000)

[NT05]   Y. Narukawa, V. Torra, Fuzzy measure and probability distributions: distorted probabilities. IEEE Trans. Fuzzy Syst. **13**, 617–629 (2005)

[O'H88]   M.O. O'Hagan, Aggregating template or rule antecedents in real-time expert systems with fuzzy set logic, in *22nd Annual IEEE Asilomar Conference on Signals, Systems, Computers*, Pacific Grove, CA, 1988, pp. 681–689

[PY06]   G. Pasi, R.R. Yager, Modeling the concept of majority opinion in group decision making. Inf. Sci. **176**, 390–414 (2006)

[Rud91]   W. Rudin, *Functional Analysis* (McGraw-Hill, New York, 1991)

[Sch83]   L.L. Schumaker, On shape preserving quadratic interpolation. SIAM J. Numer. Anal. **20**, 854–864 (1983)

[Tor97]   V. Torra, The weighted OWA operator. Int. J. Intell. Syst. **12**, 153–166 (1997)

[Tor98]   V. Torra, On some relationships between WOWA operator and the Choquet integral, in *8th International Conference on Information Processing and Management of Uncertainty*, Paris, 1998, pp. 818–824

[Tor00]   V. Torra, TheWOWA operator and the interpolation function W*: Chen and Otto's interpolation revisited. Fuzzy Sets Syst. **113**, 389–396 (2000)

[Tor02]   V. Torra, Learning weights for the quasi-weighted means. IEEE Trans. Fuzzy Syst. **10**, 653–666 (2002)

[Tor04]   V. Torra, OWA operators in data modeling and reidentification. IEEE Trans. Fuzzy Syst. **12**, 652–660 (2004)

[TN07]   V. Torra, Y. Narukawa, *Modeling Decisions. Information Fusion and Aggregation Operators.* (Springer, Berlin, 2007)

[TY05a]   L. Troiano, R.R. Yager A meaure of dispersion for OWA operators, in *Proceedings of the 11th IFSA World Congress*, eds. by Y. Liu, G. Chen, M. Ying (Tsinghua University Press and Springer, Beijing, 2005), pp. 82–87

[TY05b]   L. Troiano, R.R. Yager, Recursive and iterative OWA operators. Int. J. Uncertainty Fuzziness Knowl. Based Syst. **13**, 579–599 (2005)

[TY06]   L. Troiano, R.R. Yager, On the relationship between the quantifier threshold and OWA operators, in *Modeling Decisions for Artificial Intelligence*, eds. by V. Torra, Y. Narukawa, A. Valls, J. Domingo-Ferrer, vol. LNAI 3885 (Springer, Heidelberg, 2006), pp. 215–226

[WP05]   Y.M. Wang, C. Parkan, A minimax disparity approach for obtaining OWA operator weights. Inf. Sci. **175**, 20–29 (2005)

[Xu05]   Z.S. Xu, An overview of methods for determining OWA weights. Int. J. Intell. Syst. **20**, 843–865 (2005)

[XD02]   Z.S. Xu, Q.L. Da, The ordered weighted geometric averaging operator. Int. J. Intell. Syst. **17**, 709–716 (2002)

[Yag88]   R.R. Yager, On ordered weighted averaging aggregation operators in multicriteria decision making. IEEE Trans. Syst. Man Cybern. **18**, 183–190 (1988)

[Yag91]   R.R. Yager, Connectives and quantifiers in fuzzy sets. Fuzzy Sets Syst. **40**, 39–76 (1991)

[Yag93]   R.R. Yager, Families of OWA operators. Fuzzy Sets Syst. **59**, 125–148 (1993)

[Yag95]   R.R. Yager, Measures of entropy and fuzziness related to aggregation operators. Inf. Sci. **82**, 147–166 (1995)

[Yag96]   R.R. Yager, Quantifier guided aggregation using OWA operators. Int. J. Intell. Syst. **11**, 49–73 (1996)

[Yag98]   R.R. Yager, Fusion of ordinal information using weighted median aggregation. Int. J. Approx. Reasoning **18**, 35–52 (1998)

[Yag02a]   R.R. Yager, The induced fuzzy integral aggregation operator. Int. J. Intell. Syst. **17**, 1049–1065 (2002)

[Yag02b]   R.R. Yager, Using fuzzy methods to model nearest neighbor rules. IEEE Trans. Syst. Man Cybern. Part B Cybern. **32**, 512–525 (2002)

[Yag04a]   R.R. Yager, Choquet aggregation using order inducing variables. Int. J. Uncertainty Fuzziness Knowl. Based Syst. **12**, 69–88 (2004)

[Yag04b]   R.R. Yager, Generalized OWA aggregation operators. Fuzzy Optim. Decis. Making **3**, 93–107 (2004)

[Yag07a]   R.R. Yager, Centered OWA operators. Soft Comput. **11**, 631–639 (2007)

[Yag07b]   R.R. Yager, Using stress functions to obtain OWA operators. IEEE Trans. Fuzzy Syst. **15**, 1122–1129 (2007)

[Yag10]   R.R. Yager, Norms induced from OWA operators. IEEE Trans. Fuzzy Syst. **18**, 57–66 (2010)

[YA14a]   R.R. Yager, N. Alajlan, A generalized framework for mean aggregation: toward the modeling of cognitive aspects. Inf. Fusion **17**, 65–73 (2014)

[YA14b]   R.R. Yager, N. Alajlan, On characterizing features of OWA aggregation operators. Fuzzy Optim. Decis. Making **13**, 1–32 (2014)

[YF94]   R.R. Yager, D. Filev, *Essentials of Fuzzy Modelling and Control* (Wiley, New York, 1994)

[YF99]   R.R. Yager, D.P. Filev, Induced ordered weighted averaging operators. IEEE Trans.
           Syst. Man Cybern. Part B Cybern. **20**, 141–150 (1999)
[YK97]   R.R. Yager, J. Kacprzyk (eds.), *The Ordered Weighted Averaging Operators. Theory
           and Applications* (Kluwer, Boston, 1997)
[YKB11]   R.R. Yager, J. Kacprzyk, G. Beliakov (eds.), *Recent Developments in the Ordered
           Weighted Averaging Operators: Theory and Practice* (Springer, Berlin, 2011)
[YK03]   R.R. Yager, V. Kreinovich, Universal approximation theorem for uninorm-based
           fuzzy systems modeling. Fuzzy Sets Syst. **140**, 331–339 (2003)
[ZSA08]   M. Zarghami, F. Szidarovszky, R. Ardakanian, Sensitivity analysis of the OWA
           operator. IEEE Trans. Syst. Man Cybern. Part B **38**, 547–552 (2008)

Chapter 4
Fuzzy Integrals

Abstract This chapter presents two main types of fuzzy integrals, the Choquet integral and the Sugeno integral. Fuzzy measures are introduced and their main properties and special cases are discussed. Various indices which characterize fuzzy measures are presented. The topic of fitting fuzzy measures to empirical data is treated in detail. Induced fuzzy integrals are also presented.

4.1 Choquet Integral

4.1.1 Semantics

In this section we present a large family of aggregation functions based on Choquet integrals. The Choquet integral generalizes the Lebesgue integral, and like it, is defined with respect to a measure. Informally, a measure is a function used to measure, in some sense, sets of objects (finite or infinite). For example, the length of an interval on the real line is an example of a measure, applicable to subsets of real numbers. The area or the volume are other examples of simple measures. A broad overview of various measures is given in [Den94, Pap02, TNS14, WK92].

We note that measures can be additive (the measure of a set is the sum of the measures of its non-intersecting subsets) or non-additive. Lengths, areas and volumes are examples of additive measures. Lebesgue integration is defined with respect to additive measures. If a measure is non-additive, then the measure of the total can be larger or smaller than the sum of the measures of its components.

Choquet integration is defined with respect to not necessarily additive monotone measures, called fuzzy measures (see Definition 4.2), or *capacities* [Cho53]. In this book we are interested only in *discrete* fuzzy measures, which are defined on finite discrete subsets. This is because our construction of aggregation functions involves a finite set of inputs. In general Choquet integrals (and also various other fuzzy integrals) are defined for measures on general sets, and we refer the reader to extensive literature, e.g., [Den94, GMS00, Pap02, TNS14, WK92].

© Springer International Publishing Switzerland 2016   145
G. Beliakov et al., *A Practical Guide to Averaging Functions*,
Studies in Fuzziness and Soft Computing 329,
DOI 10.1007/978-3-319-24753-3_4

The main purpose of Choquet integral-based aggregation is to combine the inputs in such a way that not only the importance of individual inputs (as in weighted means), or of their magnitude (as in OWA), are taken into account, but also of their groups (or coalitions). For example, a particular input may not be important by itself, but become very important in the presence of some other inputs. In medical diagnosis, for instance, some symptoms by themselves may not be really important, but may become key factors in the presence of other signs.

A discrete fuzzy measure allows one to assign importances to all possible groups of criteria, and thus offers a much greater flexibility for modeling aggregation. It also turns out that weighted arithmetic means and OWA are special cases of Choquet integrals with respect to additive and symmetric fuzzy measures respectively. Thus we deal with a much broader class of aggregation functions. The uses of Choquet integrals as aggregation functions are documented in [Gra96, GKM08, GL00, Mar00a, Mar02a, Yag15].

Example 4.1 [Gra96] Consider the problem of evaluating students in a high school with respect to three subjects: mathematics (M), physics (P) and literature (L). Usually this is done by using a weighted arithmetic mean, whose weights are interpreted as importances of different subjects. However, students that are good at mathematics are usually also good at physics and vice versa, as these disciplines present some overlap. Thus evaluation by a weighted arithmetic mean will be either overestimated or underestimated for students good at mathematics and/or physics, depending on the weights.

Consider three students a, b and c whose marks on the scale from 0 to 20 are given by

Student	M	P	L
a	18	16	10
b	10	12	18
c	14	15	15

Suppose that the school is more scientifically oriented, so it weights M and P more than L, with the weights $w_M = w_P > w_L$. If the school wants to favor well equilibrated students, then student c should be considered better than a, who has weakness in L. However, there is no weighting vector $\mathbf{w}$, such that $w_M = w_P > w_L$, and $M_{\mathbf{w}}(c_M, c_P, c_L) > M_{\mathbf{w}}(a_M, a_P, a_L)$.

By aggregating scores using Choquet integral, it is possible (see Example 4.11) to construct such a fuzzy measure, that the weights of individual subjects satisfy the requirement $w_M = w_P > w_L$, but the weight attributed to the pair (M,P) is less that the sum $w_M + w_P$, and the well equilibrated student c is favored.

4.1.2 Definitions and Properties

Definition 4.2 (*Fuzzy measure*) Let $\mathcal{N} = \{1, 2, \ldots, n\}$. A discrete fuzzy measure is a set function[1] $v : 2^{\mathcal{N}} \to [0, 1]$ which is monotonic (i.e. $v(\mathcal{A}) \leq v(\mathcal{B})$ whenever $\mathcal{A} \subset \mathcal{B}$) and satisfies $v(\emptyset) = 0$ and $v(\mathcal{N}) = 1$.

Fuzzy measures are interpreted from various points of view, and are used, in particular, to model uncertainty [Den94, DP00, GMS00, WK92]. In the context of aggregation functions, we are interested in another interpretation, the importance of a coalition, which is used in game theory and in multi-criteria decision making. In the Definition 4.2, a subset $\mathcal{A} \subseteq \mathcal{N}$ can be considered as a *coalition*, so that $v(\mathcal{A})$ gives us an idea about the importance or the weight of this coalition. The monotonicity condition implies that adding new elements to a coalition does not decrease its weight.

Example 4.3 The weakest and the strongest fuzzy measures are , respectively,

1. $v(\mathcal{A}) = \begin{cases} 1, & \text{if } \mathcal{A} = \mathcal{N}, \\ 0 & \text{otherwise;} \end{cases}$

2. $v(\mathcal{A}) = \begin{cases} 0, & \text{if } \mathcal{A} = \emptyset, \\ 1 & \text{otherwise.} \end{cases}$

Example 4.4 The Dirac measure is given for any $\mathcal{A} \subseteq \mathcal{N}$ by

$$v(\mathcal{A}) = \begin{cases} 1, & \text{if } x_0 \in \mathcal{A}, \\ 0, & \text{if } x_0 \notin \mathcal{A}, \end{cases}$$

where x_0 is a fixed element in $\mathcal{N}$.

Example 4.5 The expression

$$v(\mathcal{A}) = \left(\frac{|\mathcal{A}|}{n} \right)^2,$$

where $|\mathcal{A}|$ is the number of elements in $\mathcal{A}$, is a fuzzy measure.

Definition 4.6 (*Möbius transformation*) Let v be a fuzzy measure.[2] The Möbius transformation of v is a set function defined for every $\mathcal{A} \subseteq \mathcal{N}$ as

$$\mathcal{M}(\mathcal{A}) = \sum_{\mathcal{B} \subseteq \mathcal{A}} (-1)^{|\mathcal{A} \setminus \mathcal{B}|} v(\mathcal{B}).$$

---

[1] A set function is a function whose domain consists of all possible subsets of $\mathcal{N}$. For example, for $n = 3$, a set function is specified by $2^3 = 8$ values at $v(\emptyset)$, $v(\{1\})$, $v(\{2\})$, $v(\{3\})$, $v(\{1, 2\})$, $v(\{1, 3\})$, $v(\{2, 3\})$, $v(\{1, 2, 3\})$.

[2] In general, this definition applies to any set function.

Möbius transformation is invertible, and one recovers v by using its inverse, called
Zeta transform,

$$v(\mathcal{A}) = \sum_{\mathcal{B} \subseteq \mathcal{A}} \mathcal{M}(\mathcal{B}) \quad \forall \mathcal{A} \subseteq \mathcal{N}.$$

Möbius transformation is helpful in expressing various quantities, like the interaction indices discussed in Sect. 4.1.4, in a more compact form [Gra97, Gra00,
Mar02a]. It also serves as an alternative representation of a fuzzy measure, called
Möbius representation. That is, one can either use v or $\mathcal{M}$ to perform calculations,
whichever is more convenient. The conditions of monotonicity of a fuzzy measure,
and the boundary conditions $v(\emptyset) = 0$, $v(\mathcal{N}) = 1$ are expressed, respectively, as

$$\sum_{\mathcal{B} \subseteq \mathcal{A} \mid i \in \mathcal{B}} \mathcal{M}(\mathcal{B}) \geq 0, \quad \text{for all } \mathcal{A} \subseteq \mathcal{N} \text{ and all } i \in \mathcal{A}, \tag{4.1}$$

$$\mathcal{M}(\emptyset) = 0 \text{ and } \sum_{\mathcal{A} \subseteq \mathcal{N}} \mathcal{M}(\mathcal{A}) = 1.$$

To represent set functions (for a small n), it is convenient to arrange their values
into an array,[3] e.g., for $n = 3$

$$\begin{array}{ccc}
 & v(\{1, 2, 3\}) & \\
v(\{1, 2\}) & v(\{1, 3\}) & v(\{2, 3\}) \\
v(\{1\}) & v(\{2\}) & v(\{3\}) \\
 & v(\emptyset) &
\end{array}$$

Example 4.7 Let v be the fuzzy measure on $\mathcal{N} = \{1, 2, 3\}$ given by

$$\begin{array}{ccc}
 & 1 & \\
0.9 & 0.5 & 0.3 \\
0.5 & 0 & 0.3 \\
 & 0 &
\end{array}$$

Its Möbius representation $\mathcal{M}$ is

$$\begin{array}{ccc}
 & 0.1 & \\
0.4 & -0.3 & 0 \\
0.5 & 0 & 0.3 \\
 & 0 &
\end{array}$$

Note 4.8 Observe that, the sum of all values of the Möbius transformation in the
above example is equal to 1, in accordance with (4.1). The values of v and $\mathcal{M}$ coincide
on singletons.

---

[3]Such an array is based on a Hasse diagram of the inclusion relation defined on the set of subsets
of $\mathcal{N}$.

There are various special classes of fuzzy measures, which we discuss in Sect. 4.1.3. We now proceed with the definition of the Choquet integral-based aggregation functions.

Definition 4.9 (*Discrete Choquet integral*) The discrete Choquet integral with respect to a fuzzy measure v is given by

$$C_v(\mathbf{x}) = \sum_{i=1}^{n} x_{(i)}[v(\{j|x_j \geq x_{(i)}\}) - v(\{j|x_j \geq x_{(i+1)}\})], \tag{4.2}$$

where $\mathbf{x}_\nearrow = (x_{(1)}, x_{(2)}, \ldots, x_{(n)})$ is a non-decreasing permutation of the input $\mathbf{x}$, and $x_{(n+1)} = \infty$ by convention.

Alternative expressions

- By rearranging the terms of the sum, (4.31) can also be written as

$$C_v(\mathbf{x}) = \sum_{i=1}^{n} \left[x_{(i)} - x_{(i-1)}\right] v(H_i). \tag{4.3}$$

where $x_{(0)} = 0$ by convention, and $H_i = \{(i), \ldots, (n)\}$ is the subset of indices of the $n - i + 1$ largest components of $\mathbf{x}$.
- The discrete Choquet integral is a linear function of the values of the fuzzy measure v. Let us define the following function. For each $\mathcal{A} \subseteq \mathcal{N}$ let

$$g_{\mathcal{A}}(\mathbf{x}) = \max(0, \min_{i \in \mathcal{A}} x_i - \max_{i \in \mathcal{N}\backslash\mathcal{A}} x_i), \tag{4.4}$$

The maximum and minimum over an empty set are taken as 0. Note that $g_{\mathcal{A}}(\mathbf{x}) = 0$ unless $\mathcal{A}$ is the subset of indices of the k largest components of $\mathbf{x}$, in which case $g_{\mathcal{A}}(\mathbf{x}) = x_{\searrow(k)} - x_{\searrow(k+1)}$. Then it is a matter of simple calculation to show that

$$C_v(\mathbf{x}) = \sum_{\mathcal{A} \subseteq \mathcal{N}} v(\mathcal{A}) g_{\mathcal{A}}(\mathbf{x}). \tag{4.5}$$

- Choquet integral can be expressed with the help of the Möbius transformation as

$$C_v(\mathbf{x}) = \sum_{\mathcal{A} \subseteq \mathcal{N}} \mathcal{M}(\mathcal{A}) \min_{i \in \mathcal{A}} x_i = \sum_{\mathcal{A} \subseteq \mathcal{N}} \mathcal{M}(\mathcal{A}) h_{\mathcal{A}}(\mathbf{x}), \tag{4.6}$$

with $h_{\mathcal{A}}(\mathbf{x}) = \min_{i \in \mathcal{A}} x_i$. By using Definition 4.6 we obtain

$$C_v(\mathbf{x}) = \sum_{\mathcal{A} \subseteq \mathcal{N}} v(\mathcal{A}) \sum_{\mathcal{B}|\mathcal{A} \subseteq \mathcal{B}} (-1)^{|\mathcal{B}\backslash\mathcal{A}|} \min_{i \in \mathcal{B}} x_i. \tag{4.7}$$

By comparing this expression with (4.5) we obtain

$$g_{\mathcal{A}}(\mathbf{x}) = \max(0, \min_{i \in \mathcal{A}} x_i - \max_{i \in \mathcal{N} \setminus \mathcal{A}} x_i) = \sum_{\mathcal{B} \mid \mathcal{A} \subseteq \mathcal{B}} (-1)^{|\mathcal{B} \setminus \mathcal{A}|} h_{\mathcal{B}}(\mathbf{x}). \qquad (4.8)$$

Main properties

- The Choquet integral is a continuous piecewise linear idempotent aggregation function;
- An aggregation function is a Choquet integral if and only if it is homogeneous, shift-invariant and *comonotone additive*, i.e., $C_v(\mathbf{x} + \mathbf{y}) = C_v(\mathbf{x}) + C_v(\mathbf{y})$ for all comonotone[4] $\mathbf{x}, \mathbf{y}$;
- The Choquet integral is uniquely defined by its values at the vertices of the unit cube $[0, 1]^n$, i.e., at the points $\mathbf{x}$, whose coordinates $x_i \in \{0, 1\}$. Note that there are 2^n such points, the same as the number of values that determine the fuzzy measure v;
- Choquet integral is Lipschitz-continuous, with the Lipschitz constant 1 in any p-norm, which means it is a kernel aggregation function, see Definition 1.65, p. 23;
- Choquet integral is a convex function if and only if the underlying fuzzy measure is submodular [Cho53, Den94], if and only if it is a subadditive function, see Note 4.20;
- The class of Choquet integrals includes weighted means and OWA functions, as well as minimum, maximum and order statistics as special cases (see Sect. 4.1.5 below);
- A linear convex combination of Choquet integrals with respect to fuzzy measures v_1 and v_2, $\alpha C_{v_1} + (1 - \alpha)C_{v_2}, \alpha \in [0, 1]$, is also a Choquet integral with respect to $v = \alpha v_1 + (1 - \alpha)v_2$[5];
- A pointwise maximum or minimum of Choquet integrals is not necessarily a Choquet integral (but it is in the bivariate case);
- The class of Choquet integrals is closed under duality;
- Choquet integrals have neutral and absorbent elements only in the limiting cases of min and max.

Other properties of Choquet integrals depend on the fuzzy measure being used. We discuss them in Sect. 4.1.3.

Calculation

Calculation of the discrete Choquet integral is performed using Eq. (4.3) using the following procedure. Consider the vector of pairs $((x_1, 1), (x_2, 2), \ldots, (x_n, n))$, where

---

[4]Two vectors $\mathbf{x}, \mathbf{y} \in \mathbb{R}^n$ are called comonotone if there exists a common permutation P of $\{1, 2, \ldots, n\}$, such that $x_{P(1)} \leq x_{P(2)} \leq \cdots \leq x_{P(n)}$ and $y_{P(1)} \leq y_{P(2)} \leq \cdots \leq y_{P(n)}$. Equivalently, this condition is frequently expressed as $(x_i - x_j)(y_i - y_j) \geq 0$ for all $i, j \in \{1, \ldots, n\}$.

[5]As a consequence, this property holds for a linear convex combination of any number of fuzzy measures.

the second component of each pair is just the index i of x_i. The second component will help keeping track of all permutations.

Calculation of $C_v(\mathbf{x})$.

1. Sort the components of $((x_1, 1), (x_2, 2), \ldots, (x_n, n))$ with respect to the first component of each pair in non-decreasing order. We obtain $((x_{(1)}, i_1), (x_{(2)}, i_2), \ldots, (x_{(n)}, i_n))$, so that $x_{(j)} = x_{i_j}$ and $x_{(j)} \le x_{(j+1)}$ for all i. Let also $x_{(0)} = 0$.
2. Let $T = \{1, \ldots, n\}$, and $S = 0$.
3. For $j = 1, \ldots, n$ do

   (a) $S := S + [x_{(j)} - x_{(j-1)}]v(T)$;
   (b) $T := T \setminus \{i_j\}$

4. Return S.

Example 4.10 Let $n = 3$, values of v be given and $\mathbf{x} = (0.8, 0.1, 0.6)$.

Step 1. We take $((0.8, 1), (0.1, 2), (0.6, 3))$.
   Sort this vector of pairs to obtain $((0.1, 2), (0.6, 3), (0.8, 1))$.
Step 2. Take $T = \{1, 2, 3\}$ and $S = 0$.
Step 3. (a) $S := 0 + [0.1 - 0]v(\{1, 2, 3\}) = 0.1 \times 1 = 0.1$;
   (b) $T = \{1, 2, 3\} \setminus \{2\} = \{1, 3\}$;

   (a) $S := 0.1 + [0.6 - 0.1]v(\{1, 3\}) = 0.1 + 0.5v(\{1, 3\})$;
   (b) $T := \{1, 3\} \setminus \{3\} = \{1\}$;

   (a) $S := [0.1 + 0.5v(\{1, 3\})] + [0.8 - 0.6]v(\{1\})$.

   Therefore $C_v(\mathbf{x}) = 0.1 + 0.5v(\{1, 3\}) + 0.2v(\{1\})$.

For computational purposes it is convenient to store the values of a fuzzy measure v in an array $\mathbf{v}$ of size 2^n, and to use the following indexing system, which provides a one-to-one mapping between the subsets $\mathcal{J} \subseteq \mathcal{N}$ and the set of integers $I = \{0, \ldots, 2^n - 1\}$, which index the elements of v. Take the binary representation of each index in I, e.g. $j = 5 = 101$ (binary). Now for a given subset $\mathcal{J} \subseteq \mathcal{N} = \{1, \ldots, n\}$ define its characteristic vector $\mathbf{c} \in \{0, 1\}^n$: $c_{n-i+1} = 1$ if $i \in \mathcal{J}$ and 0 otherwise. For example, if $n = 5$, $\mathcal{J} = \{1, 3\}$, then $\mathbf{c} = (0, 0, 1, 0, 1)$. Put the value $v(\mathcal{J})$ into correspondence with v_j, so that the binary representation of j corresponds to the characteristic vector of $\mathcal{J}$. In our example $v(\{1, 3\}) = v_5$.

Such an ordering of the subsets of $\mathcal{N}$ is called binary ordering:

$$\emptyset, \{1\}, \{2\}, \{1, 2\}, \{3\}, \{1, 3\}, \{2, 3\}, \{1, 2, 3\}, \{4\}, \ldots, \{1, 2, \ldots, n\}.$$

The values of v are mapped to the elements of vector $\mathbf{v}$ as follows

$v_0 = v_{(0000)}$	$v_1 = v_{(0001)}$	$v_2 = v_{(0010)}$	$v_3 = v_{(0011)}$	$v_4 = v_{(0100)}$	$v_5 = v_{(0101)}$	$\cdots$
$v(\emptyset)$	$v(\{1\})$	$v(\{2\})$	$v(\{1, 2\})$	$v(\{3\})$	$v(\{1, 3\})$	$\cdots$

Using (4.5) and the above indexing system, we can write

$$C_v(\mathbf{x}) = \sum_{j=0}^{2^n-1} v_j g_j(\mathbf{x}) = \langle \mathbf{g}(\mathbf{x}), \mathbf{v} \rangle, \tag{4.9}$$

where as earlier, functions $g_j, j = 0, \ldots, 2^n - 1$ are defined by

$$g_j(\mathbf{x}) = \max(0, \min_{i \in \mathcal{J}} x_i - \max_{i \in \mathcal{N} \setminus \mathcal{J}} x_i), \tag{4.10}$$

and the characteristic vector of the set $\mathcal{J} \subseteq \mathcal{N}$ corresponds to the binary representation of j.

An alternative ordering of the values of v is based on set cardinality:

$$\emptyset, \underbrace{\{1\}, \{2\}, \ldots, \{n\}}_{n \text{ singletons}}, \underbrace{\{1, 2\}, \{1, 3\}, \ldots, \{1, n\}, \{2, 3\}, \ldots, \{n-1, n\}}_{\binom{n}{2} \text{ pairs}}, \{1, 2, 3\}, \ldots$$

Such an ordering is useful when dealing with k-additive fuzzy measures (see Definition 4.50 and Proposition 4.63), as it allows one to group non-zero values $\mathcal{M}(\mathcal{A})$ (in Möbius representation) at the beginning of the array. We shall discuss these orderings in Sect. 4.1.6.

Example 4.11 [Gra96] We continue Example 4.1 on p. 146. Let the fuzzy measure v be given as

$$
\begin{array}{ccc}
& 1 & \\
0.5 & 0.9 & 0.9 \\
0.45 & 0.45 & 0.3 \\
& 0 &
\end{array}
$$

so that the ratio of weights $w_M : w_P : w_L = 3 : 3 : 2$, but since mathematics and physics overlap, the weight of the pair $v(\{M, P\}) = 0.5 < v(\{M\}) + v(\{P\})$. On the other hand, weights attributed to $v(\{M, L\})$ and $v(\{P, L\})$ are greater than the sum of individual weights.

Using Choquet integral, we obtain the global scores[6] $C_v(a_M, a_P, a_L) = 13.9$, $C_v(b_M, b_P, b_L) = 13.6$ and $C_v(c_M, c_P, c_L) = 14.6$, so that the students are ranked as $b \prec a \prec c$ as required. Student b has the lowest rank as requested by the scientific tendency of the school.

---

[6]Since Choquet integral is a homogeneous aggregation function, we can calculate it directly on $[0, 20]^n$ rather than scaling the inputs to $[0, 1]^n$.

4.1.3 Types of Fuzzy Measures

The properties of the Choquet integral depend on the fuzzy measure v being used. There are various generic types of fuzzy measures, which lead to specific features of Choquet integral-based aggregation, and to several special cases, such as weighted arithmetic means, OWA and WOWA discussed earlier in this chapter (see also Sect. 4.1.5). In this section we present the most important definitions and classes of fuzzy measures.

Definition 4.12 (*Dual fuzzy measure*) Given a fuzzy measure v, its dual fuzzy measure v^* is defined by

$$v^*(\mathcal{A}) = 1 - v(\mathcal{A}^c), \text{ for all } \mathcal{A} \subseteq \mathcal{N},$$

where $\mathcal{A}^c = \mathcal{N} \setminus \mathcal{A}$ is the complement of $\mathcal{A}$ in $\mathcal{N}$.

Definition 4.13 (*Self-dual fuzzy measure*) A fuzzy measure v is self-dual if it is equal to its dual $v*$, i.e.,

$$v(\mathcal{A}) + v(\mathcal{A}^c) = 1, \text{ holds for all } A \subseteq \mathcal{N}.$$

Definition 4.14 (*Submodular and supermodular fuzzy measure*) A fuzzy measure v is called submodular if for any $\mathcal{A}, \mathcal{B} \subseteq \mathcal{N}$

$$v(\mathcal{A} \cup \mathcal{B}) + v(\mathcal{A} \cap \mathcal{B}) \leq v(\mathcal{A}) + v(\mathcal{B}). \tag{4.11}$$

It is called supermodular if

$$v(\mathcal{A} \cup \mathcal{B}) + v(\mathcal{A} \cap \mathcal{B}) \geq v(\mathcal{A}) + v(\mathcal{B}). \tag{4.12}$$

Two weaker conditions which are frequently used are called sub- and super-additivity. These are special cases of sub- and supermodularity for disjoint subsets.

Definition 4.15 (*Subadditive and superadditive fuzzy measure*) A fuzzy measure v is called subadditive if for any two nonintersecting subsets $\mathcal{A}, \mathcal{B} \subset \mathcal{N}, \mathcal{A} \cap \mathcal{B} = \emptyset$:

$$v(\mathcal{A} \cup \mathcal{B}) \leq v(\mathcal{A}) + v(\mathcal{B}). \tag{4.13}$$

It is called superadditive if

$$v(\mathcal{A} \cup \mathcal{B}) \geq v(\mathcal{A}) + v(\mathcal{B}). \tag{4.14}$$

Note 4.16 Clearly sub-(super-) modularity implies sub-(super-) additivity.

Note 4.17 A fuzzy measure is supermodular if and only if its dual is submodular. However the dual of a subadditive fuzzy measure is not necessarily superadditive and vice versa.

Note 4.18 A general fuzzy measure may be submodular only with respect to specific pairs of subsets $\mathcal{A}$, $\mathcal{B}$, and supermodular with respect to other pairs.

Note 4.19 The discrete Choquet integral can be written as the maximum of linear functions (and hence is a convex function)

$$C_v(\mathbf{x}) = \max_{i=1,\ldots,n!} \{\langle \mathbf{w}_i, \mathbf{x} \rangle\}, \mathbf{w}_i \in \mathbb{R}^n_+,$$

if and only if v is submodular [BJL11].

Note 4.20 The discrete Choquet integral is superadditive (i.e., $C_v(\mathbf{x} + \mathbf{y}) \geq C_v(\mathbf{x}) + C_v(\mathbf{y})$) if and only if v is supermodular [KMM11]. If this is the case, C_v is a concave function which can be written as a minimum of linear functions. By using duality, the respective result is obtained for subadditive Choquet integrals, which are therefore convex.

Definition 4.21 (*Additive (probability) measure*) A fuzzy measure v is called **additive** if for any $\mathcal{A}, \mathcal{B} \subset \mathcal{N}, \mathcal{A} \cap \mathcal{B} = \emptyset$:

$$v(\mathcal{A} \cup \mathcal{B}) = v(\mathcal{A}) + v(\mathcal{B}). \tag{4.15}$$

An additive fuzzy measure is called a **probability** measure.

Note 4.22 A fuzzy measure is both sub- and supermodular if and only if it is additive. A fuzzy measure is both sub- and superadditive if and only if it is additive.

Note 4.23 For an additive fuzzy measure clearly $v(\mathcal{A}) = \sum_{i \in \mathcal{A}} v(\{i\})$.

Note 4.24 Additivity implies that for any subset $\mathcal{A} \subseteq \mathcal{N} \setminus \{i, j\}$

$$v(\mathcal{A} \cup \{i, j\}) = v(\mathcal{A} \cup \{i\}) + v(\mathcal{A} \cup \{j\}) - v(\mathcal{A}).$$

Definition 4.25 (*Boolean measure*) A fuzzy measure v is called a boolean fuzzy measure or $\{0, 1\}$-measure if it holds:

$$v(\mathcal{A}) = 0 \text{ or } v(\mathcal{A}) = 1,$$

for all $\mathcal{A} \subseteq \mathcal{N}$.

Definition 4.26 (*Balanced measure*) A fuzzy measure v is called balanced if it holds:

$$|\mathcal{A}| < |\mathcal{B}| \Longrightarrow v(\mathcal{A}) \leq v(\mathcal{B}), \text{ for all } \mathcal{A}, \mathcal{B} \subseteq \mathcal{N}.$$

Definition 4.27 (*Symmetric fuzzy measure*) A fuzzy measure v is called symmetric if the value $v(\mathcal{A})$ depends only on the cardinality of the set $\mathcal{A}$, i.e., for any $\mathcal{A}, \mathcal{B} \subseteq \mathcal{N}$,

$$\text{if } |\mathcal{A}| = |\mathcal{B}| \text{ then } v(\mathcal{A}) = v(\mathcal{B}).$$

Alternatively, one can say that a fuzzy measure v is symmetric if for any $\mathcal{A} \subseteq \mathcal{N}$ it is

$$v(\mathcal{A}) = Q\left(\frac{|\mathcal{A}|}{n}\right), \qquad (4.16)$$

for some monotone non-decreasing function $Q : [0, 1] \to [0, 1], Q(0) = 0$ and $Q(1) = 1$.

Example 4.28 The following fuzzy measure is additive

$$
\begin{array}{ccc}
 & 1 & \\
0.4 & 0.7 & 0.9 \\
0.1 & 0.3 & 0.6 \\
 & 0 &
\end{array}
$$

The following fuzzy measure is symmetric

$$
\begin{array}{ccc}
 & 1 & \\
0.7 & 0.7 & 0.7 \\
0.2 & 0.2 & 0.2 \\
 & 0 &
\end{array}
$$

The following fuzzy measure is superadditive but not submodular

$$
\begin{array}{ccc}
 & 1 & \\
0.6 & 0.5 & 0.6 \\
0.3 & 0.1 & 0.2 \\
 & 0 &
\end{array}
$$

The following fuzzy measure is subadditive and symmetric

$$
\begin{array}{ccc}
 & 1 & \\
0.5 & 0.5 & 0.5 \\
0.5 & 0.5 & 0.5 \\
 & 0 &
\end{array}
$$

Example 4.29 Let v be $\{0, 1\}$-fuzzy measure on $\mathcal{N} = \{1, 2, 3\}$

$$
\begin{array}{ccc}
 & 1 & \\
1 & 1 & 0 \\
0 & 0 & 0 \\
 & 0 &
\end{array}
$$

This measure is superadditive but its dual fuzzy measure v^*, given by

$$
\begin{array}{ccc}
 & 1 & \\
1 & 1 & 1 \\
1 & 0 & 0 \\
 & 0 &
\end{array}
$$

is not subadditive, because, for instance, $v^*(\{2, 3\}) = 1$ and $v^*(\{2\}) + v^*(\{3\}) = 0$, nor is it superadditive, because $v^*(\{1, 2, 3\}) < v^*(\{1\}) + v^*(\{2, 3\})$.

Definition 4.30 (*Possibility and necessity measures*) A fuzzy measure is called a possibility, *Pos*, if for all $\mathcal{A}, \mathcal{B} \subseteq \mathcal{N}$ it satisfies

$$Pos(\mathcal{A} \cup \mathcal{B}) = \max\{Pos(\mathcal{A}), Pos(\mathcal{B})\}.$$

A fuzzy measure is called a necessity, *Nec*, if for all $\mathcal{A}, \mathcal{B} \subseteq \mathcal{N}$ it satisfies

$$Nec(\mathcal{A} \cap \mathcal{B}) = \min\{Nec(\mathcal{A}), Nec(\mathcal{B})\}.$$

Note 4.31 Possibility and necessity measures are dual to each other in the sense of Definition 4.12, that is, for all $\mathcal{A} \subseteq \mathcal{N}$

$$Nec(\mathcal{A}) = 1 - Pos(\mathcal{A}^c).$$

A possibility measure is subadditive. A necessity measure is superadditive.

Possibility and necessity measures are the basis of the theory of possibility [DP88, WK92, Zad78].

Example 4.32 The following fuzzy measure v is a possibility measure

$$
\begin{array}{ccc}
 & 1 & \\
1 & 0.3 & 1 \\
0.3 & 1 & 0.2 \\
 & 0 &
\end{array}
$$

Example 4.33 The following fuzzy measure v is a necessity measure, dual to the possibility measure in the previous example

$$
\begin{array}{ccc}
 & 1 & \\
0.8 & 0 & 0.7 \\
0 & 0.7 & 0 \\
 & 0 &
\end{array}
$$

Definition 4.34 (*Belief Measure*) A *belief measure Bel* : $2^{\mathcal{N}} \rightarrow [0, 1]$ is a fuzzy measure that satisfies the following condition: for all $m > 1$

$$Bel(\bigcup_{i=1}^{m} A_i) \geq \sum_{\emptyset \neq I \subset \{1,...,m\}} (-1)^{|I|+1} Bel(\bigcap_{i \in I} A_i),$$

where $\{A_i\}_{i \in \{1,...,m\}}$, is any finite family of subsets of $\mathcal{N}$.[7]

Definition 4.35 (*Plausibility measure*) A *plausibility measure Pl* : $2^{\mathcal{N}} \rightarrow [0, 1]$ is a fuzzy measure that satisfies the following condition: for all $m > 1$

$$Pl(\bigcap_{i=1}^{m} A_i) \leq \sum_{\emptyset \neq I \subset \{1,...,m\}} (-1)^{|I|+1} Pl(\bigcup_{i \in I} A_i),$$

where $\{A_i\}_{i \in \{1,...,m\}}$ is any finite family of subsets of $\mathcal{N}$.

Belief and *plausibility* measures constitute the basis of Dempster and Shafer Evidence Theory [Sha76]. Belief measures are related to (and sometimes defined through) *basic probability assignments*, which are the values of the Möbius transformation. We refer the reader to the literature in this field, e.g., [GMS00, WK92].

Note 4.36 A set function Pl : $2^{\mathcal{N}} \rightarrow [0, 1]$ is a plausibility measure if its dual set function is a belief measure, i.e., for all $A \subseteq \mathcal{N}$

$$Pl(A) = 1 - Bel(A^c).$$

Any belief measure is superadditive. Any plausibility measure is subadditive.

Note 4.37 A fuzzy measure is both a belief and a plausibility measure if and only if it is additive.

Note 4.38 A possibility measure is a plausibility measure and a necessity measure is a belief measure.

Note 4.39 The set of all fuzzy measures (for a fixed $\mathcal{N}$) is convex.[8] The sets of subadditive, superadditive, submodular, supermodular, subadditive, superadditive, additive, belief and plausibility fuzzy measures are convex. However the sets of possibility and necessity measures are not convex.

---

[7]For a fixed $m \geq 1$ this condition is called m-monotonicity (simple monotonicity for $m = 1$), and if it holds for all $m \geq 1$, it is called total monotonicity [Den00, Gra00]. For a fixed m, condition in Definition 4.35 is called m-alternating monotonicity. 2-monotone fuzzy measures are called supermodular (see Definition 4.14), also called convex, whereas 2-alternating fuzzy measures are called submodular. If a fuzzy measure is m-monotone, its dual is m-alternating and vice versa.

[8]A set E is convex if $\alpha x + (1 - \alpha)y \in E$ for all $x, y \in E, \alpha \in [0, 1]$.

λ-fuzzy measures

Additive and symmetric fuzzy measures are two examples of very simple fuzzy measures, whereas general fuzzy measures are sometimes too complicated for applications. Next we examine some fuzzy measures with intermediate complexity, which are powerful enough to express interactions among the variables, yet require much less than 2^n parameters to express them.

As a way of reducing the complexity of a fuzzy measure Sugeno [Sug74] introduced λ-fuzzy measures (also called Sugeno measures).

Definition 4.40 (*λ-fuzzy measure*) Given a parameter $\lambda \in]-1, \infty[$, a λ-fuzzy measure is a fuzzy measure v that for all $\mathcal{A}, \mathcal{B} \subseteq \mathcal{N}, \mathcal{A} \cap \mathcal{B} = \emptyset$ satisfies

$$v(\mathcal{A} \cup \mathcal{B}) = v(\mathcal{A}) + v(\mathcal{B}) + \lambda v(\mathcal{A})v(\mathcal{B}). \tag{4.17}$$

Under these conditions, all the values $v(\mathcal{A})$ are immediately computed from n independent values $v(\{i\}), i = 1, \ldots, n$, by using the explicit formula

$$v(\bigcup_{i=1}^{m}\{i\}) = \frac{1}{\lambda}\left(\prod_{i=1}^{m}(1 + \lambda v(\{i\})) - 1\right), \quad \lambda \neq 0.$$

If $\lambda = 0$, λ-fuzzy measure becomes a probability measure. The coefficient λ is determined from the boundary condition $v(\mathcal{N}) = 1$, which gives

$$\lambda + 1 = \prod_{i=1}^{n}(1 + \lambda v(\{i\})), \tag{4.18}$$

which can be solved on $(-1, 0)$ or $(0, \infty)$ numerically (note that $\lambda = 0$ is always a solution). Thus a λ-fuzzy measure is characterized by n independent values $v(\{i\}), i = 1, \ldots, n$.

A λ-fuzzy measure v is related to a probability measure P through the relation

$$P(\mathcal{A}) = \frac{\log(1 + \lambda v(\mathcal{A}))}{1 + \lambda},$$

and, using $g(t) = ((1 + \lambda)^t - 1)/\lambda$ for $\lambda > -1, \lambda \neq 0$, and $g(t) = t$ for $\lambda = 0$,

$$g(P(\mathcal{A})) = v(\mathcal{A}).$$

Note 4.41 The set of all λ-fuzzy measures is not convex.

A λ-fuzzy measure is an example of a distorted probability measure.

Definition 4.42 (*Distorted probability measure*) A fuzzy measure v is a distorted probability measure if there exists some non-decreasing function $g : [0, 1] \to [0, 1]$, $g(0) = 0$, $g(1) = 1$, and a probability measure P, such that for all $\mathcal{A} \subset \mathcal{N}$:

$$v(\mathcal{A}) = g(P(\mathcal{A})).$$

We remind that Weighted OWA functions (see p. 109) are equivalent to Choquet integrals with respect to distorted probabilities. Distorted probabilities and their extension, m-dimensional distorted probabilities, have been recently studied in [NT05].

A λ-fuzzy measure is also an example of a decomposable fuzzy measure [GNW95].

Definition 4.43 (*Decomposable fuzzy measure*) A decomposable fuzzy measure v is a fuzzy measure which for all $\mathcal{A}, \mathcal{B} \subseteq \mathcal{N}, \mathcal{A} \cap \mathcal{B} = \emptyset$ satisfies

$$v(\mathcal{A} \cup \mathcal{B}) = f(v(\mathcal{A}), v(\mathcal{B})) \tag{4.19}$$

for some function $f : [0, 1]^2 \to [0, 1]$ known as the decomposition function.

Note 4.44 It turns out that to get $v(\mathcal{N}) = 1, f$ must necessarily be a t-conorm (see Chap. 1). In the case of λ-fuzzy measures, f is an Archimedean t-conorm with an additive generator $h(t) = \frac{\ln(1+\lambda t)}{\ln(1+\lambda)}, \lambda \neq 0$, which is a Sugeno-Weber t-conorm,.

Note 4.45 Additive measures are decomposable with respect to the Łukasiewicz t-conorm $S_L(x, y) = \min(1, x + y)$. But not every S_L-decomposable fuzzy measure is a probability. Possibility measures are decomposable with respect to the maximum t-conorm $S_{max}(x, y) = \max(x, y)$. Every S_{max}-decomposable discrete fuzzy measure is a possibility measure.

Note 4.46 A λ-fuzzy measure is either sub- or supermodular, when $-1 < \lambda \leq 0$ or $\lambda \geq 0$ respectively.

Note 4.47 When $-1 < \lambda \leq 0$, a λ-fuzzy measure is a plausibility measure, and when $\lambda \geq 0$ it is a belief measure.

Note 4.48 Dirac measures (Example 4.4) are λ-fuzzy measures for all $\lambda \in]-1, \infty[$.

Note 4.49 For a given t-conorm S and fixed $\mathcal{N}$, the set of all S-decomposable fuzzy measures is not always convex.

k-additive fuzzy measures

Another way to reduce complexity of aggregation functions based on fuzzy measures is to impose various linear constraints on their values. Such constraints acquire an interesting interpretation in terms of interaction indices discussed in the next section. One type of constraints leads to k-additive fuzzy measures.

Definition 4.50 (*k-additive fuzzy measure*) A fuzzy measure v is called k-additive ($1 \leq k \leq n$) if its Möbius transformation verifies

$$\mathcal{M}(\mathcal{A}) = 0$$

for any subset $\mathcal{A}$ with more than k elements, $|\mathcal{A}| > k$, and there exists a subset $\mathcal{B}$ with k elements such that $\mathcal{M}(\mathcal{B}) \neq 0$.

An alternative definition of k-additivity (which is also applicable to fuzzy measures on more general sets than $\mathcal{N}$) was given by Mesiar in [Mes99a, Mes99b]. It involves a weakly monotone[9] additive set function v_k defined on subsets of $\mathcal{N}^k = \mathcal{N} \times \mathcal{N} \times \cdots \times \mathcal{N}$. A fuzzy measure v is k-additive if $v(\mathcal{A}) = v_k(\mathcal{A}^k)$ for all $\mathcal{A} \subseteq \mathcal{N}$.

4.1.4 Interaction, Importance and Other Indices

When dealing with multiple criteria, it is often the case that these are not independent, and there is some interaction (positive or negative) among the criteria. For instance, two or more criteria may point essentially to the same concept, for example criteria such as "learnability" and "memorability" that are used to evaluate software user interface [SGBC03]. If the criteria are combined by using, e.g., weighted means, their scores will be double counted. In other instances, contribution of one criterion to the total score by itself may be small, but sharply rise when taken in conjunction with other criteria (i.e., in a "coalition").[10]

Thus to measure such concepts as the importance of a criterion and interaction among the criteria, we need to account for contribution of these criteria in various coalitions. To do this we will use the concepts of Shapley value, which measures the importance of a criterion i in all possible coalitions, and the interaction index, which measures the interaction of a pair of criteria i, j in all possible coalitions [Gra97, Gra00].

Definition 4.51 (*Shapley value*) Let v be a fuzzy measure. The Shapley index for every $i \in \mathcal{N}$ is

$$\phi(i) = \sum_{\mathcal{A} \subseteq \mathcal{N} \setminus \{i\}} \frac{(n - |\mathcal{A}| - 1)! |\mathcal{A}|!}{n!} [v(\mathcal{A} \cup \{i\}) - v(\mathcal{A})].$$

The Shapley value is the vector $\boldsymbol{\phi}(v) = (\phi(1), \ldots, \phi(n))$.

---

[9] Weakly monotone refers here to this property: $\forall \mathcal{A}, \mathcal{B} \subseteq \mathcal{N}, \mathcal{A} \subseteq \mathcal{B}$ implies $v_k(\mathcal{A}^k) \leq v_k(\mathcal{B}^k)$.

[10] Such interactions are well known in game theory. For example, contributions of the efforts of workers in a group can be greater or smaller than the sum of their separate contributions (if working independently).

Note 4.52 It is informative to write the Shapley index as

$$\phi(i) = \frac{1}{n} \sum_{t=0}^{n-1} \frac{1}{\binom{n-1}{t}} \sum_{A \subseteq \mathcal{N} \setminus \{i\}, |A|=t} [v(A \cup \{i\}) - v(A)].$$

Note 4.53 For an *additive* fuzzy measure we have $\phi(i) = v(\{i\})$.

The Shapley value is interpreted as a kind of average value of the contribution of each criterion alone in all coalitions.

Definition 4.54 (*Interaction index*) Let v be a fuzzy measure. The interaction index for every pair $i, j \in \mathcal{N}$ is

$$I_{ij} = \sum_{A \subseteq \mathcal{N} \setminus \{i,j\}} \frac{(n - |A| - 2)! |A|!}{(n-1)!} [v(A \cup \{i,j\}) - v(A \cup \{i\}) - v(A \cup \{j\}) + v(A)].$$

The interaction indices verify $I_{ij} < 0$ as soon as i, j are positively correlated (negative synergy, redundancy). Similarly $I_{ij} > 0$ for negatively correlated criteria (positive synergy, complementarity). $I_{ij} \in [-1, 1]$ for any pair i, j.

Proposition 4.55 *For a submodular fuzzy measure v, all interaction indices verify $I_{ij} \leq 0$. For a supermodular fuzzy measure, all interaction indices verify $I_{ij} \geq 0$.*

The issue of submodularity of fuzzy measures arose in [BJL11] when defining a Choquet integral based norm, see Note 4.74.

The Definition 4.54 due to Murofushi and Soneda was extended by Grabisch for any coalition A of the criteria (not just pairs) [Gra97].

Definition 4.56 (*Interaction index for coalitions*) Let v be a fuzzy measure. The interaction index for every set $A \subseteq \mathcal{N}$ is

$$I(A) = \sum_{B \subseteq \mathcal{N} \setminus A} \frac{(n - |B| - |A|)! |B|!}{(n - |A| + 1)!} \sum_{C \subseteq A} (-1)^{|A \setminus C|} v(B \cup C).$$

Note 4.57 Clearly $I(A)$ coincides with I_{ij} if $A = \{i, j\}$, and coincides with $\phi(i)$ if $A = \{i\}$. Also $I(A)$ satisfies the dummy criterion axiom: If i is a dummy criterion, i.e., $v(B \cup \{i\}) = v(B) + v(\{i\})$ for any $B \subset \mathcal{N} \setminus \{i\}$, then for every such $B \neq \emptyset$, $I(B \cup \{i\}) = 0$. A dummy criterion does not interact with other criteria in any coalition.

Note 4.58 An alternative single-sum expression for $I(A)$ was obtained in [GMR00]:

$$I(A) = \sum_{B \subseteq \mathcal{N}} \frac{(-1)^{|A \setminus B|}}{(n - |A| + 1) \binom{n - |A|}{|B \setminus A|}} v(B).$$

An alternative to the Shapley value is the Banzhaf index [Ban65]. It measures the same concept as the Shapley index, but weights the terms $[v(\mathcal{A} \cup \{i\}) - v(\mathcal{A})]$ in the sum equally.

Definition 4.59 (*Banzhaf Index*) Let v be a fuzzy measure. The Banzhaf index b_i for every $i \in \mathcal{N}$ is

$$b_i = \frac{1}{2^{n-1}} \sum_{\mathcal{A} \subseteq \mathcal{N} \setminus \{i\}} [v(\mathcal{A} \cup \{i\}) - v(\mathcal{A})].$$

This definition has been generalized by Roubens [Rou96].

Definition 4.60 (*Banzhaf interaction index for coalitions*) Let v be a fuzzy measure. The Banzhaf interaction index between the elements of $\mathcal{A} \subseteq \mathcal{N}$ is given by

$$J(\mathcal{A}) = \frac{1}{2^{n-|\mathcal{A}|}} \sum_{\mathcal{B} \subseteq \mathcal{N} \setminus \mathcal{A}} \sum_{\mathcal{C} \subseteq \mathcal{A}} (-1)^{|\mathcal{A} \setminus \mathcal{C}|} v(\mathcal{B} \cup \mathcal{C}).$$

Note 4.61 An alternative single-sum expression for $J(\mathcal{A})$ was obtained in [GMR00]:

$$J(\mathcal{A}) = \frac{1}{2^{n-|\mathcal{A}|}} \sum_{\mathcal{B} \subseteq \mathcal{N}} (-1)^{|\mathcal{A} \setminus \mathcal{B}|} v(\mathcal{B}).$$

Möbius transformation help one to express the indices mentioned above in a more compact form [Gra97, Gra00, GMR00, Mar02a], namely

$$\phi(i) = \sum_{\mathcal{B} \mid i \in \mathcal{B}} \frac{1}{|\mathcal{B}|} \mathcal{M}(\mathcal{B}),$$

$$I(\mathcal{A}) = \sum_{\mathcal{B} \mid \mathcal{A} \subseteq \mathcal{B}} \frac{1}{|\mathcal{B}| - |\mathcal{A}| + 1} \mathcal{M}(\mathcal{B}),$$

$$J(\mathcal{A}) = \sum_{\mathcal{B} \mid \mathcal{A} \subseteq \mathcal{B}} \frac{1}{2^{|\mathcal{B}| - |\mathcal{A}|}} \mathcal{M}(\mathcal{B}).$$

Example 4.62 Let v be the fuzzy measure defined as follows:

$$
\begin{array}{ccc}
& 1 & \\
0.5 & 0.6 & 0.7 \\
0.1 & 0.2 & 0.3 \\
& 0 &
\end{array}
$$

Then the Shapley indices are $\phi(1) = 0.7/3$, $\phi(2) = 1/3$, $\phi(3) = 1.3/3$.

The next result due to Grabisch [Gra97, Gra00] establishes a fundamental property of k-additive fuzzy measures, which justifies their use in simplifying interactions between the criteria in multiple criteria decision making.

Proposition 4.63 *Let v be a k-additive fuzzy measure, $1 \le k \le n$. Then*

- $I(\mathcal{A}) = 0$ *for every* $\mathcal{A} \subseteq \mathcal{N}$ *such that* $|\mathcal{A}| > k$;
- $I(\mathcal{A}) = J(\mathcal{A}) = \mathcal{M}(\mathcal{A})$ *for every* $\mathcal{A} \subseteq \mathcal{N}$ *such that* $|\mathcal{A}| = k$.

Thus k-additive measures acquire an interesting interpretation. These are fuzzy measures that limit interaction among the criteria to groups of size at most k. For instance, for 2-additive fuzzy measures, there are pairwise interactions among the criteria but no interactions in groups of 3 or more. By limiting the class of fuzzy measures to k-additive measures, one reduces their complexity (the number of values) by imposing linear equality constraints. The total number of linearly independent values is reduced from $2^n - 1$ to $\sum_{i=1}^{k} \binom{n}{i} - 1$.

Orness value

We recall the definition of the measure of orness of an aggregation function on p. 163. By using the Möbius transform one can calculate the orness of a Choquet integral C_v with respect to a fuzzy measure v as follows.

Proposition 4.64 (Orness of Choquet integral) [Mar04] *For any fuzzy measure v the orness of the Choquet integral with respect to v is*

$$orness(C_v) = \frac{1}{n-1} \sum_{\mathcal{A} \subseteq \mathcal{N}} \frac{n - |\mathcal{A}|}{|\mathcal{A}| + 1} \mathcal{M}(\mathcal{A}),$$

where $\mathcal{M}(\mathcal{A})$ is the Möbius representation of $\mathcal{A}$. In terms of v the orness value is

$$orness(C_v) = \frac{1}{n-1} \sum_{\mathcal{A} \subseteq \mathcal{N}} \frac{(n - |\mathcal{A}|)! |\mathcal{A}|!}{n!} v(\mathcal{A}).$$

Another (simplified) criterion which measures positive interaction among pairs of criteria is based on the degree of substitutivity.[11]

Definition 4.65 (*Substitutive criteria*) Let i, j be two criteria, and let $v_{ij} \in [0, 1]$ be the degrees of substitutivity. The fuzzy measure v is called substitutive with respect to the criteria i, j if for any subset $\mathcal{A} \subseteq \mathcal{N} \setminus \{i, j\}$

$$v(\mathcal{A} \cup \{i, j\}) \le v(\mathcal{A} \cup \{i\}) + (1 - v_{ij}) v(\mathcal{A} \cup \{j\}),$$
$$v(\mathcal{A} \cup \{i, j\}) \le v(\mathcal{A} \cup \{j\}) + (1 - v_{ij}) v(\mathcal{A} \cup \{i\}). \tag{4.20}$$

---

[11] See discussion in [GNW95], p. 318.

Note 4.66 When $v_{ij} = 1$, and in view of the monotonicity of fuzzy measures, we obtain the equalities $v(\mathcal{A} \cup \{i, j\}) = v(\mathcal{A} \cup \{i\}) = v(\mathcal{A} \cup \{j\})$, i.e., fully substitutive (identical) criteria. One of these criteria can be seen as dummy.

Note 4.67 If $v_{ij} = 0$, i.e., the criteria are not positively substitutive, then $v(\mathcal{A} \cup \{i, j\}) \leq v(\mathcal{A} \cup \{i\}) + v(\mathcal{A} \cup \{j\})$, which is a weaker version of (4.13). It does not imply independence, as the criteria may have negative interaction. Also note that it does not imply subadditivity, as $v(\mathcal{A}) \geq 0$, and it only applies to one particular pair of criteria, i, j, not to all pairs. On the other hand subadditivity implies (4.20) for all pairs of criteria, and with some $v_{ij} > 0$, i.e., all the criteria are substitutive to some degree.

Entropy

The issue of the entropy of Choquet integrals was treated in [KMR05, Mar02b].

Definition 4.68 The entropy of a fuzzy measure v is

$$H(v) = \sum_{i \in \mathcal{N}} \sum_{\mathcal{A} \subseteq \mathcal{N} \setminus \{i\}} \frac{(n - |\mathcal{A}| - 1)! |\mathcal{A}|!}{n!} h(v(\mathcal{A} \cup \{i\}) - v(\mathcal{A})),$$

with $h(t) = -t \log t$, if $t > 0$ and $h(0) = 0$.

This definition coincides with the definition of weights dispersion (Definition 2.38) used for weighted arithmetic mean and OWA functions, when v is additive or symmetric (see Proposition 4.72). The maximal value of H is $\log n$ and is achieved if and only if v is an additive symmetric fuzzy measure, i.e., $v(\mathcal{A}) = \frac{|\mathcal{A}|}{n}$ for all $\mathcal{A} \subseteq \mathcal{N}$. The minimal value 0 is achieved if and only if v is a Boolean fuzzy measure. Also H is a strictly concave function of v, which is useful when maximizing H over a convex subset of fuzzy measures, as it leads to a unique global maximum.

4.1.5 Special Cases of the Choquet Integral

Let us now study special cases of Choquet integral with respect to fuzzy measures with specific properties.

Proposition 4.69 *If v^* is a fuzzy measure dual to a fuzzy measure v, the Choquet integrals C_v and C_{v^*} are dual to each other. If v is self-dual, then C_v is a self-dual aggregation function.*

Proposition 4.70 *The Choquet integral with respect to an additive fuzzy measure v is the weighted arithmetic mean $M_{\mathbf{w}}$ with the weights $w_i = v(\{i\})$.*

Proposition 4.71 *The Choquet integral with respect to a symmetric fuzzy measure* v *defined by means of a quantifier* Q *as in* (4.16) *is the OWA function* $OWA_{\mathbf{w}}$ *with the weights*[12] $w_i = Q(\frac{i}{n}) - Q(\frac{i-1}{n})$.

The values of the fuzzy measure v, associated with an OWA function with the weighting vector $\mathbf{w}$, are also expressed as

$$v(\mathcal{A}) = \sum_{i=n-|\mathcal{A}|+1}^{n} w_i.$$

If a fuzzy measure is symmetric and additive at the same time, we have

Proposition 4.72 *The Choquet integral with respect to a symmetric additive fuzzy measure is the arithmetic mean* M, *and the values of* v *are given by*

$$v(\mathcal{A}) = \frac{|\mathcal{A}|}{n}.$$

Proposition 4.73 *The Choquet integral with respect to a boolean fuzzy measure* v *coincides with the Sugeno integral (see Sect. 4.2).*

2- and 3-additive symmetric fuzzy measures

These are two special cases of symmetric fuzzy measures, for which we can write down explicit formulas for determination of the values of the fuzzy measure. By Proposition 4.71, Choquet integral with respect to any symmetric fuzzy measure is an OWA function with the weights $w_i = Q(\frac{i}{n}) - Q(\frac{i-1}{n})$. If v is additive, then $Q(t) = t$, and the Choquet integral becomes the arithmetic mean M.

Let us now determine function Q for less restrictive 2- and 3-additive fuzzy measures. These fuzzy measures allow interactions of pairs and triples of criteria, but not in bigger coalitions.

It turns out that in these two cases, the function Q is necessarily quadratic or cubic, as given in Proposition 3.30, p. 125, namely, for 2-additive symmetric fuzzy measure

$$Q(t) = at^2 + (1-a)t \text{ for some } a \in [-1, 1],$$

---

[12]We remind that in the definition OWA, we used a non-increasing permutation of the components of $\mathbf{x}$, $\mathbf{x}_{\searrow}$, whereas in Choquet integral we use a non-decreasing permutation $\mathbf{w}_{\nearrow}$. Then OWA is expressed as

$$C_v(\mathbf{x}) = \sum_{i=1}^{n} x_{\nearrow(i)} \left(Q(\frac{n-i+1}{n}) - Q(\frac{n-i}{n}) \right) = \sum_{i=1}^{n} x_{\searrow(i)} \left(Q(\frac{i}{n}) - Q(\frac{i-1}{n}) \right).$$

We also remind that $Q : [0, 1] \to][0, 1], Q(0) = 0, Q(1) = 1$ is a RIM quantifier which determines values of v as $v(\mathcal{A}) = Q\left(\frac{|\mathcal{A}|}{n}\right)$.

which implies that OWA weights are equidistant (i.e., $w_{i+1} - w_i = const$ for all $i = 1, \ldots, n - 1$). In the 3-additive case we have

$$Q(t) = at^3 + bt^2 + (1 - a - b)t \text{ for some } a \in [-2, 4],$$

such that

- if $a \in [-2, 1]$ then $b \in [-2a - 1, 1 - a]$;
- if $a \in]1, 4]$ then $b \in [-3a/2 - \sqrt{3a(4 - a)/4}, -3a/2 + \sqrt{3a(4 - a)/4}]$.

4.1.6 Fitting Fuzzy Measures

Identification of the $2^n - 2$ values from the data (two are given explicitly as $v(\emptyset) = 0$, $v(N) = 1$) involves the least squares or least absolute deviation problems

$$\text{minimize} \sum_{k=1}^{K} (C_v(x_{1k}, \ldots, x_{nk}) - y_k)^2, \text{ or}$$

$$\text{minimize} \sum_{k=1}^{K} |C_v(x_{1k}, \ldots, x_{nk}) - y_k|,$$

subject to the conditions of monotonicity of the fuzzy measure. They translate into a number of linear constraints. In Mobius representation the monotonicity constraints on the fuzzy measure are expressed as

$$\sum_{B \subseteq A | i \in B} \mu_B \geq 0, \text{ for all } A \subseteq \mathcal{N}, |A| > 1 \text{ and all } i \in A, \tag{4.21}$$

$$\sum_{A \subseteq \mathcal{N}} \mu_A = 1.$$

If we use the indexing system outlined on p. 151, the conditions of monotonicity $v(\mathcal{T}) \leq v(\mathcal{S})$ whenever $\mathcal{T} \subseteq \mathcal{S}$ can be written as $v_i \leq v_j$ if $i \leq j$ and $AND(i, j) = i$ (AND is the usual bitwise operation, applied to the binary representations of i, j, which sets k-th bit of the result to 1 if and only if the k-th bits of i and j are 1).

Thus, in the least squares case we have the optimization problem

$$\text{minimize} \sum_{k=1}^{K} (< \mathbf{g}(x_{1k}, \ldots, x_{nk}), \mathbf{v} > -y_k)^2, \tag{4.22}$$

$$\text{s.t. } v_j - v_i \geq 0, \text{ for all } i < j \text{ such that } AND(i, j) = i,$$

$$v_j \geq 0, j = 1, \ldots, 2^n - 2, v_{2^n-1} = 1,$$

which is clearly a QP problem. In the least absolute deviation case we obtain

$$\text{minimize} \quad \sum_{k=1}^{K} |< \mathbf{g}(x_{1k}, \dots, x_{nk}), \mathbf{v} > -y_k|, \tag{4.23}$$

$$\text{s.t.} \quad v_j - v_i \geq 0, \text{ for all } i < j \text{ such that } AND(i, j) = i,$$

$$v_j \geq 0, j = 1, \dots, 2^n - 2, v_{2^n-1} = 1,$$

which is converted into an LP problem (see in Sect. 1.7.1 how LAD is converted into LP). These are two standard problem formulations that are solved by standard QP and LP methods.

Note that in formulations (4.22) and (4.23) most monotonicity constraints will be redundant. It is sufficient to include only constraints such that $AND(i, j) = i$, i and j differ by only one bit (i.e., the cardinalities of the corresponding subsets satisfy $|\mathcal{J}| - |\mathcal{I}| = 1$). There will be $n(2^{n-1} - 1)$ non-redundant constraints. Explicit expression of the constraints in matrix form is complicated, but they are easily specified by using an incremental algorithm for each n. Further, many non-negativity constraints will be redundant as well (only $v(\{i\}) \geq 0, i = 1, \dots, n$ are needed), but since they form part of a standard LP problem formulation anyway, we will keep them.

However, the main difficulty in these problems is the large number of unknowns, and typically a much smaller number of data [GKM08, MG99]. While modern LP and QP methods handle well the resulting (degenerate) problems, for $n \geq 10$ one needs to take into account the sparse structure of the system of constraints. For larger n (e.g., $n = 15$, $2^n - 1 = 32767$) QP methods are not as robust as LP, which can handle millions of variables. It is also important to understand that if the number of data $K \ll 2^n$, there will be multiple optimal solutions, i.e., (infinitely) many fuzzy measures that fit the data.

As discussed in [MG99], multiple solutions sometimes lead to counterintuitive results, because many values of v will be near 0 or 1. It was proposed to use a heuristic, which, in the absence of any data, chooses the "most additive" fuzzy measure, i.e., converts Choquet integral into the arithmetic mean. Grabisch [GKM08, GMS00] has developed a heuristic least mean squares algorithm.

It is not difficult to incorporate the above mentioned heuristic into QP or LP problems like (4.22) and (4.23). Firstly, one may use the variables $\tilde{v}_j = v_j - \bar{v}_j$, instead of v_j ($\bar{v}_j$ denote the values of the additive symmetric fuzzy measure $v(\mathcal{A}) = \frac{|\mathcal{A}|}{n}$). In this case the default 0 values of the fitted $\tilde{v}_j$ (in the absence of data) will result in $v_j = \bar{v}_j$. On the other hand, it is possible to introduce penalty terms into the objective functions in (4.22) and (4.23) by means of artificial data, e.g., to replace the objective function in (4.22) with

$$\sum_{k=1}^{K} (< \mathbf{g}(x_{1k}, \dots, x_{nk}), \mathbf{v} > -y_k)^2 + \frac{p}{2^n} \sum_{k=1}^{2^n} (< \mathbf{g}(a_{1k}, \dots, a_{nk}), \mathbf{v} > -b_k)^2,$$

with p being a small penalty parameter, $\mathbf{a}_k$ being the vertices of the unit cube and
$b_k = \frac{\sum_{i=1}^{n} a_{ik}}{n}$.

There are also alternative heuristic methods for fitting discrete Choquet integrals to empirical data. An overview of exact and approximate methods is provided in [GKM08].

Other requirements on fuzzy measures

There are many other requirements that can be imposed on the fuzzy measure from the problem specifications. Some conditions are aimed at reducing the complexity of the fitting problem (by reducing the number of parameters), whereas others have direct meaningful interpretation.

Importance and interaction indices

The interaction indices defined in Sect. 4.1.4 are all linear functions of the values of the fuzzy measure. Conditions involving these functions can be expressed as linear equations and inequalities.

One can specify given values of importance (Shapley value) and interaction indices $\phi(i)$, I_{ij} (see pp. 160, 161) by adding linear equality constraints to the problems (4.22) and (4.23). Of course, these values may not be specified exactly, but as intervals, say, for Shapley value we may have $a_i \le \phi(i) \le b_i$. In this case we obtain a pair of linear inequalities.

Substitutive criteria

For substitutive criteria i, j we add (see p. 163)

$$v(\mathcal{A} \cup \{i,j\}) \le v(\mathcal{A} \cup \{i\}) + (1 - v_{ij})v(\mathcal{A} \cup \{j\}),$$
$$v(\mathcal{A} \cup \{i,j\}) \le v(\mathcal{A} \cup \{j\}) + (1 - v_{ij})v(\mathcal{A} \cup \{i\}).$$

for all subsets $\mathcal{A} \subseteq \mathcal{N} \setminus \{i,j\}$, where $v_{ij} \in [0, 1]$ is the degree of substitutivity. These are also linear inequality constraints added to the quadratic or linear programming problems.

k-additivity

Recall that Definition 4.50 specifies k-additive fuzzy measures through their Möbius transform

$$\mathcal{M}(\mathcal{A}) = 0$$

for any subset $\mathcal{A}$ with more than k elements. Since Möbius transform is a linear combination of values of v, we obtain a set of linear equalities. By using interaction indices, we can express k-additivity as (see Proposition 4.63) $I(\mathcal{A}) = 0$ for every $\mathcal{A} \subseteq \mathcal{N}, |\mathcal{A}| > k$, which is again a set of linear equalities.

All of the mentioned conditions on the fuzzy measures do not reduce the complexity of quadratic or linear programming problems (4.22) and (4.23). They only add a number of equality and inequality constraints to these problems. The aim of introducing these conditions is not to simplify the problem, but to better fit a fuzzy measure to the problem and data at hand, especially when the number of data is small.

Let us now consider simplifying assumptions, which do reduce the complexity. First recall that adding the symmetry makes Choquet integral an OWA function. In this case we only need to determine n (instead of $2^n - 2$) values. Thus we can use the techniques for fitting OWA weights, discussed in Sect. 3.4.1. In the case of 2- and 3-additive symmetric fuzzy measures we can fit generating functions Q, as discussed in Sect. 3.4.4.

Sub- and super-modularity

Next we present a convenient formulation of the submodularity constraints. First, we note that the interaction indices satisfy $I_{ij} \leq 0$ for submodular fuzzy measures. Then, if we limit ourselves to 2-additive fuzzy measures where $I_A = 0$ when $|A| > 2$, the additional linear constraints are simply $I_{ij} \leq 0$.

For general fuzzy measures, submodularity constraints can be expressed in various forms [BGW08, CJ89, Gra00]. Here we take the submodulatiry condition in the Möbius representation, $\sum_{C \subseteq A \cup B} \mu_C + \sum_{C \subseteq A \cap B} \mu_C - \sum_{C \subseteq A} \mu_C - \sum_{C \subseteq B} \mu_C \leq 0$. Let us rewrite this sum as

$$\sum_{C \subseteq A \cap B} \mu_C + \mu_C - \mu_C - \mu_C + \sum_{C \subseteq A, C \not\subseteq B} \mu_C - \mu_C$$
$$+ \sum_{C \subseteq B, C \not\subseteq A} \mu_C - \mu_C + \sum_{C \subseteq A \cup B, C \not\subseteq A, C \not\subseteq B} \mu_C \leq 0,$$

which simplifies to

$$\sum_{C \subseteq A \cup B, C \not\subseteq A, C \not\subseteq B} \mu_C \leq 0, \quad \forall A, B \subseteq \mathcal{N}. \tag{4.24}$$

To eliminate the redundant constraints, we eliminate all subsets $A \subseteq B$ (including the cases $A = \emptyset$ and $B = \mathcal{N}$, as this will always give 0 for the above constraint), and of course, half of those remaining due to the symmetry with respect to A and B. We also note that in the sum we necessarily have $|C| \geq 2$. There are $\frac{2^n(2^n+1)}{2}$ combinations (A, B) (this eliminates symmetry, but includes the cases $A = B$ and $A, B = \emptyset, \mathcal{N}$). Each set of size $k = 0, \ldots, n$ has 2^k subsets, and there are $\frac{n!}{k!(n-k)!}$ subsets of size k for any given number of inputs. We therefore reduce the number of constraints by $\sum_{k=0}^{n} \frac{n!2^k}{k!(n-k)!} = 3^n$. This leaves $\frac{2^n(2^n+1)}{2} - 3^n$ constraints.

An alternative expression for the subadditivity constraints is due to [CJ89],

$$\sum_{C \subseteq B \subseteq A,} \mu_B \leq 0, \quad \forall A \text{ and } \forall C \subseteq A \subseteq \mathcal{N} \text{ such that } |C| = 2. \tag{4.25}$$

Here we only require combinations of every subset $\mathcal{A}$, $|\mathcal{A}| \geq 2$ with each of its subset pairs, $|\mathcal{C}| = 2$. There are $\frac{n!}{k!(n-k)!}$ subsets $\mathcal{A}$ of size k, and for each such subset there are $\frac{k(k-1)}{2}$ pairs. This gives a total of

$$\sum_{k=2}^{n} \frac{n!k(k-1)}{2 \cdot k!(n-k)!} = \frac{1}{8}2^n n(n-1)$$

constraints. This number is drastically smaller than the number of constraints in the alternative expression (4.24). Table 4.1 provides the number of monotonicity constraints and submodularity constraints required as n increases. An algorithmic implementation of both (4.24) and (4.25) is quite efficient, because the operations verifying whether $\mathcal{A} \subseteq \mathcal{B}$ for any given $\mathcal{A}$, $\mathcal{B}$ can be implemented with complexity $O(1)$ using binary operations.

The supermodularity constraints can be derived by using duality.

Note 4.74 We note that submodularity constraints are important when using Choquet integrals to define metrics (or norms) [BJL11, BGW08, Nar07]. It was shown in [BJL11] that a discrete Choquet integral defines a norm if and only if v is submodular (i.e., if and only if C_v is a convex function). In this case the Choquet integral is a subadditive and convex function, see Note 4.20. In particular, when v is a symmetric fuzzy measure and therefore the Choquet integral becomes an OWA function $OWA_{\mathbf{w}}$, the weights $\mathbf{w}$ are decreasing. Hence OWA functions define norms [Yag10] if and only if the weights are decreasing (such OWA are called the buoyancy measure). The methods of learning Choquet-integral based metrics from data are presented in [BJL11].

λ-fuzzy measures

Fitting λ-fuzzy measures also involves a reduced set of parameters. Recall (p. 158) that λ-fuzzy measures are specified by n parameters $v(\{i\})$, $i = 1, \ldots, n$. The other values are determined from

Table 4.1 Number of required constraints for $3 \leq n \leq 8$

n	Monotonicity	Submodularity	
	From Eq. (4.21)	From Eq. (4.24)	From Eq. (4.25)
3	9	9	6
4	28	55	24
5	75	285	80
6	186	1351	240
7	441	6069	672
8	1016	26335	1792

$$v(\mathcal{A}) = \frac{1}{\lambda}\left(\prod_{i\in\mathcal{A}}(1 + \lambda v(\{i\})) - 1\right)$$

with the help of the parameter λ, which itself is computed from

$$\lambda + 1 = \prod_{i=1}^{n}(1 + \lambda v(\{i\})).$$

The latter is a non-linear equation, which is solved numerically on $(-1, 0)$ or $(0, \infty)$. This means that the Choquet integral C_v becomes a *nonlinear* function of parameters $v(\{i\})$, and therefore problems (4.22) and (4.23) become difficult non-linear programming problems. The problem of fitting these fuzzy measures was studied in [CW01, IS85, KWH97, LL95, WWK98, WLW00] and the methods are based on genetic algorithms.

Representation based on Möbius transformation

In this section we reformulate fitting problem for general k-additive fuzzy measures based on Möbius representation. Our goal is to use this representation to reduce the complexity of problems (4.22) and (4.23). We remind that this is an invertible linear transformation such that:

-
$$\mathcal{M}(\mathcal{A}) = \sum_{\mathcal{B}\subseteq\mathcal{A}} (-1)^{|\mathcal{A}\setminus\mathcal{B}|} v(\mathcal{B}) \text{ and}$$

$$v(\mathcal{A}) = \sum_{\mathcal{B}\subseteq\mathcal{A}} \mathcal{M}(\mathcal{B}), \quad \text{for all } \mathcal{A} \subseteq \mathcal{N}.$$

- Choquet integral is expressed as

$$C_v(\mathbf{x}) = \sum_{\mathcal{A}\subseteq\mathcal{N}} \mathcal{M}(\mathcal{A}) \min_{i\in\mathcal{A}} x_i.$$

- k-additivity holds when

$$\mathcal{M}(\mathcal{A}) = 0$$

for any subset $\mathcal{A}$ with more than k elements.

As the variables we will use $m_j = m_\mathcal{A} = \mathcal{M}(\mathcal{A})$ such that $|\mathcal{A}| \le k$ in some appropriate indexing system, such as the one based on cardinality ordering on p. 152. This is a much reduced set of variables ($\sum_{i=1}^{k} \binom{n}{i} - 1$ compared to $2^n - 2$). Now, monotonicity of a fuzzy measure, expressed as

$$v(\mathcal{A} \cup \{i\}) - v(\mathcal{A}) \ge 0, \quad \forall \mathcal{A}|i \notin \mathcal{A}, i = 1, \ldots, n,$$

converts into (4.1), and using k-additivity, into

$$\sum_{\mathcal{B}\subseteq\mathcal{A}|i\in\mathcal{B},|\mathcal{B}|\leq k} m_{\mathcal{B}} \geq 0, \quad \text{for all } \mathcal{A} \subseteq \mathcal{N} \text{ and all } i \in \mathcal{A}.$$

The (non-redundant) set of non-negativity constraints $v(\{i\}) \geq 0, i = 1, \dots, n$, is a special case of the previous formula when $\mathcal{A}$ is a singleton, which simply become (see Note 4.8)

$$\sum_{\mathcal{B}=\{i\}} m_{\mathcal{B}} = m_{\{i\}} \geq 0, \ i = 1, \dots, n.$$

Finally, condition $v(\mathcal{N}) = 1$ becomes $\displaystyle\sum_{\mathcal{B}\subseteq\mathcal{N}||\mathcal{B}|\leq k} m_{\mathcal{B}} = 1$.

Then the problem (4.22) is translated into a simplified QP problem

$$\text{minimize} \ \sum_{j=1}^{K} \left(\sum_{\mathcal{A}| \, |\mathcal{A}|\leq k} h_{\mathcal{A}}(\mathbf{x}_j)m_{\mathcal{A}} - y_j \right)^2, \tag{4.26}$$

$$\text{s.t.} \ \sum_{\mathcal{B}\subseteq\mathcal{A}|i\in\mathcal{B},|\mathcal{B}|\leq k} m_{\mathcal{B}} \geq 0,$$

$$\text{for all } \mathcal{A} \subseteq \mathcal{N}, |\mathcal{A}| > 1, \text{ and all } i \in \mathcal{A},$$

$$m_{\{i\}} \geq 0, \ i = 1, \dots, n,$$

$$\sum_{\mathcal{B}\subseteq\mathcal{N}||\mathcal{B}|\leq k} m_{\mathcal{B}} = 1,$$

where $h_{\mathcal{A}}(\mathbf{x}) = \min_{i\in\mathcal{A}} x_i$. Note that only the specified $m_{\mathcal{B}}$ are non-negative, others are unrestricted. The number of monotonicity constraints is the same for all k-additive fuzzy measures for $k = 2, \dots, n$. Similarly, a simplified LP problem is obtained from (4.23), with the same set of constraints as in (4.26).

Software

A software package Kappalab provides a number of tools to calculate various quantities using fuzzy measures, such as the Möbius transform, interaction indices, k-additive fuzzy measures in various representations, and also allows one to fit values of fuzzy measures to empirical data. This package is available from http://cran.r-project. org/web/packages/kappalab/index.html and it works under R environment [KG05]. Another C library called FMTools, with R and Matlab interface [BVL14] is available from http://www.deakin.edu.au/~gleb/fmtools.html. A set of C++ algorithms for the same purpose is available from http://www.deakin.edu.au/~gleb/aotool.html.

4.1.7 Generalized Choquet Integral

Mesiar [Mes95] has proposed a generalization of the Choquet integral, called Choquet-like integral.

Definition 4.75 (*Generalized discrete Choquet integral*) Let $g : [0, 1] \rightarrow [-\infty, \infty]$ be a continuous strictly monotone function. A generalized Choquet integral with respect to a fuzzy measure v is the function

$$C_{v,g}(\mathbf{x}) = g^{-1}\left(C_v(g(\mathbf{x}))\right),$$

where C_v is the discrete Choquet integral with respect to v and $g(\mathbf{x}) = (g(x_1), \ldots, g(x_n))$.

A special case of this construction was presented in [Yag04].

$$C_{v,q}(\mathbf{x}) = \left(\sum_{i=1}^{n} x_{(i)}^q [v(H_i) - v(H_{i+1})]\right)^{1/q}, \quad q \in \mathbb{R}. \tag{4.27}$$

It is not difficult to see that this is equivalent to

$$C_{v,q}(\mathbf{x}) = \left(\sum_{i=1}^{n} \left[x_{(i)}^q - x_{(i-1)}^q\right] v(H_i)\right)^{1/q}. \tag{4.28}$$

The generalized Choquet integral depends on the properties of the fuzzy measure v, discussed in this Chapter. For additive fuzzy measures it becomes a weighted quasi-arithmetic mean with the generating function g, and for symmetric fuzzy measures, it becomes a generalized OWA function, with the generating function g. Continuity of $C_{v,g}$ holds if $Ran(g) \neq [-\infty, \infty]$.

Fitting generalized Choquet integral to empirical data involves a modification of problems (4.22), (4.23) or (4.26), which consists in applying g to the components of $\mathbf{x}_k$ and y_k, (i.e., using the data $\{g(\mathbf{x}_k), g(y_k)\}$) provided g is known, or is fixed. In the case of fitting both g and v to the data, we use bi-level optimization, similar to that in Sect. 2.3.7.

The Choquet integral, as well as the Sugeno integral treated in the next section, are special cases of more general integrals with respect to a fuzzy measure. The interested reader is referred to [Mes95, RA80, Web84].

4.2 Sugeno Integral

4.2.1 Definition and Properties

Similarly to the Choquet integral, Sugeno integral is also frequently used to aggregate inputs, such as preferences in multicriteria decision making [DMPRS01, Mar00b]. Various important classes of aggregation functions, such as medians, weighted minimum and weighted maximum are special cases of Sugeno integral.

Definition 4.76 (*Discrete Sugeno integral*) The Sugeno integral with respect to a fuzzy measure v is given by

$$S_v(\mathbf{x}) = \max_{i=1,\ldots,n} \min\{x_{(i)}, v(H_i)\}, \tag{4.29}$$

where $\mathbf{x}_{\nearrow} = (x_{(1)}, x_{(2)}, \ldots, x_{(n)})$ is a non-decreasing permutation of the input $\mathbf{x}$, and $H_i = \{(i), \ldots, (n)\}$.

Sugeno integrals can be expressed, for arbitrary fuzzy measures, by means of the Median function (see Sect. 3.6.1 below) in the following way:

$$S_v(\mathbf{x}) = Med(x_1, \ldots, x_n, v(H_2), v(H_3), \ldots, v(H_n)).$$

Let us denote max by $\vee$ and min by $\wedge$ for compactness. We denote by $\mathbf{x} \vee \mathbf{y} = \mathbf{z}$ the componentwise maximum of $\mathbf{x}, \mathbf{y}$ (i.e., $z_i = \max(x_i, y_i)$), and by $\mathbf{x} \wedge \mathbf{y}$ their componentwise minimum.

Main properties

- Sugeno integral is a continuous idempotent aggregation function;
- An aggregation function is a Sugeno integral if and only if it is *min-homogeneous*, i.e., $S_v(x_1 \wedge r, \ldots, x_n \wedge r) = S_v(x_1, \ldots, x_n) \wedge r$ and *max-homogeneous*, i.e., $S_v(x_1 \vee r, \ldots, x_n \vee r) = S_v(x_1, \ldots, x_n) \vee r$ for all $\mathbf{x} \in [0, 1]^n, r \in [0, 1]$ (See [Mar00a], Theorem 4.3. There are also alternative characterizations);
- Sugeno integral is *comonotone maxitive* and *comonotone minimitive*, i.e., $S_v(\mathbf{x} \vee \mathbf{y}) = S_v(\mathbf{x}) \vee S_v(\mathbf{y})$ and $S_v(\mathbf{x} \wedge \mathbf{y}) = S_v(\mathbf{x}) \wedge S_v(\mathbf{y})$ for all comonotone[13] $\mathbf{x}, \mathbf{y} \in [0, 1]^n$;
- Sugeno integral is Lipschitz-continuous, with the Lipschitz constant 1 in any p-norm, which means it is a kernel aggregation function, see Definition 1.65, p. 23;
- The class of Sugeno integrals is closed under duality.

Calculation

Calculation of the discrete Sugeno integral is performed using Eq. (4.29) similarly to calculating the Choquet integral on p. 175. We take the vector of pairs

---

[13]See footnote 4 on p. 150.

$((x_1, 1), (x_2, 2), \ldots, (x_n, n))$, where the second component of each pair is just the index i of x_i. The second component will help keeping track of all permutations.

Calculation of $S_v(\mathbf{x})$.

1. Sort the components of $((x_1, 1), (x_2, 2), \ldots, (x_n, n))$ with respect to the first component of each pair in non-decreasing order. We obtain $((x_{(1)}, i_1), (x_{(2)}, i_2), \ldots, (x_{(n)}, i_n))$, so that $x_{(j)} = x_{i_j}$ and $x_{(j)} \le x_{(j+1)}$ for all i.
2. Let $\mathcal{T} = \{1, \ldots, n\}$, and $S = 0$.
3. For $j = 1, \ldots, n$ do

    (a) $S := \max(S, \min(x_{(j)}, v(\mathcal{T})))$;
    (b) $\mathcal{T} := \mathcal{T} \setminus \{i_j\}$

4. Return S.

Example 4.77 Let $n = 3$, let the values of v be given by

$$v(\emptyset) = 0, \ v(\mathcal{N}) = 1, \ v(\{1\}) = 0.5, v(\{2\}) = v(\{3\}) = 0,$$

$$v(\{1, 2\}) = v(\{2, 3\}) = 0.5, \ v(\{1, 3\}) = 1.$$

and $\mathbf{x} = (0.8, 0.1, 0.6)$.

Step 1. We take $((0.8, 1), (0.1, 2), (0.6, 3))$.
    Sort this vector of pairs to obtain $((0.1, 2), (0.6, 3), (0.8, 1))$.
Step 2. Take $\mathcal{T} = \{1, 2, 3\}$ and $S = 0$.
Step 3. (a) $S := \max(0, \min(0.1, v(\{1, 2, 3\}))) = 0.1$;
    (b) $\mathcal{T} = \{1, 2, 3\} \setminus \{2\} = \{1, 3\}$;
    (a) $S := \max(0.1, \min(0.6, v(\{1, 3\}))) = 0.6$;
    (b) $\mathcal{T} := \{1, 3\} \setminus \{3\} = \{1\}$;
    (a) $S := \max(0.6, \min(0.8, v(\{1\}))) = 0.6$.
    Therefore $S_v(\mathbf{x}) = 0.6$.

4.2.2 Special Cases

- The Sugeno integral with respect to a symmetric fuzzy measure given by $v(\mathcal{A}) = v(|\mathcal{A}|)$ is the Median $Med(x_1, \ldots, x_n, v(n-1), v(n-2), \ldots, v(1))$.
- Weighted maximum (max-min) $WMAX_{\mathbf{w}}(\mathbf{x}) = \max_{i=1,\ldots,n} \min\{w_i, x_i\}$. An aggregation function is a weighted maximum if and only if it is the Sugeno integral with respect to a possibility measure;
- Weighted minimum (min-max) $WMIN_{\mathbf{w}}(\mathbf{x}) = \min_{i=1,\ldots,n} \max\{w_i, x_i\}$. An aggregation function is a weighted minimum if and only if it is the Sugeno integral with respect to a necessity measure;

- Ordered weighted maximum $OWMAX_{\mathbf{w}}(\mathbf{x}) = \max\limits_{i=1,\ldots,n} \min\{w_i, x_{(i)}\}$ with a non-increasing weighting vector $1 = w_1 \geq w_2 \geq \ldots \geq w_n$. An aggregation function is an ordered weighted maximum if and only if it is the Sugeno integral with respect to a symmetric fuzzy measure. It can be expressed by means of the Median function as

$$OWMAX_{\mathbf{w}}(\mathbf{x}) = Med(x_1, \ldots, x_n, w_2, \ldots, w_n);$$

- Ordered weighted minimum $OWMIN_{\mathbf{w}}(\mathbf{x}) = \min\limits_{i=1,\ldots,n} \max\{w_i, x_{(i)}\}$ with a non-increasing weighting vector $w_1 \geq w_2 \geq \ldots \geq w_n = 0$. An aggregation function is an ordered weighted minimum if and only if it is the Sugeno integral with respect to a symmetric fuzzy measure. It can be expressed by means of the Median function as

$$OWMIN_{\mathbf{w}}(\mathbf{x}) = Med(x_1, \ldots, x_n, w_1, \ldots, w_{n-1});$$

- The Sugeno integral coincides with the Choquet integral if v is a boolean fuzzy measure.

Note 4.78 The weighting vectors in weighted maximum and minimum do not satisfy $\sum w_i = 1$, but $\max(\mathbf{w}) = 1$ and $\min(\mathbf{w}) = 0$ respectively.

Note 4.79 For the weighted maximum $WMAX_{\mathbf{w}}$ (v is a possibility measure) it holds $WMAX_{\mathbf{w}}(\mathbf{x} \vee \mathbf{y}) = WMAX_{\mathbf{w}}(\mathbf{x}) \vee WMAX_{\mathbf{w}}(\mathbf{y})$ for *all* vectors $\mathbf{x}, \mathbf{y} \in [0, 1]^n$ (a stronger property than comonotone maxitivity). For the weighted minimum $WMIN_{\mathbf{w}}$ (v is a necessity measure) it holds $WMIN_{\mathbf{w}}(\mathbf{x} \wedge \mathbf{y}) = WMIN_{\mathbf{w}}(\mathbf{x}) \wedge WMIN_{\mathbf{w}}(\mathbf{y})$ for all $\mathbf{x}, \mathbf{y} \in [0, 1]^n$.

Note 4.80 Ordered weighted maximum and minimum functions are related through their weights as follows: $OWMAX_{\mathbf{w}}(\mathbf{x}) = OWMIN_{\mathbf{u}}(\mathbf{x})$ if and only if $u_i = w_{i+1}, i = 1, \ldots, n - 1$ [Mar00a].

Example 4.81 Let v be a fuzzy measure on $\{1, 2\}$ defined by $v(\{1\}) = a \in [0, 1]$ and $v(\{2\}) = b \in [0, 1]$. Then

$$S_v(\mathbf{x}) = \begin{cases} x_1 \vee (b \wedge x_2), & \text{if } x_1 \leq x_2, \\ (a \wedge x_1) \vee x_2, & \text{if } x_1 \geq x_2. \end{cases}$$

Example 4.82 Another interesting example of the Sugeno integral is the famous Hirsch h-index (and some other related bibliometric indices). Recall that a scientist's h-index is his/her number of papers h with at least h citations each. Torra [TN08] wrote the h-index as

$$Hindex = \max_i \min(i, x_{(i)}),$$

where x_i is the number of citations of the ith paper, and $\mathbf{x}$ is ordered in decreasing order. We immediately see h-index as the weighted maximum function $WMAX_{\mathbf{w}}$ where the weighting vector has components $w_i = i$. This was further explored in

[CG13a, CG13b, CG15, Gag13, GM13, GM14]. Hence the h-index is a special case of the Sugeno integral with respect to counting fuzzy measure.

The only issue not previously picked up is that the above expression for the h-index coincides with WMAX only if $x_i \leq n$ (on $[0, n]^n$). Recall that the weighting vector here satisfies max $w_i \leq n$. If some $x_i > n$ we no longer have (a scaled version of) the WMAX function because the idempotency (and internality) is lost: compute, for example, the h-index of $\mathbf{x} = (5, 5, 5)$, which is 3 rather than 5. The correct expression is the following:

$$Hindex = WMAX_{\mathbf{w}}(\min(x_1, n), \min(x_2, n), \ldots, \min(x_n, n)). \tag{4.30}$$

4.3 Induced Fuzzy Integrals

It is well known that the OWA function is generalized by the Choquet integral, which also has a reordering step in its calculation. In [Yag02], Yager made extensions to the induced Sugeno integral and in [YK03], the induced Choquet integral was presented.

Definition 4.83 (*Induced Choquet integral*) The induced Choquet integral with respect to a fuzzy measure v and an order inducing variable $\mathbf{z}$ is given by

$$IC_v(\langle \mathbf{x}, \mathbf{z} \rangle) = \sum_{i=1}^{n} x_{\theta(i)}[v(\{j | z_j \geq z_{\theta(i)}\}) - v(\{j | z_j \geq z_{\theta(i+1)}\})], \tag{4.31}$$

where the $\theta(.)$ notation denotes the inputs $\langle x_1, z_1 \rangle$ reordered such that $z_{\theta(1)} \leq z_{\theta(2)} \leq \cdots \leq z_{\theta(n)}$ and $z_{\theta(n+1)} = \infty$ by convention.

The same convention can be used for ties, and note again that the calculation of the Choquet integral is usually performed by firstly arranging the arguments in non-decreasing order—in this case with respect to the inducing variable $\mathbf{z}$.

As the Choquet integral generalizes the OWA, motivation for extensions to the induced Choquet integral occur naturally in application. The fuzzy measure used in calculation still plays the same role, giving a weight to each coalition and allowing for interaction. Consider the final component in calculation of the standard Choquet integral, $x_{\tau(n)} v(j | x_j = x_{\tau(n)})$. The highest input is multiplied by the value of its corresponding singleton. This means that if $v(j)$ is high, a high score for x_j will be sufficient for an overall large score. In the case of the induced Choquet integral, however, this final term will be $x_{\theta(n)} v(j | z_j = z_{\theta(n)})$. If $v(j)$ is high in this instance, a high score for z_j results in the output being heavily influenced by the input x_j, whether it is small or large, i.e. $f(\langle \mathbf{x}, \mathbf{z} \rangle) \approx x_j$. Such examples arise naturally in nearest-neighbor rules as we will discuss later.

The Sugeno integral is similar to the Choquet integral, in that it is defined by a fuzzy measure v and is calculated with a reordering of the input vector. The Sugeno integral is based on operations of min and max and hence is capable of handling non-numerical inputs, provided they are taken from a finite ordinal scale.

Definition 4.84 (*Induced Sugeno integral*) The induced Sugeno integral with respect to a fuzzy measure v and an inducing variable $\mathbf{z}$ is given by

$$IS_v(\langle \mathbf{x}, \mathbf{z} \rangle) = \max_{i=1,\dots,n} \min\{x_{\theta(i)}, v(H_{i,\mathbf{z}})\}, \tag{4.32}$$

where the $\theta(.)$ notation denotes the inputs $\langle x_1, z_1 \rangle$ reordered such that $z_{\theta(1)} \leq z_{\theta(2)} \leq \cdots \leq z_{\theta(n)}$, with $H_{i,\mathbf{z}} = \{\theta(i), \theta(i+1), \dots, \theta(n)\}$. If q of the z_i are tied, i.e. $z_{\theta(i)} = z_{\theta(i+1)} = \cdots = z_{\theta(i+q-1)}$,

$$x_j \leq x_k \Rightarrow z_{\theta(j)} \leq z_{\theta(k)} \forall j, k = \theta(i), \dots, \theta(i+q-1).$$

Note here that a different convention is employed for ties, namely that if the inducing variables are tied, the arguments within ties are reordered according to the relative values of x_i.

It might be noted that the circumstances under which an induced OWA could be applied may also allow for the application of the induced Choquet integral. The induced Sugeno integral, however, may be applied to situations which are quite different, where the semantics of an OWA function might not make sense. In particular, Sugeno integrals are capable of operating on values expressed linguistically e.g. $\{very\ good, good, fair, poor, very\ poor\}$. This is one reason why ties are dealt with differently, since with linguistic variables it might not make sense to take the average. The induced Sugeno integral has a tighter bound than that expressed by the averaging property. Yager noted in [Yag02] that the bounds on IS_v are

$$\min\{\mathbf{x}\} \leq IS_v(\langle \mathbf{x}, \mathbf{z} \rangle) \leq x_{\theta(n)}.$$

In other words, the Sugeno integral is bounded from above by the input associated with the largest inducing variable input $z_{\theta(n)}$, as well as from below by the minimum argument value of $\mathbf{x}$.

References

[Ban65]   J.F. Banzhaf, Weight voting doesn't work: a mathematical analysis. Rutgers Law Rev. **19**, 317–343 (1965)
[BJL11]   G. Beliakov, S. James, G. Li, Learning Choquet-integralbased metrics for semisupervised clustering. IEEE Trans. Fuzzy Syst. **19**, 562–574 (2011)
[BVL14]   G. Beliakov, Q.H. Vu, G. Li, A Choquet integral toolbox and its application in customer preference analysis, in *Data Mining Applications with R* eds. by Y. Zhao, Y. Cen (Academic Press, Waltham, 2014), pp. 247–272
[BGW08]   J. Bolton, P. Gader, J.N. Wilson, Discrete Choquet integral as a distance metric. IEEE Trans. Fuzzy Syst. **16**, 1107–1110 (2008)

[CG13a] A. Cena, M. Gagolewski, OM3: Ordered Maxitive, Minitive, and Modular aggrega-
         tion operators. A simulation study (II), in *Aggregation Functions in Theory and in
         Practise*, eds. by H. Bustince, J. Fernandez, R. Mesiar, T. Calvo. Advances in Intel-
         ligent Systems and Computing, vol. 228 (Springer, Heidelberg, 2013), pp. 105–115
[CG13b] A. Cena, M. Gagolewski, OM3: Ordered Maxitive, Minitive, and Modular aggrega-
         tion operators: axiomatic analysis under arity-dependence (I), in *Aggregation Func-
         tions in Theory and in Practise*, eds. by H. Bustince, J. Fernandez, R. Mesiar, T.
         Calvo. Advances in Intelligent Systems and Computing, vol. 228 (Springer, Heidel-
         berg, 2013), pp. 93–103
[CG15]   A. Cena, M. Gagolewski, OM3: ordered maxitive, minitive, and modular aggregation
         operators axiomatic and probabilistic properties in an arity-monotonic setting. Fuzzy
         Sets Syst. **264**, 138–159 (2015)
[CJ89]   A. Chateauneuf, J.Y. Jaffray, Some characterizations of lower probabilities and other
         monotone capacities through the use of Möbius inversion. Math. Soc. Sci. **17**, 263–
         283 (1989)
[CW01]   T.-Y. Chen, J.-C. Wang, Identification of λ-fuzzy measures using sampling design
         and genetic algorithms. Fuzzy Sets Syst. **123**, 321–341 (2001)
[Cho53] G. Choquet, Theory of capacities. Ann. Inst. Fourier **5**, 131–295 (1953)
[Den94] D. Denneberg, *Non Additive Measure and Integral* (Kluwer Academic Publishers,
         Dordorecht, 1994)
[Den00] D. Denneberg, Non-additive measure and integral, basic concepts and their role for
         applications, in *Fuzzy Measures and Integrals. Theory and Applications*, eds. by M.
         Grabisch, T.Murofushi, M. Sugeno (Physica-Verlag, Heidelberg, 2000), pp. 42–69
[DP88]   D. Dubois, H. Prade, *Possibility Theory* (Plenum Press, New York, 1988)
[DP00]   D. Dubois, H. Prade, *Fundamentals of Fuzzy Sets* (Kluwer, Boston, 2000)
[DMPRS01] D. Dubois, J.-L. Marichal, H. Prade, M. Roubens, R. Sabbadin, The use of the discrete
         Sugeno integral in decision making: a survey. Int. J. Uncertainty Fuzziness Knowl.-
         Based Syst. **9**, 539–561 (2001)
[Gag13] M. Gagolewski, On the relationship between symmetric maxitive, minitive, and mod-
         ular aggregation operators. Inf. Sci. **221**, 170–180 (2013)
[GM13]   M. Gagolewski, R. Mesiar, Aggregating different paper quality measures with a gen-
         eralized h-index. J. Informetrics **6**, 566–579 (2013)
[GM14]   M. Gagolewski, R. Mesiar, Monotone measures and universal integrals in a uniform
         framework for the scientific impact assessment problem. Inf. Sci. **263**, 166–174 (2014)
[Gra96] M. Grabisch, The applications of fuzzy integrals in multicriteria decision making.
         Euro. J. Oper. Res. **89**, 445–456 (1996)
[Gra97] M. Grabisch, k-order additive discrete fuzzy measures and their representation. Fuzzy
         Sets Syst. **92**, 167–189 (1997)
[Gra00] M. Grabisch, The interaction and Möbius representation of fuzzy measures on finite
         spaces, k-additive measures: a survey, in *Fuzzy Measures and Integrals. Theory and
         Applications*. eds. by M. Grabisch, T. Murofushi, M. Sugeno (Physica-Verlag, Hei-
         delberg, 2000), pp. 70–93
[GL00]   M. Grabisch, C. Labreuche, To be symmetric or asymmetric? A dilemma in decision
         making, in *Preferences and Decisions under Incomplete Knowledge*, eds. by J. Fodor,
         B. De Baets, P. Perny (Physica-Verlag, Heidelberg, 2000)
[GNW95] M. Grabisch, H.T. Nguyen, E.A. Walker, *Fundamentals of Uncertainty Calculi, with
         Applications to Fuzzy Inference* (Kluwer, Dordrecht, 1995)
[GMR00] M. Grabisch, J.-L. Marichal, M. Roubens, Equivalent representations of set functions.
         Math. Oper. Res. **25**, 157–178 (2000)
[GMS00] M. Grabisch, T. Murofushi, M. Sugeno (eds.), *Fuzzy Measures and Integrals, Theory
         and Applications* (Physica-Verlag, Heidelberg, 2000)
[GKM08] M. Grabisch, I. Kojadinovic, P. Meyer, A review of methods for capacity identification
         in Choquet integral based multiattribute utility theory. Euro. J. Oper. Res. **186**, 766–
         785 (2008)

[IS85]   K. Ishii, M. Sugeno, A model of human evaluation process using fuzzy measure. Int. J. Man-Mach. Stud. **22**, 19–38 (1985)

[KMM11]   E.P. Klement, M. Manzi, R. Mesiar, Ultramodular aggregation functions. Inf. Sci. **181**, 4101–4111 (2011)

[KWH97]   G. Klir, Z. Wang, D. Harmanec, Constructing fuzzy measures in expert systems. Fuzzy Sets Syst. **92**, 251–264 (1997)

[KG05]   I. Kojadinovic, M. Grabisch, *Non additive measure and integral manipulation functions, R package version 0.4-6* (2005), http://cran.r-project.org/web/packages/kappalab/index.html

[KMR05]   I. Kojadinovic, J.-L. Marichal, M. Roubens, An axiomatic approach to the definition of the entropy of a discrete Choquet capacity. Inf. Sci. **172**, 131–153 (2005)

[LL95]   K.M. Lee, H. Leekwang, Identification of λ-fuzzy measure by genetic algorithms. Fuzzy Sets Syst. **75**, 301–309 (1995)

[Mar00a]   J.-L. Marichal, On Choquet and Sugeno integrals as agregation functions, in *Fuzzy Measures and Integrals. Theory and Applications*, eds. by M. Grabisch, T. Murofushi, M. Sugeno (Physica-Verlag, Heidelberg, 2000), pp. 247–272

[Mar00b]   J.-L. Marichal, On Sugeno integral as an aggregation function. Fuzzy Sets Syst. **114**, 347–365 (2000)

[Mar02a]   J.-L. Marichal, Aggregation of interacting criteria by means of the discrete Choquet integral, in *Aggregation Operators. New Trends and Applications*, eds. by T. Calvo, G. Mayor, R. Mesiar (Physica-Verlag, Heidelberg, 2002), pp. 224–244

[Mar02b]   J.-L. Marichal, Entropy of discrete Choquet capacities. Euro. J. Oper. Res. **137**, 612–624 (2002)

[Mar04]   J.-L. Marichal, Tolerant or intolerant character of interacting criteria in aggregation by the Choquet integral. Euro. J. Oper. Res. **155**, 771–791 (2004)

[Mes95]   R. Mesiar, Choquet-like integrals. J. Math. Anal. Appl. **194**, 477–488 (1995)

[Mes99a]   R. Mesiar, Generalizations of k-order additive discrete fuzzy measures. Fuzzy Sets Syst. **102**, 423–428 (1999)

[Mes99b]   R. Mesiar, k-order additive measures. Int. J. Uncertainty Fuzziness Knowl.-Based Syst. **7**, 561–568 (1999)

[MG99]   P. Miranda, M. Grabisch, Optimization issues for fuzzy measures. Int. J. Uncertainty, Fuzziness Knowl.-Based Syst. **7**, 545–560 (1999)

[Nar07]   Y. Narukawa, Distances defined by Choquet integral, in *FUZZIEEE*, London, U.K., July 2007, pp. 511–516

[NT05]   Y. Narukawa, V. Torra, Fuzzy measure and probability distributions: distorted probabilities. IEEE Trans. Fuzzy Syst. **13**, 617–629 (2005)

[Pap02]   E. Pap (ed.), *Handbook of Measure Theory* (North Holland/Elsevier, Amsterdam, 2002)

[RA80]   D. Ralescu, G. Adams, The fuzzy integral. J. Math. Anal. Appl. **86**, 176–193 (1980)

[Rou96]   M. Roubens, Interaction between criteria and definition of weights in MCDA problems, in *44th Meeting of the European Working Group Multicriteria Aid for Decisions*. Brussels, Belgium, 1996

[Sha76]   G. Shafer, *A Mathematical Theory of Evidence* (Princeton University Press, Princeton, 1976)

[SGBC03]   M.A. Sicilia, E. García Barriocanal, T. Calvo, An inquirybased method for Choquet integral-based aggregation of interface usability parameters, in *Kybernetika*, vol. 39 (2003), pp. 601–614

[Sug74]   M. Sugeno, Theory of Fuzzy Integrals and Applications. Ph.D. thesis. Tokyo Institute of Technology, 1974

[TN08]   V. Torra, Y. Narukawa, The h-index and the number of citations: two fuzzy integrals. IEEE Trans. Fuzzy Syst. **16**, 795–797 (2008)

[TNS14]   V. Torra, Y. Narukawa, M. Sugeno (eds.), *Non-Additive Measures* (Springer, Berlin, 2014)

[WK92]   Z. Wang, G. Klir, *Fuzzy Measure Theory* (Plenum Press, New York, 1992)

[WWK98] W. Wang, Z. Wang, G.J. Klir, Genetic algorithms for determining fuzzy measures from data. J. Intell. Fuzzy Syst. **6**, 171–183 (1998)

[WLW00] Z. Wang, K.S. Leung, J. Wang, Determining nonnegative monotone set functions based on Sugeno's integral: an application of genetic algorithms. Fuzzy Sets Syst. **112**, 155–164 (2000)

[Web84] S. Weber, Measures of fuzzy sets and measures on fuzziness. Fuzzy Sets Syst. **13**, 114–138 (1984)

[Yag02] R.R. Yager, The induced fuzzy integral aggregation operator. Int. J. Intell. Syst. **17**, 1049–1065 (2002)

[Yag04] R.R. Yager, Generalized OWA aggregation operators. Fuzzy Optim. Decis. Making **3**, 93–107 (2004)

[Yag10] R.R. Yager, Norms induced from OWA operators. IEEE Trans. Fuzzy Syst. **18**, 57–66 (2010)

[Yag15] R.R. Yager, Multicriteria decision-making using fuzzy measures. Cybern. Syst. **46**, 150–171 (2015)

[YK03] R.R. Yager, V. Kreinovich, Universal approximation theorem for uninorm-based fuzzy systems modeling. Fuzzy Sets Syst. **140**, 331–339 (2003)

[Zad78] L. Zadeh, Fuzzy sets as a basis for a theory of possibility. Fuzzy Sets Syst. **1**, 3–28 (1978)

Chapter 5
Penalty Based Averages

Abstract Every averaging function can be looked at from the perspective of minimizing some sort of a penalty, a price paid for the deviation of the output from the inputs. Penalty functions are formally defined and their special classes are presented. The quasi-arithmetic means, OWA and Choquet integral are obtained as special cases of the penalty based averages. Other classes of averages such as the deviation and entropic means are also presented. This chapter introduces new penalty based averages and relates the averages to the maximum likelihood principle.

5.1 Motivation and Definitions

The aggregated value is often interpreted as some sort of representative, or consensus value of the inputs. A measure of deviation from this value, or a penalty for not having a consensus, has been studied by many authors [BS38, BTCT89, CMY04, Gin58, Jac21, Mes07, MKKC08, MSV08, Ped09, TN07, YR97]. It is known that the weighted arithmetic and geometric means, the median and the mode are functions that minimize some simple penalty functions; these examples are discussed in detail in Sect. 5.3. There is a related class of deviation (and quasi-deviation) means [Bul03]: we will show in Sect. 5.3.2 that deviation means are a subclass of penalty-based aggregation functions.

In this chapter we present a general view of the penalty based aggregation, show how the existing families of averaging aggregation functions arise as special cases, and design new aggregation functions based on problem specific information. We follow a systematisation of penalty based averages in [CB10].

Penalty based representation of the arithmetic mean and the median (see special cases in Sect. 5.3) were already known to Laplace (quoted from [TN07], p. 15), see also [Gin58]. The main motivation is the following. Let $\mathbf{x}$ be the inputs and y the output. If all the inputs coincide $x = x_1 = \cdots = x_n$, then the output is $y = x$, and we have a unanimous vote. If some input $x_i \neq y$, then we impose a "penalty" for this disagreement. The larger the disagreement, and the more inputs disagree with the output, the larger (in general) is the penalty. We look for an aggregated value

© Springer International Publishing Switzerland 2016

G. Beliakov et al., *A Practical Guide to Averaging Functions*,
Studies in Fuzziness and Soft Computing 329,
DOI 10.1007/978-3-319-24753-3_5

which minimizes the penalty; in some sense we look for a consensus value which minimizes the disagreement.

Thus we need to define a suitable measure of disagreement, or dissimilarity. We start with a very broad definition of penalties, and then particularize it and obtain many known aggregation functions as special cases. Let us consider a vector of inputs **x** and the vector $\mathbf{y} = (y, y, \ldots, y)$.

Definition 5.1 (*Penalty function*) The function $P : \mathbb{I}^{n+1} \rightarrow \bar{\mathbb{R}}_+ = [0, \infty]$ is a penalty function if it satisfies:

  (i) $P(\mathbf{x}, y) \geq 0$ for all **x**, y;
  (ii) $P(\mathbf{x}, y) = 0$ if and only if $\mathbf{x} = \mathbf{y}$;
  (iii) For every fixed **x**, the set of minimizers of $P(\mathbf{x}, y)$ is either a singleton or an interval.

Definition 5.2 (*Penalty based function*) Given a penalty function P, the penalty based function f is

$$f(\mathbf{x}) = \arg \min_y P(\mathbf{x}, y), \qquad\qquad (5.1)$$

if y is the unique minimizer, and $y = \frac{a+b}{2}$ if the set of minimizers is the interval (a, b) (open or closed).

The first two conditions have useful interpretations: no penalty is imposed if there is full agreement, and no negative penalties are allowed.

Note 5.3 The "only if" part in the condition ii) is essential for the idempotency of the penalty based function. Indeed, let us construct the following penalty $P(\mathbf{x}, y) = 0$ if $y \in [0, \bar{x}]$ and 1 otherwise, where $\bar{x}$ denotes the arithmetic mean of the inputs. Then P satisfies the conditions of the Definition 5.1 safe the "only if" part. Then the value of the penalty based function f is $\frac{\bar{x}}{2}$ for any input **x**, so f is clearly not idempotent.

Since adding a constant to P does not change its minimizers, technically condition (ii) can be relaxed: P just needs to reach its absolute minimum if and only if $\mathbf{x} = \mathbf{y}$. Condition (iii) ensures that the function f is well defined (see Note 5.15 for a counterexample). If P is quasiconvex in y, then (iii) is automatically satisfied. We should also note that a penalty based function is necessarily idempotent, but it is not always monotone.

Note 5.4 We remind that a function f is quasiconvex, if its level sets $S = \{\mathbf{x} : f(\mathbf{x}) \leq C\}$ are convex for any $C \in Ran(f)$. Quasiconvex functions have a unique minimum, but may have many minimizers that form a convex set (an interval in our case). In that case a convention how to choose the minimizer is needed, and we take the midpoint of the interval of minimizers.

Note 5.5 It is not necessary to explicitly state f, provided a suitable penalty function is given and the optimization problem solvable. Subsequently it is sufficient to solve (5.1) to obtain the aggregate $y = f(\mathbf{x})$.

Definition 5.6 (*Penalty based aggregation*) A penalty based function f, which is monotone increasing in all components of $\mathbf{x}$ is called a penalty based aggregation function.

Penalty based aggregation functions are necessarily averaging due to monotonicity and idempotency.

Next we establish a few general results.

Proposition 5.7 *Let f be a penalty based aggregation function on $\mathbb{I}^n$. Let h be a continuous strictly monotone function $X \to \mathbb{I}$. Then $f_h(\mathbf{x}) = h^{-1}(f(h(\mathbf{x})))$ is also a penalty based aggregation function on X^n, with $P_h(\mathbf{x}, y) = P(h(\mathbf{x}), h(y))$.*

Proof The first assertion follows from Proposition 1.88. Considering $f(\mathbf{x}) = \arg\min_y P(\mathbf{x}, y)$, we have that $f(h(\mathbf{x})) = \arg\min_y P(h(\mathbf{x}), v)$, with $v = h(y)$ or equivalently $y = h^{-1}(v)$. Furthermore, the minimum value of y is equal to $h^{-1}(v) = h^{-1}(f(h(\mathbf{x})))$. $\qquad\square$

Proposition 5.8 *Let f be a penalty based aggregation function. Let P satisfy the following condition $P(r\mathbf{x}, ry) = p_1(r)P(\mathbf{x}, y)$ for all $r \in \mathbb{R}_+$ such that $(r\mathbf{x}, ry) \in Dom(P)$, where p_1 is some function. Then f is homogeneous.*

Proof We have $f(r\mathbf{x}) = \arg\min_t P(r\mathbf{x}, t) = r \arg\min_y P(\mathbf{x}, y)$ with $t = ry$, so we have $f(r\mathbf{x}) = ry = rf(\mathbf{x})$. $\qquad\square$

Note 5.9 Positively homogeneous functions of degree α are the functions which satisfy $f(r\mathbf{x}) = r^\alpha f(\mathbf{x}), r > 0$.

Proposition 5.10 [BB87] *Let f be a homogeneous averaging aggregation function on $\mathbb{I}^n$, and h be a continuous strictly monotone function $X \to \mathbb{I}$. Then $f_h(\mathbf{x}) = h^{-1}(f(h(\mathbf{x})))$ is also homogeneous if $h(t) = t^p$, $p \neq 0$, or $h(t) = -\log(t)$.*

Proposition 5.11 *Let a penalty based function f be defined by the minimization of the penalty*

$$P(\mathbf{x}, y) = \sum G_i(\mathbf{x})D(x_i - y),$$

where functions G_i are constant under the translation $\mathbf{x} \to \mathbf{x} + c\mathbf{1}$ and D some function. Then f is shift-invariant.

Proof Since under the translation $\mathbf{x} \to \mathbf{x} + c\mathbf{1}$ all G_i remain constant then the minimum of $P(\mathbf{x} + c\mathbf{1}, y)$ will be achieved at $y = y^* + c$, where y^* is the minimizer of $P(\mathbf{x}, y)$. Properties of D do not matter here, as its value is not affected. $\qquad\square$

Let us now consider the following questions: (a) can any averaging function be represented as a penalty based function? and (b) are penalty based functions always agregation functions?

First we prove that any averaging aggregation function can be expressed as a penalty based function. This result does not lead by itself to new aggregation functions. It just shows that this class is as powerful as the class of averaging aggregation functions itself, and that in principle it is possible to express any averaging aggregation function with the help of an appropriately chosen penalty function.

Theorem 5.12 [CB10] *Let* $f : \mathbb{I}^n \to \mathbb{I}$ *be an idempotent function. Then there exists a penalty function* $P : \mathbb{I}^{n+1} \to \bar{\mathbb{R}}_+$, *such that*

$$f(\mathbf{x}) = \arg\min_y P(\mathbf{x}, y).$$

Proof The function $P(\mathbf{x}, y) = (f(\mathbf{x}) - y)^2$ is one such penalty function. In fact, any strictly convex (or quasi-convex) univariate function of $t = f(\mathbf{x}) - y$ can serve as such a penalty function. □

Corollary 5.13 *Any averaging aggregation function* f *can be expressed as a penalty based aggregation function.*

We have stated earlier that not every penalty based function is monotone (see Example 5.42). But for some types of penalty based functions we can establish when monotonicity holds. The next result applies to twice differentiable penalties, whereas another condition for a particular class of penalties will be given in Sect. 5.2.1.

Theorem 5.14 [CB10] *If the penalty function* P *is twice continuously differentiable, then the necessary and sufficient condition for* f *to be an aggregation function is that the ratio of partial derivatives at all* $y = f(\mathbf{x})$,

$$\frac{P_{yx_i}}{P_{yy}} \leq 0,$$

for all $i = 1, \dots, n$.

Proof Consider an implicit function $y = f(\mathbf{x})$ given by $F(\mathbf{x}, y) = 0$, where $F = P_y$, which is the necessary condition of the minimum of P at $y = f(\mathbf{x})$. Taking the derivatives,

$$\frac{\partial y}{\partial x_i} = -\frac{F_{x_i}}{F_y} = -\frac{P_{yx_i}}{P_{yy}} \geq 0,$$

from which the assertion of the Theorem follows. □

5.2 Types of Penalty Functions

5.2.1 Faithful Penalty Functions

A special class of penalty functions was considered in [CMY04]. Let P be given as

$$P(\mathbf{x}, y) = \sum_{i=1}^{n} w_i\, p(x_i, y), \tag{5.2}$$

where $p : \mathbb{I}^2 \to \mathbb{R}_+$ is a dissimilarity function with the properties

(1) $p(t, s) = 0$ if and only if $t = s$, and

(2) $p(t_1, s) \geq p(t_2, s)$ whenever $t_1 \geq t_2 \geq s$ or $t_1 \leq t_2 \leq s$,

and $\mathbf{w}$ is a weighting vector. Note that the condition (2) is weaker than that in [YR97], which is $p(t_1, s) \geq p(t_2, s)$ if $|t_1 - s| > |t_2 - s|$.

The resulting penalty based function, if it exists, is idempotent, but it need not be monotone, see Chap. 7.

Note 5.15 Under conditions (1) and (2) above, P is not a penalty function in the sense of Definition 5.1 (condition iii) is not assumed). There is no guarantee that f is well defined. For example, function $p(x, y) = 0$ if $x = y$ and $p(x, y) = \sin(y)$ otherwise, satisfies (1) and (2), but it has several separate minimizers.

To ensure that y^* is unique, and f is an aggregation function, the authors in [CMY04] use the so called "faithful" penalty function.

Definition 5.16 (*Faithful penalty function*) The function $p : \mathbb{I}^2 \to \mathbb{R}_+$ is called faithful penalty function, if it satisfies 1) and can be represented as $p(t, s) = K(h(t), h(s))$, where $h : \mathbb{I} \to \mathbb{R}$ is some continuous monotone function (scaling function) and $K : \mathbb{R}^2 \to \mathbb{R}_+$ is convex.

Definition 5.17 (*Faithful penalty based function*) Let the penalty function P be given by (5.2), where $p : \mathbb{I}^2 \to \mathbb{R}_+$ is a faithful penalty function. The function

$$f(\mathbf{x}) = y^* = \arg\min_y P(\mathbf{x}, y)$$

is a faithful penalty based aggregation function.

Note 5.18 Condition (2) is satisfied automatically by the faithful penalty functions, because p is quasi-convex in t. Also P given in (5.2) is quasi-convex in y in this case (since the sum of convex functions K is always convex), and hence it has a unique minimum, but possibly many minimizers. In the latter case y^* is taken as the midpoint of the set of minimizers.

A special class of faithful penalty based functions was considered in [Mes07, MSV08] (dissimilarity functions). The (faithful) penalties p are expressed as

$$p(t, s) = K(h(t) - h(s)), \tag{5.3}$$

where $K : \mathbb{R}^2 \to \mathbb{R}$ is convex (shape function) with the unique minimum $K(0) = 0$, and h is the scaling function.[1] In this case the following result holds.

---

[1] In [Mes07] the author considers different penalties p_i for each argument x_i, all with different shape and scaling functions K_i and h_i. We shall deal with this case in Sect. 5.4.

Theorem 5.19 [Mes07] *The penalty based function with the penalty expressed in* (5.2) *and* (5.3) *is an aggregation function.*

Proof The proof in [Mes07] is based on the following. Take $h = Id$ and consider $y^* = f(\mathbf{x})$, and take $\mathbf{x}' = (x_1, \ldots, x_{i-1}, x_i + \varepsilon, x_{i+1}, \ldots, x_n)$. Denote $y^{**} = \arg\min_y P(\mathbf{x}', y)$. We need to show that for $\varepsilon > 0$, $y^{**} \geq y^*$. This follows from the fact that for a convex function K, $K(u) - K(v) \leq K(u + w) - K(v + w)$ for $u \geq v$ and all $w \geq 0$, and consequently $p(x_i + \varepsilon, y^*) - p(x_i, y^*) \leq p(x_i + \varepsilon, y) - p(x_i, y)$ for $y < y^*$ (taking $u = x_i + \varepsilon - y^*$, $v = x_i - y^*$ and $w = y^* - y$). Then, from $P(\mathbf{x}, y^*) \leq P(\mathbf{x}, y)$ we have

$$P(\mathbf{x}', y^*) = \sum_{j \neq i} p(x_j, y^*) + p(x_i + \varepsilon, y^*) \leq \sum_{j \neq i} p(x_j, y) + p(x_i + \varepsilon, y) = P(\mathbf{x}', y)$$

for all $y < y^*$. Hence $y^{**} \geq y^*$. The result for $h \neq Id$ follows from Proposition 5.7. $\qquad\qquad\square$

5.2.2 Restricted Dissimilarity Functions

In [BBP08] the concept of *restricted dissimilarity function* is introduced and different theorems of construction and characterization are presented. In this section we fix the interval $\mathbb{I} = [0, 1]$.

Definition 5.20 (*Restricted dissimilarity function*) A function $d_R : \mathbb{I}^2 \to \mathbb{I}$ is a restricted dissimilarity function if:

1. $d_R(x, y) = d_R(y, x)$ for every $x, y \in \mathbb{I}$;
2. $d_R(x, y) = 1$ if and only if $x = 0$ and $y = 1$ or $x = 1$ and $y = 0$; that is, if $\{x, y\} = \{0, 1\}$;
3. $d_R(x, y) = 0$ if and only if $x = y$;
4. For any $x, y, z \in \mathbb{I}$, if $x \leq y \leq z$, then $d_R(x, y) \leq d_R(x, z)$ and $d_R(y, z) \leq d_R(x, z)$.

We will say that d_R is a strict restricted dissimilarity function if for any $x, y, z \in \mathbb{I}$, if $x < y < z$, then $d_R(x, y) < d_R(x, z)$ and $d_R(y, z) < d_R(x, z)$.

Example 5.21 The function $d_R(x, y) = |x - y|$ provides a simple example of a restricted dissimilarity function which is strict. On the other hand, as an example of non-strict restricted dissimilarity function we present the following. Take $c \in]0, 1[$. Then

$$d_R(x, y) = \begin{cases} 1, & \text{if } \{x, y\} = \{0, 1\} \\ 0, & \text{if } x = y; \\ c & \text{otherwise,} \end{cases}$$

is a restricted dissimilarity function. Observe that this function is not continuous.

Since penalty functions need to be quasi-convex in order to satisfy item (iii) in Definition 5.1, we focus on the set of restricted dissimilarity functions which are convex or quasi-convex following [BJPMB13].

Theorem 5.22 *Let $d_R : \mathbb{I}^2 \to \mathbb{I}$ be a restricted dissimilarity function. Then d_R is quasi-convex in one variable; that is, for all $x, y_1, y_2, \lambda \in \mathbb{I}$*

$$d_R(x, \lambda y_1 + (1 - \lambda)y_2) \leq \max(d_R(x, y_1), d_R(x, y_2)). \qquad (5.4)$$

Because of symmetry of d_R it is quasi-convex in each of its variables. But it is not quasi-convex in both variables, as the following example illustrates.

Example 5.23 Take $d_R(x, y) = (|x^2 - y^2|)^{\frac{1}{2}}, \lambda = \frac{1}{2}, x_1^2 = \frac{1}{4}, x_2^2 = 0, y_1^2 = \frac{3}{4}$ and $y_2^2 = \frac{1}{2}$. Then calculation shows that $d_R(\lambda x_1 + (1 - \lambda)y_1, \lambda x_2 + (1 - \lambda)y_2) = \frac{1}{2}|\frac{1}{2} + \sqrt{3}|^{\frac{1}{2}} > \frac{1}{2}$ whereas $d_R(x_1, x_2) = \frac{1}{2} = d_R(y_1, y_2)$.

Proposition 5.24 [BBP08] *If φ_1, φ_2 are two automorphisms of the unit interval, then*

$$d(x, y) = \varphi_1(|\varphi_2(x) - \varphi_2(y)|)$$

is a restricted dissimilarity function.

Proposition 5.25 *Let $d_R : \mathbb{I}^2 \to \mathbb{I}$ be a symmetric quasi-convex function in each of its variables such that*

1. $d_R(x, y) = 0$ *if and only if $x = y$;*
2. $d_R(x, y) = 1$ *if and only if $\{x, y\} = \{0, 1\}$.*

Then d_R is a restricted dissimilarity function.

Corollary 5.26 *Let d_R be a symmetric and strictly convex function such that*

1. $d_R(x, x) = 0$ *for all $x \in \mathbb{I}$;*
2. $d_R(x, y) = 1$ *if and only if $\{x, y\} = \{0, 1\}$.*

Then, d_R is a strict restricted dissimilarity function.

Some other methods of constructing quasi-convex restricted dissimilarity functions from implication functions and negations can be found in [BJPMB13]. Similarly to faithful penalty functions, faithful restricted dissimilarity functions can be expressed by using a convex function K with the unique minimum $K(0) = 0$ and a strictly monotone continuous function $h : \mathbb{I} \to \mathbb{R}$ by using $p(x, y) = K(h(x) - h(y))$.

Next we construct the suitable penalty functions by using the following results from [BJPMB13].

Proposition 5.27 *Take $d_R : \mathbb{I}^2 \to \mathbb{I}$. Then the following items are equivalent.*

 (i) d_R is a faithful restricted dissimilarity function.
(ii) There exists a convex automorphism K and a bijection φ on $\mathbb{I}$ such that

$$d_R(x, y) = K(|\varphi(x) - \varphi(y)|)$$

   for all $x, y \in \mathbb{I}$.

Example 5.28   1. If $K(x) = x^2$, then $d_R(x, y) = (\varphi(x) - \varphi(y))^2$,
2. If $K(x) = |x|$, then $d_R(x, y) = |\varphi(x) - \varphi(y)|$.

Theorem 5.29 *Let $M : \mathbb{I}^n \to \mathbb{I}$ be an aggregation function verifying:*
A1 : $M(x_1, \cdots, x_n) = 0$ if and only if $x_1 = \cdots = x_n = 0$,
and $d_R : \mathbb{I}^2 \to \mathbb{I}$ a faithful restricted dissimilarity function. Then

$$P : \mathbb{I}^{n+1} \to \mathbb{I} \text{ given by}$$

$$P(\mathbf{x}, y) = \overset{n}{\underset{i=1}{M}} d_R(x_i, y) \tag{5.5}$$

is a penalty function.

Corollary 5.30 *In the setting of Theorem 5.29, if $M : \mathbb{I}^n \to \mathbb{I}$ is an aggregation function such that (A1) holds, then*

1. *the penalty based function*

$$f(\mathbf{x}) = \arg\min_{y} P(\mathbf{x}, y) = \arg\min_{y} \overset{n}{\underset{i=1}{M}} d_R(x_i, y) \tag{5.6}$$

   is an idempotent function.
2. *If M fulfills property:*

   A2 : $M(x_1, \ldots, x_n) = 1$ if and only if $x_1 = \cdots = x_n = 1$,

   then $P(\mathbf{x}, y) = 1$ if and only if $\{x_i, y\} = \{0, 1\}$ for all $i = 1, \ldots, n$;
3. *If $\mathbf{x} = (1, \ldots, 1)$, then $P(\mathbf{x}, y) = \overset{n}{\underset{i=1}{M}} N(y)$ with $N(y) = d_R(1, y)$. If M is idempotent then $P(\mathbf{x}, y) = N(y)$;*
4. *$P(\mathbf{x}, 1) = \overset{n}{\underset{i=1}{M}} n(x_i)$ with $n(x_i) = d_R(1, x_i)$ for all $i = 1, \ldots, n$;*
5. *If φ is a concave automorphism, then*

$$P(\mathbf{x}, y) = \overset{n}{\underset{i=1}{M}} \varphi^{-1}(|\varphi(x_i) - \varphi(y)|) \tag{5.7}$$

   is a penalty function.

Example 5.31 Let $\varphi(x) = x^{\frac{1}{2}}$

- If we take $M(x_1, \ldots, x_n) = \frac{1}{n}\sum_{i=1}^{n} x_i$, then

$$P(\mathbf{x}, y) = \frac{1}{n}\sum_{i=1}^{n} (|x_i^{\frac{1}{2}} - y^{\frac{1}{2}}|)^2. \tag{5.8}$$

- If we take the *3-max operator*,

$$M(x_1, \ldots, x_n) = \frac{\max(x_1, \ldots, x_n)}{\max(x_1, \ldots, x_n) + \max(1 - x_1, \ldots, 1 - x_n)},$$

we have

$$P(\mathbf{x}, y) = \frac{\max\limits_{i=1,\ldots,n} (|x_i^{\frac{1}{2}} - y^{\frac{1}{2}}|)^2}{\max\limits_{i=1,\ldots,n} (|x_i^{\frac{1}{2}} - y^{\frac{1}{2}}|)^2 + \max\limits_{i=1,\ldots,n} (1 - (|x_i^{\frac{1}{2}} - y^{\frac{1}{2}}|)^2)}. \tag{5.9}$$

- If we take a *convex linear combinations* of min and of max with $\lambda \in (0, 1)$,

$$M(x_1, \ldots, x_n) = \lambda \min(x_1, \ldots, x_n) + (1 - \lambda) \max(x_1, \ldots, x_n)$$

then

$$P(\mathbf{x}, y) = \lambda \min\limits_{i=1,\ldots,n} (|x_i^{\frac{1}{2}} - y^{\frac{1}{2}}|)^2 + (1 - \lambda) \max\limits_{i=1,\ldots,n} (|x_i^{\frac{1}{2}} - y^{\frac{1}{2}}|)^2. \tag{5.10}$$

Theorem 5.32 *Let $M : \mathbb{I}^n \to \mathbb{I}$ be a weighted quasi-arithmetic mean with strictly positive weights. Let $d_R : \mathbb{I}^2 \to \mathbb{I}$ be a faithful restricted dissimilarity function. Then, the mapping*

$$F(x_1, \ldots, x_n) = \arg\min_{y} M(d_R(x_1, y), \ldots, d_R(x_n, y)) \tag{5.11}$$

is an averaging aggregation function.

5.2.3 Minkowski Gauge Based Penalties

Definition 5.33 (*Minkowski gauge*) Let S be a bounded star-shaped set in $\mathbb{R}^n$ that includes the origin in its interior, i.e., S is compact and if $\mathbf{x} \in S$ and $\lambda \in [0, 1]$, then $\lambda \mathbf{x} \in S$ [Rub00]. Minkowski gauge is the function

$$\mu_P(\mathbf{x}) = \inf\{\lambda > 0 : \mathbf{x} \in \lambda S\}.$$

It enjoys several interesting properties, as reported in [DR95, Rub00], in particular it is defined on $\mathbb{R}^n$, is non-negative and positively homogeneous of degree one. If S is convex, then $\mu_S(\mathbf{x})$ is also convex and sublinear. That is we have [DR95, Rub00]

(1) $\mu_S(\lambda\mathbf{x}) = \lambda\mu_S(\mathbf{x})$, $\forall \mathbf{x} \in \mathbb{R}^n$, $\forall \lambda > 0$.
(2) $\mu_S(\mathbf{x} + \mathbf{y}) \leq \mu_S(\mathbf{x}) + \mu_S(\mathbf{y})$, $\forall \mathbf{x}, \mathbf{y} \in \mathbb{R}^n$ (for a convex S).

The function $D_S(\mathbf{x}, \mathbf{y}) = \mu_S(\mathbf{x} - \mathbf{y})$ is a distance function defined with the help of Minkowski gauge. It verifies $D(\mathbf{x}, \mathbf{x}) = 0$ for all $\mathbf{x}$, and the triangular inequality (if S is convex), but D needs not be symmetric. D becomes a metric if S is symmetric with respect to the origin.

The set S is interpreted as the unit "sphere" in such a distance function. All the standard metrics are special cases of the Minkowski gauge. Other examples include polyhedral distances, when S is a polytope.

Definition 5.34 (*Minkowski gauge distance*) The Minkowski gauge based distance is

$$D_S(\mathbf{x}, \mathbf{y}) = \mu_S(\mathbf{x} - \mathbf{y}).$$

Definition 5.35 (*Minkowski gauge penalty based function*) Let $P : \mathbb{R}^{n+1} \to \mathbb{R}_+$ be a penalty function defined by $P(\mathbf{x}, y) = D_S(\mathbf{x}, \mathbf{y})$. The Minkowski gauge penalty based aggregation function is

$$f(\mathbf{x}) = \arg\min_y P(\mathbf{x}, y).$$

Note 5.36 f is obviously a homogeneous function. However it is not necessarily an aggregation function, since monotonicity condition may fail. This condition depends on the set S, and even convexity of S does not guarantee monotonicity.

Note 5.37 By using a scaling function h, we can define averaging aggregation functions $f_h(\mathbf{x}) = h^{-1}(f(\mathbf{x}))$. From Proposition 5.10 it follows that f_h is a homogeneous function for $h(x) = x^p$ for any $p \in \mathbb{R}$, $p \neq 0$ and $h(x) = -\log(x)$ for $p = 0$.

Example 5.38 Let S be the square centered at the origin, with side length 2, so that μ_S is the ∞-norm. Then $P(\mathbf{x}, y) = \max_i |x_i - y|$. The corresponding penalty based aggregation function is the mid-range, i.e., $f(\mathbf{x}) = \frac{1}{2}(\max x_i + \min x_i)$. One can apply a scaling function h and use Proposition 1.88 to obtain a scaled version of the mid-range.

5.3 Examples

5.3.1 Quasi-arithmetic Means, OWA and Choquet Integral

We shall now state how some of the averages we already discussed in the previous chapters can be represented in penalty based form. Consider the following faithful penalty based aggregation functions with P given by (5.2).

1. Let $p(t, s) = (t - s)^2$. The corresponding faithful penalty based aggregation function is the weighted arithmetic mean.
2. Let $p(t, s) = |t - s|$. The corresponding faithful penalty based aggregation function is the weighted median.
3. Let $p(t, s) = (h(t) - h(s))^2$. The corresponding faithful penalty based aggregation function is the weighted quasi-arithmetic mean with the generator h.
4. Let $p(t, s) = |h(t) - h(s)|$. The corresponding faithful penalty based aggregation function is the weighted quasi-median with the generator h, defined as $f(\mathbf{x}) = h^{-1}(Med_\mathbf{w}(h(\mathbf{x})))$.
5. Let $P(\mathbf{x}, y) = \sum_{i=1}^{n} w_i p(x_{(i)}, y)$, where $x_{(i)}$ is the i-th largest component of $\mathbf{x}$. We obtain the ordered weighted counterparts of the means in the previous examples, namely the OWA, ordered weighted median and generalized OWA.
6. Let $p(t, s) = |t - s|^r, r \geq 1$. Then in general no closed form solution exists, but the minimum in (5.2) can be found numerically; this is discussed in Sect. 5.4. Interestingly, the limiting case $r \to 1$ does not correspond to the median when $n = 2k$ (see [Jac21]), but to a solution of the following equation,

$$(y - x_{(1)})(y - x_{(2)}) \ldots (y - x_{(k)}) = (y - x_{(k+1)}) \ldots (y - x_{(n)}).$$

   For example, for $n = 4$ we have

$$f(\mathbf{x}) = \frac{x_{(1)} x_{(2)} - x_{(3)} x_{(4)}}{(x_{(1)} + x_{(2)}) - (x_{(3)} + x_{(4)})},$$

   whereas the standard definition of median gives $f(\mathbf{x}) = \frac{x_{(2)} + x_{(3)}}{2}$. Weighted medians were considered in [BS38].
7. Let $P(\mathbf{x}, y) = \sum_{i=1}^{n} w_i(\mathbf{x})(x_{(i)} - y)^2$, where $x_{(i)}$ is the i-th smallest component of $\mathbf{x}$, and $w_i(\mathbf{x}) = v(H_i) - v(H_{i+1})$, as in Definition 1.9. We obtain the Choquet integral with respect to v. Note that the weights depend on the ordering of the components of $\mathbf{x}$.
8. We obtain the generalized Choquet integral $C_{v,h}(\mathbf{x}) = h^{-1}(C_v(h(\mathbf{x})))$ by minimizing $P(\mathbf{x}, y) = \sum_{i=1}^{n} w_i(\mathbf{x})(h(x_{(i)}) - h(y))^2$.
9. Let $c \geq 0$ and

$$p(t, s) = \begin{cases} t - s, & \text{if } s \leq t, \\ c(s - t), & \text{if } s > t. \end{cases}$$

Then f is the α-quantile operator, with $\alpha = c/(1+c)$ [CMY04]. To obtain the i-th order statistic, we take $c = \frac{i-1/2}{n-i+1/2}$.

5.3.2 Deviation Means

Next consider the class of deviation means, see [Bul03], p. 316.

Definition 5.39 (*Deviation mean*) Let $d : \mathbb{I}^2 \to \mathbb{R}$ be a continuous function strictly increasing with respect to the second argument, and satisfying $d(t,t) = 0$ for all $t \in \mathbb{I}$, and $\mathbf{w}$ is a weighting vector. The equation

$$\sum_{i=1}^{n} w_i d(x_i, y) = 0 \qquad (5.12)$$

has the unique solution $y*$, which is the value of the function $f(\mathbf{x})$ called the deviation mean.

Note 5.40 The function d is called the deviation function. In general, one can use a family of deviation functions $d_i, i = 1, \ldots, n$ and solve the equation $\sum_{i=1}^{n} d_i(x_i, y) = 0$, so that Definition 5.39 becomes a special case. Deviation means were recovered in [BC09] as lever aggregation functions.

Note 5.41 If $d(t,s) = h(s) - h(t)$ for some continuous strictly monotone function h, one recovers the class of weighted quasi-arithmetic means with the generator h.

Example 5.42 Let $d_i(x,y) = w_i(x)(h(y) - h(x))$. We obtain the Bajraktarevic mean (2.28) (see [Baj58, Baj63, Bul03]), defined by

$$f(\mathbf{x}) = h^{-1}\left(\frac{\sum_{i=1}^{n} w_i(x_i) h(x_i)}{\sum_{i=1}^{n} w_i(x_i)}\right).$$

Note that this mean is not always monotone, see Chap. 7; some conditions ensuring monotonicity were studied in [MPR03].

Theorem 5.43 *Let the penalty function P be defined as*

$$P(\mathbf{x}, y) = \sum_{i=1}^{n} w_i d(x_i, y)^2,$$

where d is a deviation function. Then the penalty based aggregation function is the deviation mean.

Proof Of course, the Eq. (5.12) is the necessary condition of a minimum, which is unique since d^2 is strictly quasiconvex with respect to y. $\square$

Hence all deviation means can be represented as penalty based functions but not vice versa (because P needs not be differentiable with respect to y, nor strictly convex in y). Therefore they form a subclass of penalty based functions.

5.3.3 Entropic Means

Another interesting special case is that of entropic means [BTCT89].

Definition 5.44 (*Entropic mean*) Let $\phi : \mathbb{R}_+ \to \mathbb{R}$ be a strictly convex differentiable function with $(0, 1] \subset dom\ \phi$ and such that $\phi(1) = \phi'(1) = 0$, and $\mathbf{w}$ is a weighting vector. The penalty d_ϕ is defined as

$$d_\phi(x, y) = x\phi(y/x).$$

The entropic mean is the function

$$f(\mathbf{x}) = y^* = \arg \min_{y \in \mathbb{R}_+} P_\phi(\mathbf{x}, y) = \arg \min_{y \in \mathbb{R}_+} \sum_{i=1}^{n} w_i d_\phi(x_i, y).$$

It turns out that $d_\phi(\alpha, \cdot)$ is strictly convex for any $\alpha > 0$, and $d_\phi(\alpha, \beta) \geq 0$ with equality if and only if $\alpha = \beta$. The differentiability assumption can be relaxed, see [BTCT89].

However the condition (1) in the Sect. 5.2.1 is not satisfied, hence the set of entropic means is not a subset of faithful penalty based aggregation functions. On the other hand, all entropic means are homogeneous aggregation functions [BTCT89].

In Table 5.1 we present several special cases of entropic means.

5.3.4 Bregman Loss Functions

Definition 5.45 (*Bregman loss function*) [Bre67] Let $\psi : \mathbb{R}^n \to \mathbb{R}$ be a strictly convex differentiable function. Then the Bregman loss function $D_\psi : \mathbb{R}^n \times \mathbb{R}^n \to \mathbb{R}$ is defined as

$$D_\psi(\mathbf{x}, \mathbf{y}) = \psi(\mathbf{x}) - \psi(\mathbf{y}) - \langle \mathbf{x} - \mathbf{y}, \nabla\psi(\mathbf{y}) \rangle,$$

where ∇ denotes the gradient and $\langle \cdot, \cdot \rangle$ is the standard inner product.

We concentrate on the univariate Bregman loss functions, defined by

$$D_\psi(x, y) = \psi(x) - \psi(y) - (x - y)\psi'(y). \tag{5.13}$$

Table 5.1 Special cases of entropic means

$\phi(t)$	Penalty $d_\phi(x, y)$	Mean
$-\log t + t - 1$	$-x \log \frac{y}{x} + y - x$	WAM
$(t-1)^2$	$\frac{(y-x)^2}{x}$	WHM
$1 - 2\sqrt{t} + t$	$x + y - 2\sqrt{xy}$	WPM, $p = \frac{1}{2}$
$t \log t - t + 1$	$x - y + y \log \frac{y}{x}$	WGM
$\frac{1}{p-1}(t^{1-p} - p) + t$	$x \left(\frac{1}{p-1} \left((\frac{y}{x})^{1-p} - p \right) + \frac{y}{x} \right)$	WPM
$\max\{0, (1-t)\}^2$	$x \cdot \max\{0, (1 - \frac{y}{x})\}^2$	max
$\frac{t^{2-p}}{2-p} - \frac{t^{1-p}}{1-p} + \frac{1}{(2-p)(1-p)}$	$x \left(\frac{(\frac{y}{x})^{2-p}}{2-p} - \frac{(\frac{y}{x})^{1-p}}{1-p} + \frac{1}{(2-p)(1-p)} \right)$, with $0 < p < 1$[1]	Lehmer
$\frac{t^{1-r}-1}{1-r} - \frac{t^{1-s}-1}{1-s}$	$x \left(\frac{(\frac{y}{x})^{1-r}-1}{1-r} - \frac{(\frac{y}{x})^{1-s}-1}{1-s} \right)$, with $s \geq 0 > r$ or $s > 0 \geq r$[1]	Gini
$t - \frac{2}{3} \log t + \frac{t^2}{3} - \frac{1}{3}$	$x \left(-\frac{2}{3} \log \frac{y}{x} + \frac{(\frac{y}{x})^2}{3} - \frac{1}{3} \right)$	Composition

[1] For these values of the parameters the mean is monotone

Definition 5.46 (*Bregman penalty based function*) Let $\psi : \mathbb{R} \to \mathbb{R}$ be a strictly convex differentiable function. Then Bregman penalty based aggregation function is

$$f(\mathbf{x}) = y^* = \arg \min_{y \in \mathbb{R}_+} \sum_{i=1}^{n} w_i D_\psi(y, x_i).$$

Note 5.47 Notice that the arguments of D_ψ are in this particular order. If instead we minimize $\sum_{i=1}^{n} w_i D_\psi(x_i, y)$, we always obtain the arithmetic mean as the result [BGW05], as can be seen below.

Consider the derivative (assuming ψ is twice differentiable)

$$\frac{\partial D_\psi(x, y)}{\partial y} = (y - x)\psi''(y).$$

Then clearly the necessary condition of the minimum of $\sum_{i=1}^{n} w_i D_\psi(x_i, y)$ is

$$\sum_{i=1}^{n} w_i(y^* - x_i)\psi''(y^*) = 0,$$

and since $\psi''(y) \neq 0$ (strict convexity) it follows that $y^*\psi''(y^*) = \psi''(y^*) \sum_{i=1}^{n} w_i x_i$, and hence y^* is the WAM.

Returning to the Definition 5.46, and taking the partial derivative of D_ψ, we have [BTCT89]:

$$\frac{\partial D_\psi(x, y)}{\partial x} = \psi'(x) - \psi'(y),$$

and the necessary condition of the minimum becomes

$$\sum_{i=1}^{n} w_i (\psi'(y^*) - \psi'(x_i)) = 0,$$

from which it follows that y^* satisfies

$$\psi'(y^*) = \sum_{i=1}^{n} w_i \psi'(x_i).$$

Because ψ is strictly convex, the minimum is unique, ψ' is strictly increasing, and $f(\mathbf{x}) = y^*$ is a weighted quasi-arithmetic mean with the generator $h(t) = \psi'(t)$.

Example 5.48 Penalties based on Bregman loss function [BTCT89]:

1. $\psi(t) = (t-1)^2$, then $D_\psi(x, y) = (x-y)^2$ and f is WAM;
2. $\psi(t) = t \log t$, then $D_\psi(x, y) = y \log x/y$, and $f = M_g$ is WGM;
3. $\psi(t) = t \log t - (1+t) \log(1+t)$, then $h(t) = \psi'(t) = \log(\frac{t}{1+t})$, $h^{-1}(t) = \frac{e^t}{1+e^t}$, and

$$f(\mathbf{x}) = \frac{\prod x_i^{w_i}}{\prod (1+x_i)^{w_i} - \prod x_i^{w_i}} = \frac{G(\mathbf{x})}{G(1+\mathbf{x}) - G(\mathbf{x})}.$$

5.4 New Penalty Based Aggregation Functions

We have provided several examples in which the penalties were symmetric with respect to the inputs, save the weights, i.e., penalties of the form (5.2). Let us now consider more general penalties, based on the formula

$$P(\mathbf{x}, y) = \sum_{i=1}^{n} p_i(x_i, y). \tag{5.14}$$

Here, in addition to the weights (absorbed in p_i), with their usual interpretation of the relative importance of the i-th input (or i-th largest input), we can vary the contribution of the i-th input based on the functional form of the corresponding penalty $p_i(x_i, y)$. This is useful in the following context. Consider the inputs of different sensors, which need to be averaged (e.g., temperature sensors). The inputs from sensors are random variables with different distributions (e.g., normal, Laplace or another member of exponential family). Then taking the weighted arithmetic mean or median is not appropriate, because sensors are heterogeneous. We can take into account the diversity of inputs errors distributions by means of different penalty functions. The following example presents the penalty suitable when the first distribution is Laplace, and the second is normal (see also Example 5.56 in the next section).

Example 5.49 [Mes07, MKKC08] Let $n = 2$ and the penalty be

$$P(\mathbf{x}, y) = |x_1 - y| + (x_2 - y)^2.$$

Solving the equation of the necessary condition for a minimum, and taking into account that P is convex, we obtain

$$f(x_1, x_2) = Med(x_1, x_2 - \frac{1}{2}, x_2 + \frac{1}{2}).$$

For a weighted penalty function

$$P(\mathbf{x}, y) = w_1|x_1 - y| + w_2(x_2 - y)^2$$

the solution is

$$f(x_1, x_2) = Med(x_1, x_2 - \frac{w_1}{2w_2}, x_2 + \frac{w_1}{2w_2}).$$

Extending Example 5.49 we have

Example 5.50 [CB10]

$$P(\mathbf{x}, y) = \sum_{i=1}^{n} w_i|x_i - y|^i.$$

We cannot provide a closed form solution in this case, but note that P is convex with respect to y, and a numerical solution is easily obtained using, e.g., the method of golden section. The result is always an aggregation function.

Note 5.51 When $p_i(x_i, y) = K_i(h_i(x_i) - h_i(y))$, where K_i is a convex function with the unique minimum at $K_i(0) = 0$ and h_i is a scaling function for all $i = 1, \ldots, n$, the resulting penalty based function is monotone, as shown in [Mes07] (Theorem 5.19 arises as a special case).

Example 5.52 An interesting concept of weighting functions was proposed in [MSV08]. The penalty is given by

$$P(\mathbf{x}, y) = \sum_{i=1}^{n} g(x_i)K(h(x_i) - h(y)),$$

where K is a convex shape function, h is a scaling function (see Sect. 5.2.1), and g : $\mathbb{I} \to \mathbb{R}_+$ is a given weighting function. When $K(t) = t^2$ we obtain the Bajraktarevic mean (cf. Example 5.42), which is not always monotone (see [MPR03]). For $K(t) = |t|$, the conditions (on g) which ensure monotonicity of the resulting penalty based function are established in [MSV08]. Under these conditions, f is always one of the medians (standard, upper, or lower).

Example 5.53 [CB10] Let $P(\mathbf{x}, y) = \max(0, y - x_{(1)}) + \max(0, y - x_{(2)}) + \sum_{i=3}^{n} |x_{(i)} - y|$, and $x_{(i)}$ is the i-th largest component of $\mathbf{x}$. The first two terms penalize solutions y exceeding the largest and the second largest inputs. As the result, we discard the two largest values of $\mathbf{x}$. The solution is equivalent to a weighted median with the weighting vector $\mathbf{w} = (0, 0, \frac{1}{n-2}, \ldots, \frac{1}{n-2})$. By changing the absolute value to the squared differences, we obtain an OWA function with the same weighting vector.

Example 5.54 [CB10] Let $P(\mathbf{x}, y) = \sum_{i=1}^{n-1} w_i (x_i - y)^2 + w_n \max(0, y - x_n)^2$. The meaning of the last term is the following. Suppose the n-th input (e.g., the n-th expert) usually underestimates the result y. Then we wish to penalize $y > x_n$ but not $y < x_n$. So the n-th input is discarded only if $y < x_n$. The resulting penalty P is a piecewise quadratic function whose minimum is easily found: it is the minimum of the weighted arithmetic means M of the first $n - 1$ and of all components of $\mathbf{x}$, $f(\mathbf{x}) = \min(M(x_1, \ldots, x_{n-1}), M(x_1, \ldots, x_n))$.

This example can be extended by changing more terms of the sum to some asymmetric functions of $y - x_i$, in which case the solution can be found numerically. The interpretation is similar: positive and negative deviations from x_i are penalized differently.

Let us now use the concept of Minkowski gauge. The resulting penalty is generally not of the form (5.14), although in many cases it is. For example, in cases 1, 2, 6 of Sect. 5.3.1 D_S is simply a 1-, 2-, and r-norm, in the Example 5.38 we have the ∞-norm, and the penalty in Example 5.49 corresponds to D_S with S being the region between two parabolas $S\{(x_1, x_2) \in \mathbb{R}^2 | x_2^2 - 1 \leq x_1 \leq 1 - x_2^2\}$.

Example 5.55 Define the simplicial distance function as follows. Take the simplex S centered at 0, which is the intersection of $n + 1$ halfspaces

$$S = \bigcap_{i=1}^{n+1} \{\mathbf{x} : \langle \mathbf{x}, \mathbf{h}_i \rangle \leq 1\}, \qquad (5.15)$$

where $\mathbf{h}_i \in \mathbb{R}^n$ are the directional vectors. Then

$$D_S(\mathbf{x}, \mathbf{y}) = \max\{\langle \mathbf{x} - \mathbf{y}, \mathbf{h}_i \rangle : 1 \leq i \leq n + 1\}.$$

Define (5.15) by vectors

$$\mathbf{h}_1 = (-1, 0, 0, \ldots),$$
$$\mathbf{h}_2 = (0, -1, 0, \ldots),$$
$$\vdots$$
$$\mathbf{h}_n = (0, \ldots, 0, -1),$$
$$\mathbf{h}_{n+1} = (1, 1, \ldots, 1).$$

The simplex S is illustrated on Fig. 5.1. Then

$$D_S(\mathbf{x}, \mathbf{y}) = \max\{\max_{i=1,\ldots,n} (y_i - x_i), \sum_{i=1}^{n} (x_i - y_i)\}.$$

The penalty P is expressed as

$$P(\mathbf{x}, y) = \max\{y - x_1, y - x_2, \ldots, y - x_n, \sum_{i=1}^{n} (x_i - y)\}.$$

Minimization of P can be performed by methods of linear programming as in [CB10]. As the result we obtain $y = \frac{1}{n+1}(x_{(1)} + \cdots + x_{(n-1)} + 2x_{(n)})$. This is the OWA function with the weighting vector $\mathbf{w} = (\frac{1}{n+1}, \ldots, \frac{1}{n+1}, \frac{2}{n+1})$.

Taking D_{-S} we obtain the dual of such an OWA function with the weights $\mathbf{w} = (\frac{2}{n+1}, \ldots, \frac{1}{n+1}, \frac{1}{n+1})$. Note that such weighting vectors correspond to Yager's andlike and orlike S-OWA functions respectively [Yag93] with the parameter $\alpha = \frac{1}{n+1}$. We remind that the weighting vectors of S-OWA functions are given by

$$w_1 = \frac{1-\alpha}{n} + \alpha, \; w_i = \frac{1-\alpha}{n}, \; i = 2, \ldots, n \text{ (orlike S-OWA)},$$

Fig. 5.1 The simplex S used in the Example 5.55 in the case of $n = 2$

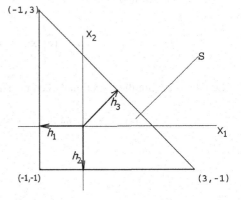

$$w_n = \frac{1-\alpha}{n} + \alpha, \, w_i = \frac{1-\alpha}{n}, i = 1, \ldots, n-1 \text{ (andlike S-OWA)}.$$

Of course, this method of solution can be used for other polyhedral distances D_S, and extended with the help of a scaling function, using Proposition 5.7.

5.5 Relation to the Maximum Likelihood Principle

Consider the following problem: given n noisy estimations $x_1, \ldots, x_n$ of a quantity y, determine the best estimate for y. For example, we can have n different temperature sensors in a room, and want to determine the best estimate for the temperature. The estimates x_i are independent random variables, characterized by their own distribution density functions ρ_i. Then the likelihood function is

$$L(y) = \prod_{i=1}^{n} \rho_i(x_i|y).$$

The best estimate of y^* is the one which maximizes L, according to the maximum likelihood principle. It is convenient to maximize the logarithm of L, when ρ_i are from the exponential family of distributions. When x_i are normally distributed with the mean y and dispersions σ_i^2, the maximum likelihood principle leads to the weighted arithmetic mean as the best estimate of y,

$$y^* = \sum_{i=1}^{n} w_i x_i, \text{ with the weights } w_i = \frac{\sigma_i^{-2}}{\sum_{j=1}^{n} \sigma_j^{-2}}.$$

Now, let the sensors have different distributions functions ρ_i from the exponential family, $\rho_i(x|y) = h_i(x)e^{p_i(x,y)}$. For instance $p_i(x, y) = -\frac{(x-y)^2}{2\sigma_i^2}$ corresponds to normal distribution and $p_i(x, y) = -\frac{|x-y|}{\sigma_i/\sqrt{2}}$ corresponds to double exponential (Laplace). The logarithm of L is

$$\log L(y) = \sum_{i=1}^{n} p_i(x_i, y) + Const.$$

Minimizing $-L(y)$ leads to the best estimate of y being the penalty based mean

$$f(\mathbf{x}) = \arg \min_{y} \sum_{i=1}^{n} -p_i(x_i, y).$$

Note that the normalization constants implicitly present in ρ_i and L (so that the integral of ρ_i over x is 1) do not play any role in the minimization, as obviously $\arg\min_y f(y) = \arg\min_y (Af(y) + B)$.

Example 5.56 Let $p_1(x, y) = -w_1|x - y|$ be Laplace and $p_2(x, y) = -w_2(x - y)^2$ be normal distributions. Then the corresponding weighted average is given in the Example 5.49.

We see that the principle of maximum likelihood naturally leads to penalty based aggregation functions, which has been noted in [CMY04]. While in special cases (e.g., normal and double exponential distributions) it leads to closed form solutions, in general this is not the case, and the value y^* should be computed numerically. If the resulting penalty P is convex, this is easily done by the golden section method, otherwise Pijavski method of global optimization can be used [Pij72] (see also [Bel08, HJ95] and [BU07] for implementation in a software package). Since y is the only variable, this is numerically quite efficient. Furthermore, the distributions ρ_i can be found experimentally and tabulated, rather than be given in a nice analytical form. The cost of numerical solution increases only marginally in this case.

Example 5.57 Consider ρ_i being truncated Laplace densities, given by $\rho_i(x, y) = e^{p_i(x,y)}$ with

$$p_i(x, y) = \begin{cases} -w_i|x - y|, & \text{if } |x - y| < r_i, \\ -\infty & \text{otherwise} \end{cases}$$

Then minimizing the penalty function $P(\mathbf{x}, y) = \sum_{i=1}^{n} -p_i(x_i, y)$ leads to the following linear programming problem

$$\begin{aligned} \text{minimize} \quad & \sum_{i=1}^{n} w_i(p_i^+ + p_i^-) \\ \text{s.t.} \quad & p_i^+ - p_i^- + y = x_i, \\ & p_i^+ + p_i^- \leq r_i \\ & p_i^+, p_i^- \geq 0, \ i = 1, \dots, n. \end{aligned}$$

Example 5.58 The following penalty function appear in robust regression as one of M-estimators [Hub03], and is due to Huber,

$$p(x, y) = \begin{cases} (x - y)^2, & \text{if } |x - y| \leq c, \\ c(2|x - y| - c) & \text{otherwise,} \end{cases}$$

c is a parameter and $P(\mathbf{x}, y) = \sum p(x_i, y)$ (the corresponding distribution density ρ is Gaussian in the middle with double exponential tails).

5.6 Representation of Averages

In this chapter we presented a general view of penalty based aggregation functions. We have shown that every averaging function can be represented as a penalty based function, and have considered many special cases, which lead to the weighted arithmetic and quasi-arithmetic means, medians, OWA and generalized OWA functions, Choquet integrals, entropic, Bregman and deviation means, Gini, Lehmer and Bajraktarevic means, composition of means, half-range, as well as new averaging functions. In those cases which do not lead to closed form solutions, efficient numerical optimization techniques, in particular linear programming and Lipschitz optimization, can be used. There is also a relation of penalty based means to the maximum likelihood principle.

An important point we would like to emphasize is that penalty functions may have a direct intuitive interpretation in terms of the problem at hand. Thus one can specify the penalties and let the resulting aggregation function be determined by these penalties by means of the optimization process (either analytically or numerically). A particularly important application is averaging noisy experimental values, where the noise has distinct distributions for each datum, e.g., distinct sensors. The maximum likelihood principle leads directly to penalty based aggregation. In the special cases of Gaussian and Laplace distributions it leads to the weighted arithmetic mean and weighted median. For other distributions from the exponential family the aggregation function is determined numerically.

These methods open a new possibility to incorporate weighting functions into the aggregation process. While the traditional weights are constant values, which relate to the parameters of the same underlying noise distribution for all the aggregated data (e.g., the dispersions for Gauss and Laplace distributions), this new method allows one to average data with distinct noise distributions. This can also be viewed from the perspective of aggregating experts opinions: the relative importance of each input is not just a fixed constant, but a function (not necessarily symmetric) which depends on the deviation of this expert from the consensus. It is possible to account for experts who typically underestimate or overestimate the result (pessimists and optimists).

Another important role of the penalty based functions is proper definitions of averages on lattices and other structures, where the traditional analytical expressions make no sense. Yet distances and distance based penalties on lattices can be defined in various ways, and minimization of such penalties result is well defined averages. We discuss this in detail in Chap. 8.

Finally, penalty based representation of averages may also help in technical proofs, and in Chap. 7 we will encounter several examples of this.

References

[Baj58]   M. Bajraktarević, Sur une équation fonctionelle aux valeurs moyennes. Glasnik Mat.-
           Fiz. i Astronom. Drustvo Mat. Fiz. Hrvatske. Ser. II **13**, 243–248 (1958)
[Baj63]   M. Bajraktarević, Sur une genéralisation des moyennes quasilineaire. Publ. Inst. Math.
           Beograd **3**(17), 69–76 (1963)
[BC09]   M.A. Ballester, T. Calvo, Lever aggregation operators. Fuzzy Sets Syst. **160**, 1984–
           1997 (2009)
[BGW05]   A. Banerjee, X. Guo, H. Wang, On the optimality of conditional expectation as a
           Bregman predictor. IEEE Trans. Inf. Theory **51**, 2664–2669 (2005)
[BS38]   J.B. Souto, El modo y otras medias, casos particulares de una misma expresion matem-
           atica. Boletin Matematico **11**, 29–41 (1938)
[Bel08]   G. Beliakov, Extended cutting angle method of global optimization. Pac. J. Optim. **4**,
           153–176 (2008)
[BU07]   G. Beliakov, J. Ugon, Implementation of novel methods of global and non-smooth
           optimization: GANSO programming library. Optimization **56**, 543–546 (2007)
[BTCT89]   A. Ben-Tal, A. Charnes, M. Teboulle, Entropic means. J. Math. Anal. Appl. **139**,
           537–551 (1989)
[BB87]   J.M. Borwein, P.B. Borwein, *PI and the AGM: A Study in Analytic Number Theory
           and Computational Complexity* (Wiley, New York, 1987)
[Bre67]   L.M. Bregman, The relaxation method of finding the common point of convex sets and
           its application to the solution of problems in convex programming. USSR Comput.
           Math. Phys. **7**, 200–217 (1967)
[Bul03]   P.S. Bullen, *Handbook of Means and Their Inequalities* (Kluwer, Dordrecht, 2003)
[BBP08]   H. Bustince, E. Barrenechea, M. Pagola, Relationship between restricted dissimilarity
           functions, restricted equivalence functions and normal EN-functions: image thresh-
           olding invariant. Pattern Recogn. Lett. **29**, 525–536 (2008)
[BJPMB13] H. Bustince, A. Jurio, A. Pradera, R. Mesiar, G. Beliakov, Generalization of the
           weighted voting method using penalty functions constructed via faithful restricted
           dissimilarity functions. Eur. J. Oper. Res. **225**, 472–478 (2013)
[CB10]   T. Calvo, G. Beliakov, Aggregation functions based on penalties. Fuzzy Sets Syst.
           161, 1420–1436 (2010)
[CMY04]   T. Calvo, R. Mesiar, R.R. Yager, Quantitative weights and aggregation. IEEE Trans.
           Fuzzy Syst. **12**, 62–69 (2004)
[DR95]   V.F. Demyanov, A.M. Rubinov, *Constructive Nonsmooth Analysis* (Peter Lang, Frank-
           furt am Main, 1995)
[Gin58]   C. Gini. Le Medie. Milan (Russian translation, Srednie Velichiny, Statistica, Moscow,
           1970): Unione Tipografico- Editorial Torinese (1958)
[HJ95]   P. Hansen, B. Jaumard, Lipschitz optimization, in *Handbook of Global Optimization*,
           ed. by R. Horstm, P. Pardalos (Dordrecht: Kluwer, 1995), pp. 407–493
[Hub03]   P.J. Huber, *Robust Statistics* (Wiley, New York, 2003)
[Jac21]   D. Jackson, Note on the median of a set of numbers. Bull. Am. Math. Soc. **27**, 160–164
           (1921)
[MPR03]   R.A.M. Pereira, R.A. Ribeiro, Aggregation with generalized mixture operators using
           weighting functions. Fuzzy Sets Syst. **137**, 43–58 (2003)
[Mes07]   R. Mesiar, Fuzzy set approach to the utility, preference relations, and aggregation
           operators. Eur. J. Oper. Res. **176**, 414–422 (2007)
[MKKC08]   R. Mesiar, M. Komornikova, A. Kolesarova, T. Calvo, Aggregation functions: a revi-
           sion, in *Fuzzy Sets and Their Extensions: Representation, Aggregation and Models*,
           ed. by H. Bustince, F. Herrera, J. Montero (Springer, Berlin, 2008)
[MSV08]   R. Mesiar, J. Spirková, L. Vavríková, Weighted aggregation operators based on min-
           imization. Inf. Sci. **178**, 1133–1140 (2008)
[Ped09]   W. Pedrycz, Statistically grounded logic operators in fuzzy sets. Eur. J. Oper. Res.
           193, 520–529 (2009)

[Pij72]   S.A. Pijavski, An algorithm for finding the absolute extremum of a function. USSR Comput. Math. Math. Phys. **2**, 57–67 (1972)

[Rub00]   A.M. Rubinov, *Abstract Convexity and Global Optimization* (Kluwer, Dordrecht, 2000)

[TN07]   V. Torra, Y. Narukawa, *Modeling Decisions. Information Fusion and Aggregation Operators* (Springer, Berlin, 2007)

[Yag93]   R.R. Yager, Families of OWA operators. Fuzzy Sets Syst. **59**, 125–148 (1993)

[YR97]   R.R. Yager, A. Rybalov, Understanding the median as a fusion operator. Int. J. Gen. Syst. **26**, 239–263 (1997)

Chapter 6
More Types of Averaging and Construction Methods

Abstract This chapter treats some advanced topics, including a number of construction methods, such as idempotization, graduation curves and flying parameter. It also presents various related functions, such as the overlap and grouping functions. The recent extensions of the Bonferroni mean are treated in detail. The issue of consistency and stability of weighted averages is presented.

6.1 Some Construction Methods

6.1.1 Idempotization

Averaging functions can be constructed from other suitable functions by inverting their diagonal.

Definition 6.1 (*Idempotizable function*) A function $g : \mathbb{I}^n \to \bar{\mathbb{R}}$ is called idempotizable if its diagonal d_g (see Definition 1.16) is strictly increasing and $ran(d_g) = ran(g)$.

Then the function $f = d_g^{-1} \circ g$ is idempotent. If g is also monotone increasing, so is f, hence we obtain an averaging aggregation function f.

The two basic examples of using the idempotization method are the arithmetic and geometric means obtained from the sum and product operations respectively. Another example refers to strict Archimedean t-norms/t-conorms, which can be idempotized by using their diagonal. For example, the Einstein sum in Example 1.83 is idempotized by using $d_g(t) = \frac{2t}{1+t^2}$, whose inverse is $d_g^{-1}(t) = \frac{1+\sqrt{1-t^2}}{t}$, and produces a quasi-arithmetic mean M_h with the generator $h(t) = \log\frac{1+t}{1-t}$, which is the same as the generator of the Einstein sum itself.

In fact, this will be true for every strict Archimedean triangular norm or conorm with the generator h. Since the diagonal can be expressed (in the bivariate case) as $d_g(t) = h^{-1}(2h(t))$, its inverse is $d_g^{-1}(t) = h^{-1}(h(t)/2)$, and therefore by idempotization method we obtain

© Springer International Publishing Switzerland 2016   207
G. Beliakov et al., *A Practical Guide to Averaging Functions*,
Studies in Fuzziness and Soft Computing 329,
DOI 10.1007/978-3-319-24753-3_6

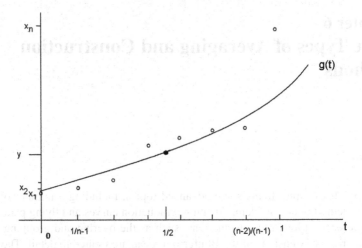

Fig. 6.1 An example of a graduation *curve*. Function $g(t)$ approximates the data $(\frac{i-1}{n-1}, x_i)$. The average is the value $y = g(\frac{1}{2})$

$$f(x, y) = h^{-1}(h(g(x, y))/2) = h^{-1}(h(h^{-1}(h(x) + h(y))))/2) \qquad (6.1)$$
$$= h^{-1}((h(x) + h(y))/2) = M_h(x, y).$$

The multivariate case is dealt with similarly.

6.1.2 Means Defined by Using Graduation Curves

The following construction from [BBF11] based on graduation curves was inspired by [Gin58]. Consider the unweighted averages first. Order the inputs in the non-decreasing order and take the points

$$(0, x_{(1)}), (\frac{1}{n-1}, x_{(2)}), \dots, (\frac{n-2}{n-1}, x_{(n-1)}), (1, x_{(n)})).$$

We can draw these points in the coordinate plane and obtain a picture presented in Fig. 6.1.

Now let us interpolate (or approximate) the resulting points with a function $g :$ $[0, 1] = \mathbb{I} \rightarrow \mathbb{R}$, whose graph is called the graduation curve.[1] To underline its dependence on the data, we will use the notation $g(t; x_1, \dots, x_n)$ where necessary. The value $f(x_1, \dots, x_n) = g(\frac{1}{2}; x_1, \dots, x_n)$ is an average of the components of **x**. We can formulate the following general results [BBF11].

---

[1] Intervals $\mathbb{I}$ other than $[0, 1]$ can be easily accommodated.

Proposition 6.2 *If g preserves the range of the data, i.e., $\min(x_1, \ldots, x_n) \leq g(t; x_1, \ldots, x_n) \leq \max(x_1, \ldots, x_n)$ for all $0 \leq t \leq 1$, then f is an averaging function.*

Proof Clearly $x_{(1)} \leq g(\frac{1}{2}) = f(\mathbf{x}) \leq x_{(n)}$.

The converse of this proposition is also true: for any averaging aggregation function f there exists a function g, such that $g(\frac{1}{2}; x_1, \ldots, x_n) = f(x_1, \ldots, x_n)$.

Corollary 6.3 *If g interpolates the data and is monotone in t, then f is an averaging function.*

Proposition 6.4 *If g is monotone increasing in $x_1, \ldots, x_n$, then f is also monotone, and hence an aggregation function.*

Proposition 6.5 *If g is scale invariant (homogeneous), i.e., $\lambda g(t; \mathbf{x}) = g(t; \lambda \mathbf{x})$, then f is homogeneous. If g is shift-invariant, i.e., $g(t; \mathbf{x}) + \lambda = g(t; \mathbf{x} + \lambda)$, then f is shift-invariant.*

Not all methods of interpolation or approximation deliver these properties of g. However several known methods do, and below we consider several examples.

Example 6.6 Let g be a piecewise linear interpolant to the data, also called broken line or linear spline. The function g is monotone in $\mathbf{x}$, scale invariant and shift-invariant. The resulting value of $g(\frac{1}{2}) = Med(\mathbf{x})$.

Example 6.7 Let g be a constant function fitted to the data in the least squares sense. In this case $g(t) = M(\mathbf{x})$ for all $t \in \mathbb{I}$, the arithmetic mean.

Example 6.8 Let g be an affine function fitted to the data, $g(t) = at + b$. From the theory of linear regression we know that if $\hat{t} = \frac{1}{n} \sum t_i$ then $g(\hat{t}) = a\hat{x} + b = \frac{1}{n} \sum x_i$. So we again obtain the arithmetic mean $y = g(\frac{1}{2}) = M(\mathbf{x})$. Note that linear regression function g is monotone in $\mathbf{x}$, homogeneous and shift-invariant.

If we use polynomial interpolation, the resulting polynomial g is not always increasing in t, nor in $\mathbf{x}$, hence this method is not suitable. Below we present three advanced monotone approximation techniques which yield suitable graduation curves.

Let us take monotone interpolating, smoothing or regression splines as g. Splines are piecewise continuous functions joined at the knots $t_i, i = 1, \ldots, n$. In our case $t_i = \frac{i-1}{n-1}$. There are several possibilities here.

Example 6.9 Function g is a monotone interpolating spline in tension [SK88, Sch66] given by

$$g(t) = \frac{h_i^2}{u_i \sinh(u_i)}[c_i \sinh(u_i z) + c_{i+1} \sinh(u_i(1-z))]$$
$$+(x_i - h_i^2 c_i / u_i^2)z + (x_{i+1} - h_i^2 c_{i+1} / u_i^2)(1-z),$$

where $t_i \leq t \leq t_{i+1}$, $z = (t_{i+1} - t)/h_i$, $t_i = ph_i$, $h_i = t_{i+1} - t_i$, $p \geq 0$ is tension parameter, and c_i are spline coefficients found by solving a linear system of equations with a tridiagonal matrix. We note that for odd n we obtain $g(\frac{1}{2}) = Med(\mathbf{x})$, because this spline is interpolating and $g(t_i) = x_i$ for all $i = 1, \ldots, n$.

Example 6.10 Function g is a monotone smoothing spline presented in [AE91]. The coefficients of the spline are found by solving a convex optimization problem. Here the value $f(\mathbf{x}) = g(\frac{1}{2})$ is different from the median for both even and odd n.

Example 6.11 Function g is a monotone regression spline presented in [Bel00, Bel02]. The coefficients of the spline are found by solving a quadratic optimization problem.

In the Examples 6.9–6.11 above, the value of the spline depends on several data, located just around the central value (the number of such data depends on the order of the spline). Polynomial splines are typically represented in the B-spline basis. Because B-splines have local support, the extreme values are excluded. In these examples g is scale and shift-invariant and so is f.

Now we look at the weighted averages. In the proposed scheme based on the graduation curves, this is achieved by changing the abscissae of the data: given a weighting vector $\mathbf{w}$, we position the data in the following way. Partition the interval $\mathbb{I}$ into n subintervals, the length of each is u_i, i.e. $t_{i+1} - t_i = u_i$, $i = 1, \ldots, n$ with $t_1 = 0$ and $t_{n+1} = 1$, and the components of vector $\mathbf{u}$ are a permutation of components of $\mathbf{w}$ induced by $\mathbf{x}$, i.e., $u_k = w_i$ if $x_i = x_{(k)}$, as in Definition 3.46. The data to be fitted is (ϵ_i, x_i) where the points $\epsilon_i \in [t_i, t_{i+1}]$ are chosen as either the centers or extremes of the respective intervals. When ϵ_i is in the center we obtain the usual median-like function, otherwise we obtain the analogues of the lower and upper weighted medians respectively. This coincides with the definition of the lower and upper median when g is a piecewise linear function interpolating the data in the Example 6.8. All the examples we presented except Example 6.7 remain valid for the weighted averages.

6.1.3 Aggregation Functions with Flying Parameter

We have studied several families of aggregation functions which depend on some parameter. For instance power means (Definition 2.12) and some others. In all cases the parameter which characterizes a specific member of each family did not depend on the input. An obvious question arises: what if the parameter is also a function of the inputs? This leads to aggregation functions with flying parameter [CKKM02], p. 43, [YF94].

Proposition 6.12 *Let f_r, $r \in \mathbb{I}$ be a family of aggregation functions, such that $f_{r_1} \leq f_{r_2}$ as long as $r_1 \leq r_2$, and let g be another aggregation function. Then the function*

$$f_g(x_1, \ldots, x_n) = f_{g(x_1,\ldots,x_n)}(x_1, \ldots, x_n)$$

is also an aggregation function. Further, if f_r is a family of conjunctive, disjunctive, averaging or mixed aggregation functions, f_g also belongs to the respective class.

The same result is valid for extended aggregation functions. Note that many families of aggregation functions are defined with respect to a parameter ranging over $[0, \infty]$ or $[-\infty, \infty]$. To apply flying parameter construction, it is possible to redefine parameterizations using some strictly increasing bijection $\varphi : [0, 1] \rightarrow [0, \infty]$ (or $\rightarrow [-\infty, \infty]$), for instance $\varphi(t) = \frac{t}{t+1}$. In this case we have $f_r = f_{\varphi(g(x_1,\ldots,x_n))}$.

Example 6.13 1. Let $f = M_{[r]}$ be the power mean and let g be the product. Then we have for $x_i > 0$

$$f_g(x_1, \ldots, x_n) = \left(\frac{1}{n} \sum_{i=1}^{n} x_i^{\prod x_j} \right)^{1/\prod x_j}.$$

2. Let $f = $ Bonferroni mean, with $q = 1$ and $p = g = M \geq 0$. Then we have

$$f_g(x_1, \ldots, x_n) = \left(\frac{1}{(n-1)n} \sum_{i,j=1,i \neq j}^{n} x_i(x_j)^{\frac{\sum x_k}{n}} \right)^{\frac{n}{n+\sum x_k}}.$$

3. Let $f = $ Lehmer mean and $g + 1 = M$. Then we have on $(0, 1]^n$

$$f_g(x_1, \ldots, x_n) = \frac{\sum_{i=1}^{n} x_i^{\frac{\sum x_k}{n}}}{\sum_{i=1}^{n} x_i^{\frac{\sum x_k}{n}-1}}.$$

4. Let $f = Med_g(x_1, \ldots, x_n)$ (g–median) and $g = M$ then

$$f_g(x_1, \ldots, x_n) = Median(x_1, \ldots, x_n, \overbrace{M(x_1, \ldots, x_n), \ldots, M(x_1, \ldots, x_n)}^{(n-1)-times}).$$

Weighting vectors of various aggregation functions, such as weighted quasi-arithmetic means and OWA can also be made dependent on the input vector $\mathbf{x}$. Examples of such functions are the Bajraktarevic mean (Definition 2.51 on p. 89), counter-harmonic means (see p. 83), mixture functions (see p. 89) and neat OWA (Definition 3.7 on p. 106). One difficulty when making the weights dependent on $\mathbf{x}$ is that the resulting function f is not necessarily monotone increasing, and hence is not an aggregation function. For mixture functions a sufficient condition for monotonicity of f was established in [MPR03], see discussion on p. 92.

6.1.4 Construction of Shift-Invariant Functions

Shift-invariant functions (Definition 1.46) are stable with respect to linear changes of the scale. The weighted arithmetic means, OWA functions and discrete Choquet integrals are shift-invariant, together with some other averages discussed in Chap. 7. In this section we construct bivariate shift-invariant functions by using one-dimensional continuous function h and the arithmetic mean.

Proposition 6.14 *Let a function* $h : [-1, 1] \rightarrow [-\frac{1}{2}, \frac{1}{2}]$ *be continuous piecewise differentiable and satisfy the following conditions (a)* $h(0) = 0$ *and (b)* $|h'(t)| \leq \frac{1}{2}$ *on the set of its differentiability. Then the function*

$$f(x, y) = \frac{x + y}{2} + h(x - y)$$

is an averaging shift-invariant aggregation function on $[0, 1]^2$. *If* h *is an even function then* f *is symmetric.*

Proof Clearly f is shift-invariant and idempotent. The partial derivatives of f (where they exist) are non-negative as the partial derivatives of the arithmetic mean are both equal to $\frac{1}{2}$, hence f is increasing and therefore is averaging. Consequently the range of f is $[0, 1]$ and it is an aggregation function. The symmetry is also clear. □

The other domains are dealt with similarly by extending the domain of h.

Example 6.15 Take $h(t) = -\frac{|t|}{2}$. Then f is the minimum function. If $h(t) = \frac{|t|}{2}$ then f is the maximum function. If $h(t) = \frac{1}{2}a|t|$, then f is an OWA function with $orness(f) = (a + 1)/2$.

Example 6.16 A piecewise linear function h and the function $h(t) = \frac{1}{s}\sin(|t|), s \geq 2$ generate shift-invariant functions f illustrated in Fig. 6.2.

6.1.5 Interpolatory Constructions

In this section we consider a method of construction based almost entirely on fitting the empirical data, yet also incorporating some important application specific properties if required. The resulting averaging function does not belong to any specific family: it is a general averaging aggregation function, which, according to Definition 1.5 is simply a monotone increasing idempotent function satisfying $f(\mathbf{0}) = 0$ and $f(\mathbf{1}) = 1$ (we fix the interval $\mathbb{I} = [0, 1]$). This function is not given by an algebraic formula, but typically as an algorithm. Yet it is an aggregation function, and for computational purposes this may be as good as an algebraic formula in terms of efficiency.

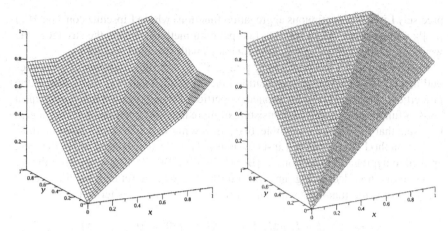

Fig. 6.2 3D plots of the shift-invariant functions in Example 6.16

Of course, having such a "black-box" function is not as transparent to the user, nor is it easy to replicate calculations with pen and paper. Still many such black-box functions are routinely used in practice, for example neural networks for pattern recognition. Hence we consider averaging functions of interpolatory type: they are based on interpolation of empirical data.

Typically the data comes in pairs $(\mathbf{x}, y)$, where $\mathbf{x} \in \mathbb{I}^n$ is the input vector and $y \in \mathbb{I}$ is the desired output. There are several pairs, which will be denoted by a subscript k: $(\mathbf{x}_k, y_k), k = 1, \ldots, K$. The data may be the result of a mental experiment: if we take the input values (x_1, x_2, x_3), what output do we expect? The data can be observed in a controlled experiment: the developer of an application could ask the domain experts to provide their opinion on the desired outputs for selected inputs. The data can also be collected in another sort of experiment, by asking a group of lay people or experts about their input and output values, but without associating these values with some aggregation rule. Finally, the data can be collected automatically by observing the responses of subjects to various stimuli. For example, by presenting a user of a computer system with some information and recording their actions or decisions.

Interpolatory constructions have been studied in [Bel02, Bel03, Bel07, BC07, BCL07, DMDB05, Gra04, KK05, KMM02]. A number of methods are presented in [BPC07], and here we focus on one particular method called Lipschitz interpolation.

The construction is formulated as the following problem. Given a set (possibly uncountable) of values of an aggregation function f, construct f, subject to a number of properties, which we discuss later. The pointwise construction method results in an algorithm whose output is a value of the aggregation function f for a given input $\mathbf{x} \in \mathbb{I}^n$.

Continuous and Lipschitz-continuous aggregation functions are of our particular interest. Lipschitz aggregation functions are very important for applications, because small errors in the inputs do not drastically affect the behavior of the system. The concept of p-stable aggregation functions was proposed in [CM01]. These are precisely Lipschitz continuous aggregation functions whose Lipschitz constant M in $||\cdot||_p$ norm is one.[2] The Lipschitz interpolation method delivers the strongest, the weakest and the optimal aggregation functions with specified conditions.

Given a data set $\mathcal{D} = \{(\mathbf{x}_k, y_k)\}_{k=1}^K$, $\mathbf{x}_k \in \mathbb{I}^n$, $y_k \in \mathbb{I}$, and a number of properties outlined below, construct an aggregation function f, such that $f(\mathbf{x}_k) = y_k$ and all the properties are satisfied. The mentioned properties of an aggregation function define a class of functions $\mathcal{F}$, typically consisting of more than just one function. Our goal is to ensure that $f \in \mathcal{F}$, and if possible, that f is in some sense the "best" element of $\mathcal{F}$.

The method of monotone Lipschitz interpolation was proposed in [Bel05] and applied to aggregation functions in [Bel07, BCL07, BC07]. Denote by *Mon* the set of monotone non-decreasing functions on $[0, 1]^n$. Then the set of general Lipschitz n-ary aggregation functions with Lipschitz constant M is characterized as

$$\mathcal{A}_{M,||\cdot||} = \{f \in Lip(M, ||\cdot||) \cap Mon : f(\mathbf{0}) = 0, f(\mathbf{1}) = 1\}.$$

We assume that the data set is consistent with the class $\mathcal{A}_{M,||\cdot||}$. If not, there are ways of smoothing the data, discussed in [Bel05]. Our goal is to determine the *best* element of $\mathcal{A}_{M,||\cdot||}$ which interpolates the data. The *best* is understood in the sense of optimal interpolation [TW80]: it is the function which minimizes the worst case error, i.e., solves the following Problem

Optimal interpolation problem

$$\min_{f \in \mathcal{A}_{M,||\cdot||}} \max_{g \in \mathcal{A}_{M,||\cdot||}} \max_{\mathbf{x} \in [0,1]^n} |f(\mathbf{x}) - g(\mathbf{x})|$$

$$\text{s.t. } f(\mathbf{x}_k) = y_k, k = 1, \ldots, K.$$

The solution to this problem will be an aggregation function f which is the "center" of the set of all possible aggregation functions in this class consistent with the data. The method of computing f is based on the following result [Bel05].

Theorem 6.17 *Let $\mathcal{D}$ be a data set compatible with the conditions $f \in Lip(M, ||\cdot||) \cap Mon$. Then for any $\mathbf{x} \in [0, 1]^n$, the values $f(\mathbf{x})$ are bounded by $\sigma_l(\mathbf{x}) \leq f(\mathbf{x}) \leq \sigma_u(\mathbf{x})$, with*

$$\sigma_u(\mathbf{x}) = \min_k \{y_k + M||(\mathbf{x} - \mathbf{x}_k)_+||\},$$
$$\sigma_l(\mathbf{x}) = \max_k \{y_k - M||(\mathbf{x}_k - \mathbf{x})_+||\}, \qquad (6.2)$$

---

[2]See p. 22 for definitions of various norms.

where $\mathbf{z}_+$ denotes the positive part of vector $\mathbf{z}$: $\mathbf{z}_+ = (\bar{z}_1, \ldots, \bar{z}_n)$, with

$$\bar{z}_i = \max\{z_i, 0\}.$$

The optimal interpolant is given by

$$f(\mathbf{x}) = \frac{1}{2}(\sigma_l(\mathbf{x}) + \sigma_u(\mathbf{x})). \tag{6.3}$$

Computation of the function f is straightforward, it requires computation of both bounds, and all the functions, σ_l, σ_u and f belong to $Lip(M, ||\cdot||) \cap Mon$. Thus, in addition to the optimal function f, one obtains as a by-product the strongest and the weakest aggregation functions from the mentioned class.

It is also useful to consider infinite data sets

$$\mathcal{D} = \{(\mathbf{t}, v(\mathbf{t})) : \mathbf{t} \in \Omega \subset [0, 1]^n, v : \Omega \to [0, 1]\}$$

in which case the bounds translate into

$$B_u(\mathbf{x}) = \inf_{\mathbf{t} \in \Omega}\{v(\mathbf{t}) + M||(\mathbf{x} - \mathbf{t})_+||\},$$
$$B_l(\mathbf{x}) = \sup_{\mathbf{t} \in \Omega}\{v(\mathbf{t}) - M||(\mathbf{t} - \mathbf{x})_+||\}. \tag{6.4}$$

The function f given in Theorem 6.17 is not yet an aggregation function, because we did not take into account the conditions $f(\mathbf{0}) = 0, f(\mathbf{1}) = 1$. By adding these conditions, we obtain the following generic construction of Lipschitz aggregation functions

$$f(\mathbf{x}) = \frac{1}{2}(\underline{A}(\mathbf{x}) + \overline{A}(\mathbf{x})). \tag{6.5}$$

$$\underline{A}(\mathbf{x}) = \max\{\sigma_l(\mathbf{x}), B_l(\mathbf{x})\}, \quad \overline{A}(\mathbf{x}) = \min\{\sigma_u(\mathbf{x}), B_u(\mathbf{x})\}, \tag{6.6}$$

where the additional bounds B_l and B_u are due to specific properties of aggregation functions. At the very least we have (because of $f(\mathbf{0}) = 0, f(\mathbf{1}) = 1$)

$$B_u(\mathbf{x}) = \min\{M||\mathbf{x}||, 1\}, \tag{6.7}$$
$$B_l(\mathbf{x}) = \max\{0, 1 - M||\mathbf{1} - \mathbf{x}||\},$$

but other conditions will tighten these bounds.

We note that as a special case of Eqs. (6.2)–(6.7) we obtain p-stable aggregation functions (Definition 1.63, p. 23), which have Lipschitz constant $M = 1$ in the norm $||\cdot||_p$. In this case the bounds (6.7) become Yager t-norm and t-conorm respectively $B_u = S_p^Y, B_l = T_p^Y$.

The additional bounds that ensure f is averaging are $\min \leq f \leq \max$. Hence we have

$$B_u(\mathbf{x}) = \min\{M||\mathbf{x}||, \max(\mathbf{x})\}, \quad B_l(\mathbf{x}) = \max\{1 - M||\mathbf{1} - \mathbf{x}||, \min(\mathbf{x})\}.$$

Preservation of symmetry

Symmetry can be imposed in a straightforward manner by ordering the inputs. Namely, consider the simplex $S = \{\mathbf{x} \in [0, 1]^n | x_1 \geq x_2 \geq \cdots \geq x_n\}$ and a function $\tilde{f} : S \rightarrow [0, 1]$. The function $f : [0, 1]^n \rightarrow [0, 1]$ defined by $f(\mathbf{x}) = \tilde{f}(\mathbf{x}_\searrow)$ is symmetric ($\mathbf{x}_\searrow$ is obtained from $\mathbf{x}$ by arranging its components in non-increasing order). Then in order to construct a symmetric f, it is sufficient to construct $\tilde{f}$.

To build $\tilde{f}$ we simply apply Eq. (6.6), with the bounds σ_u, σ_l modified as

$$\sigma_u(\mathbf{x}) = \min_k\{y_k + M||(\mathbf{x} - \mathbf{x}_{\searrow k})_+||\},$$
$$\sigma_l(\mathbf{x}) = \max_k\{y_k - M||(\mathbf{x}_{\searrow k} - \mathbf{x})_+||\},$$

i.e., we order the abscissae of each datum in non-increasing order. There is no need to modify any of the subsequent formulae for B_u, B_l, as long as the conditions which define these bounds are consistent with the symmetry themselves (B_u, B_l will be automatically symmetric).

6.2 Other Types of Aggregation and Properties

6.2.1 Bi-capacities

The concept of a fuzzy measure was extended to set functions on a product $2^{\mathcal{N}} \times 2^{\mathcal{N}}$, $\mathcal{N} = \{1, \ldots, n\}$, called bi-capacities [GL05a, GL05b]. Formally, let

$$Q(\mathcal{N}) = \{(\mathcal{A}, \mathcal{B}) \in 2^{\mathcal{N}} \times 2^{\mathcal{N}} | \mathcal{A} \cap \mathcal{B} = \emptyset\}.$$

A discrete bi-capacity is the mapping $v : Q(\mathcal{N}) \rightarrow [-1, 1]$, non-decreasing with respect to set inclusion in the first argument and non-increasing in the second, and satisfying:

$$v(\emptyset, \emptyset) = 0, v(\mathcal{N}, \emptyset) = 1, v(\emptyset, \mathcal{N}) = -1.$$

Bi-capacities are useful for aggregation on bipolar scales (on the interval $[-1, 1]$), and are used to define a generalization of the Choquet integral as an aggregation function. Bi-capacities are represented by 3^n coefficients. The Möbius transformation, interaction indices and other quantities have been defined for bi-capacities as well

[GL05a, Koj07, KM07, XG07], see also [Sam04, SM03]. Recent extensions involve multi-polar averaging functions, in particular multipolar weighted means, OWA and Choquet integrals [MZA15].

6.2.2 Linguistic Aggregation Functions

In [HHVV96] the authors proposed the linguistic OWA function, based on the definition of a convex combination of linguistic variables [DVV93]. They subsequently extended this approach to linguistic OWG functions[3] [HHV03], see also [HHV97, HHV00, HHVV98].

Linguistic variables are variables, whose values are labels of fuzzy sets [Zad75a, Zad75b, Zad75c]. The arguments of an aggregation function, such as linguistic OWA or linguistic OWG, are the labels from a totally ordered universal set S, such as $S = \{Very\ low, Low, Medium, High, Very\ high\}$. For example, such arguments could be the assessments of various alternatives by several experts, which need to be aggregated, e.g., $f(L, M, H, H, VL)$. The result of such aggregation is a label from S.

6.2.3 Multistage Aggregation

Double aggregation functions (see also discussison on p. 217) were introduced by Calvo and Pradera [CP04] with the purpose to model multistage aggregation process. They are defined as

$$f(\mathbf{x}) = a(g(\mathbf{y}), h(\mathbf{z})),$$

where a, g, h are aggregation functions, and $\mathbf{y} \in [0, 1]^k$, $\mathbf{z} \in [0, 1]^m$, $k + m = n$ and $\mathbf{x} = \mathbf{y}|\mathbf{z}$. "$\cdot|\cdot$" denotes concatenation of two vectors. A typical application of such operators is when the information contained in $\mathbf{y}$ and $\mathbf{z}$ is of different nature, and is aggregated in different ways. Function a may have more than two arguments.

Double aggregation functions can be used to model the following logical constructions *If* (*A* **AND** *B* **AND** *C*) **OR** (*D* **AND** *E*) *then ...*

In this process some input values are combined using one aggregation function, other arguments are combined using a different function, and at the second stage the outcomes are combined with a third function. While the resulting function f is an ordinary aggregation function, it has certain structure due to the properties of the functions a, g and h, such as right- and left-symmetry, and so forth [CP04].

Multi-step Choquet integrals have been treated in [MV99, NT05, TN07].

---

[3]The standard OWA and OWG functions were defined on pp. 101, 133.

6.2.4 Migrativity

The concept of α-migrativity was introduced by Durante and Sarkoci [DS08] for a class of bivariate operations having a property previously presented by Mesiar and Novak in ([MN96], Problem 1.8(b)), as Fodor and Rudas acknowledge in [FR07].

Definition 6.18 (α-*Migrativity*) Let $\alpha \in [0, 1]$. A bivariate function $f : \mathbb{I}^2 \rightarrow \mathbb{I}$ is said to be α-migrative if we have

$$f(\alpha x, y) = f(x, \alpha y) \text{ for all } x, y \in \mathbb{I}.$$

Observe that any function f is 1-migrative, whereas 0-migrativity means that

$$f(x, 0) = f(0, y) = f(0, 0)$$

for any $x, y \in \mathbb{I}$. Also notice that in Definition 6.18 migrativity refers to a fixed, predetermined α. In [BMM09] the concept of α-migrativity was generalized as follows.

Definition 6.19 (*Migrativity*) A function $f : \mathbb{I}^2 \rightarrow \mathbb{I}$ is called *migrative* if and only if

$$f(\alpha x, y) = f(x, \alpha y), \quad \text{for all } x, y \in \mathbb{I}$$

and every $\alpha \in [0, 1]$.

In [BMM09] an in-depth study of the migrativity property was carried out, even allowing α to take values greater than 1.

Lemma 6.20 *A function $f : \mathbb{I}^2 \rightarrow \mathbb{I}$ is migrative if and only if $f(x, y) = f(1, xy)$, for all $x, y \in \mathbb{I}$*

Lemma 6.21 *A function $f : \mathbb{I}^2 \rightarrow \mathbb{I}$ is migrative if and only if there exists $g : \mathbb{I} \rightarrow \mathbb{I}$ such that $f(x, y) = g(xy)$, for all $x, y \in \mathbb{I}$.*

In particular this implies that the only migrative idempotent function is the geometric mean.

6.3 Overlap and Grouping Functions

A common problem in many fields is to assign a given object to one out of several available classes. If the separation between classes is not clear, the expert may not be sure about how to assign objects to a specific class. Perhaps these classes are fuzzy in nature and the expert realizes that elements are simply in between several classes. Or it could be that the boundary between classes is sharp in reality but not clearly perceived by the expert. In any of these situations, the concept of overlap arises [AGMB01].

The concept of overlap as a bivariate aggregation function was introduced in [BFMMO10] (see also [BBPS06]) to measure the degree of overlap in a fuzzy classification system with two classes. This concept has been applied to some interesting problems such as the image segmentation problem described in [JBPPY13] (in which it is necessary to discriminate between objects and the background), or in the framework of preference relations [Bus+12].

The concept of overlap has been also extended to a more general situation with three or more classes involved [GRMBB15]. In this section, we review the results in the bivariate case and then we consider the general n-dimensional case.

6.3.1 Definition of Overlap Functions and Basic Properties

Definition 6.22 (*Overlap function*) A mapping $G_O : [0, 1]^2 \to [0, 1]$ is an overlap function if it satisfies the following conditions: $(G_O 1)$ G_O is symmetric.

$(G_O 2)$ $G_O(x, y) = 0$ if and only if $xy = 0$.
$(G_O 3)$ $G_O(x, y) = 1$ if and only if $xy = 1$.
$(G_O 4)$ G_O is increasing.
$(G_O 5)$ G_O is continuous.

There are many examples of overlap functions and not all of them are averaging aggregation functions. For instance, $G_O(x, y) = (\min(x, y))^p$ for $p > 1$ or $G_O(x, y) = xy$ are such kind of overlap functions. However, there also exist many examples of overlap functions which are averaging, as it is the case for $G_O(x, y) = \sqrt{xy}$ or $G_O(x, y) = (\min(x, y))$.

Let $\mathcal{G}$ denote the set of all overlap functions.

Theorem 6.23 [BFMMO10] $(\mathcal{G}, \leq_\mathcal{G})$ with the ordering $\leq_\mathcal{G}$ defined for $G_1, G_2 \in \mathcal{G}$ by

$$G_1 \leq_\mathcal{G} G_2 \text{ if and only if } G_1(x, y) \leq G_2(x, y)$$

for all $x, y \in [0, 1]$, is a lattice.

The lattice $(\mathcal{G}, \leq_\mathcal{G})$ is not complete (no top neither bottom elements, for example). On the other hand, it is closed with respect to appropriate aggregation functions.

Theorem 6.24 [BFMMO10] Let $M : [0, 1]^2 \to [0, 1]$ be a function. For $G_1, G_2 \in \mathcal{G}$, define the mapping $\mathcal{M}(G_1, G_2) : [0, 1]^2 \to [0, 1]$ as

$$\mathcal{M}(G_1, G_2)(x, y) = M(G_1(x, y), G_2(x, y)) \text{ for all } x, y \in [0, 1] .$$

Then, $\mathcal{M}(G_1, G_2) \in \mathcal{G}$ for any $G_1, G_2 \in \mathcal{G}$ if and only if there is a continuous aggregation function $M^* : [0, 1]^2 \to [0, 1]$ with no zero divisors and such that also its dual $(M^*)_d$ has no zero divisors (i.e., if $M^*(x, y) = 1$ then necessarily either $x = 1$ or $y = 1$) so that $M|_E = M^*|_E$, where $E =]0, 1[^2 \cup \{(0, 0), (1, 1)\}$.

Note 6.25 If we assume that M is continuous,

1. The lattice $\mathcal{G}$ is closed under the product operator.
2. The lattice $\mathcal{G}$ is not closed under the Łukasiewicz t-norm $T_L(x, y) = \max(x + y - 1, 0)$. For instance,

$$M(x, y) = T_L(\min(x, y), \min(x, y))$$

   has zero divisors (take $x = 1/4$ and $y = 1$), thus clearly it is not an overlap function.

As an important particular case we have the following result.

Corollary 6.26 *Let $G_1, \ldots, G_m$ be overlap functions and $w_1, \ldots, w_m$ be non-negative weights with $\sum w_i = 1$. Then the convex combination $G = \sum w_i G_i$ is also an overlap function.*

6.3.2 Characterization of Overlap Functions

In [BFMMO10] a characterization result for overlap functions was provided as follows.

Theorem 6.27 *The mapping $G_O : [0, 1]^2 \rightarrow [0, 1]$ is an overlap function if and only if*

$$G_O(x, y) = \frac{f(x, y)}{f(x, y) + h(x, y)}$$

for some $f, h : [0, 1]^2 \rightarrow [0, 1]$ such that
(1) f and h are symmetric;
(2) f is increasing and h is decreasing;
(3) $f(x, y) = 0$ if and only if $xy = 0$;
(4) $h(x, y) = 0$ if and only if $xy = 1$;
(5) f and h are continuous functions.

Corollary 6.28 *Under the conditions of Theorem 6.27 the following items hold:*

(1) $G_O(x, x) = x$ for some $x \in (0, 1)$ if and only if

$$f(x, x) = \frac{x}{1 - x} h(x, x) .$$

(2) The function h cannot be homogeneous of any order.
(3) If f is homogeneous of order k, then G_O is homogeneous of the same order k if and only if $f + h$ is constant, i.e., $G_O(x, y) = \frac{f(x,y)}{h(0,0)}$.

Theorem 6.27 allows defining interesting families of overlap functions.

Corollary 6.29 *Let f and h be two functions in the setting of the previous theorem. Then, for $k_1, k_2 \in]0, \infty[$, the mappings*

$$G_S^{k_1,k_2}(x,y) = \frac{f^{k_1}(x,y)}{f^{k_1}(x,y) + h^{k_2}(x,y)}$$

define a parametric family of overlap functions.

Corollary 6.30 *In the same setting of Theorem 6.27, let us assume that G_O can be expressed in two different ways:*

$$G_O(x,y) = \frac{f_1(x,y)}{f_1(x,y) + h_1(x,y)} = \frac{f_2(x,y)}{f_2(x,y) + h_2(x,y)}$$

for any $x,y \in [0,1]$ and let M be a bivariate continuous aggregation function that is homogeneous of order one. Then, if we define $f(x,y) = M(f_1(x,y), f_2(x,y))$ and $h(x,y) = M(h_1(x,y), h_2(x,y))$ it also holds that

$$G_O(x,y) = \frac{f(x,y)}{f(x,y) + h(x,y)} \; .$$

Theorem 6.31 *Let G_O be an associative overlap function. Then G_O is a t-norm.*

6.3.3 Homogeneous Overlap Functions

From the general characterization result on homogeneous functions, we have the following result.

Theorem 6.32 *G_O is a homogeneous overlap function of order $k > 0$ if and only if there exists a continuous mapping*

$$\psi : [0, \pi/2] \to [0,1]$$

with $\psi(0) = 0$, $\psi(\frac{\pi}{4}) = 2^{-\frac{k}{2}}$, ψ non-decreasing in $[0, \pi/4]$ and $\psi(\theta) = \psi(\frac{\pi}{2} - \theta)$ for any $\theta \in [0, \pi/4]$ such that

$$G_O(x,y) = (x^2 + y^2)^{\frac{k}{2}} \psi(\arctan(\tfrac{\min(x,y)}{\max(x,y)}))$$

for any $x,y \in [0,1]$.

Proof If G_O is homogeneous of order $k > 0$, then, for any $x,y \in [0,1]$ we have

$$G_O(x,y) = (x^2 + y^2)^{\frac{k}{2}} G_O(\frac{x}{(x^2 + y^2)^{\frac{1}{2}}}, \frac{y}{(x^2 + y^2)^{\frac{1}{2}}})$$

and hence, as we are reduced to consider values on the unit sphere, the result follows. The converse is obvious. $\qquad\square$

Proposition 6.33 *Let G_1 and G_2 be overlap functions which are homogeneous of order k_1 and k_2 respectively. Then*

(i) $G_1G_2(x, y) = G_1(x, y)G_2(x, y)$ is an overlap function homogeneous of order $k_1 + k_2$;

(ii) $G_1^{\frac{1}{k_1}}(x, y) = (G_1(x, y))^{\frac{1}{k_1}}$ is an overlap function homogeneous of order 1.

Proposition 6.34 *Let M be a continuous aggregation function homogeneous of order $k > 0$. If G_1 and G_2 are two overlap functions homogeneous of the same order l, then $\mathcal{M}(G_1, G_2)$ is also an overlap function homogeneous of order kl.*

Example 6.35 Let $G_O : [0, 1]^2 \to [0, 1]$ be an overlap function which is homogeneous of order k. Then, the following items hold:

(i) $G_{S1}(x, y) = \frac{xG_O(x,y)+yG_O(x,y)}{2}$ is an overlap function homogeneous of order $k+1$
    such that $G_{S1}(x, x) = x^{k+1}$ for all $x \in [0, 1]$.

(ii) $G_{S2}(x, y) = \left(\frac{xG_O(x,y)+yG_O(x,y)}{2} \right)^{\frac{1}{k+1}}$ is an idempotent overlap function homo-
    geneous of order 1.

Observe that G_{S1} and G_{S2} are not migrative. Obviously $G_{S1}(\alpha x, y) \neq G_{S1}(x, \alpha y)$ for all $\alpha \geq 0$. The same can be seen for G_{S2} in a similar way.

We can identify homogeneous migrative overlap functions as follows.

Theorem 6.36 *The only migrative homogeneous of order $k > 0$ function $G :$ $[0, 1]^2 \to [0, 1]$ such that $G(1, 1) = 1$ is $G(x, y) = (xy)^{\frac{k}{2}}$*

Corollary 6.37 *The only migrative homogeneous of order 1 function $G_O : [0, 1]^2 \to [0, 1]$ such that $G_O(1, 1) = 1$ is the geometric mean.*

Corollary 6.38 *The only migrative homogeneous of order 2 function $G : [0, 1]^2 \to [0, 1]$ such that $G(1, 1) = 1$ is the product.*

6.3.4 k-Lipschitz Overlap Functions

First of all we write the definition of k-Lipschitzicianity for overlap functions.

Definition 6.39 *(k-Lipschitz overlap function)* Let $k \geq 1$. An overlap function G_O is k-Lipschitz if for any $x, y, z, t \in [0, 1]$ it holds

$$|G_O(x, y) - G_O(z, t)| \leq k(|x - z| + |y - t|) . \tag{6.8}$$

This is the usual definition of k-Lipschitz functions, and it is valid for any function, allowing any value of k greater than zero. But, in the case of overlap functions, just by taking $x = y = z = 1$ and $t = 0$ the restriction to $k \geq 1$ becomes justified.

The set of k-Lipschitz overlap functions with respect to the ordering $\leq_{\mathcal{G}}$ is bounded, as the next result shows.

Theorem 6.40 *Let $k \geq 1$. Then the supremum of the set of k-Lipschitz overlap functions is given by the mapping $\min(kx, ky, 1)$, whereas the infimum is given by $\max(kx + ky - 2k + 1, 0)$. That is, for any k-Lipschitz overlap function G_O the inequality*

$$\max(kx + ky - 2k + 1, 0) \leq G_O(x, y) \leq \min(kx, ky, 1)$$

holds for all $x, y \in [0, 1]$.

The mapping $\max(kx + ky - 2k + 1, 0)$ is never an overlap function. On the contrary, although in general, the mapping $\min(kx, ky, 1)$ for $k > 1$ and $x, y \in [0, 1]$ such that $kx, ky \in [0, 1]$ does not define an overlap function (since by taking $x = y = \frac{1}{k}$ we see that it does not fulfill condition (G_O3)), $\min(x, y)$ is an overlap function, so we have the following corollary.

Corollary 6.41 *The mapping $\min(x, y)$ is the strongest 1-Lipschitz overlap function, in the sense that for any other 1-Lipschitz overlap function G_O the inequality*

$$G_O(x, y) \leq \min(x, y)$$

holds for any $x, y \in [0, 1]$.

In Table 6.1, which has been extracted from [BFMMO10], we present a summary of the properties fulfilled by some instances of overlap functions. For the sake of completeness we include the product even if this is not an averaging function.

Table 6.1 Properties of selected overlap functions

Expression	t-norm T_M	t-norm T_P	Geometric mean
	$\min(x, y)$	xy	$\sqrt{xy}$
G_O1	Yes	Yes	Yes
G_O2	Yes	Yes	Yes
G_O3	Yes	Yes	Yes
G_O4	Yes	Yes	Yes
G_O5	Yes	Yes	Yes
Migrativity	No	Yes	Yes
Homogeneity of order 1	Yes	No	Yes
Homogeneity of order 2	No	Yes	No
Lipschitzicianity	Yes	Yes	No

6.3.5 n-dimensional Overlap Functions

Since overlap functions are not assumed to be associative, its extensions to the n-dimensional case with $n > 2$ requires a specific definition. From now on we follow the developments in [GRMBB15].

Definition 6.42 (*n-dimensional overlap function*) An n-dimensional aggregation function $G_O : [0, 1]^n \longrightarrow [0, 1]$ is an n-dimensional overlap function if and only if:

1. G_O is symmetric.
2. $G_O(x_1, \ldots, x_n) = 0$ if and only if $\prod_{i=1}^{n} x_i = 0$.
3. $G_O(x_1, \ldots, x_n) = 1$ if and only if $x_i = 1$ for all $i \in \{1, \ldots, n\}$.
4. G_O is increasing.
5. G_O is continuous.

Note that, taking into account this definition, an object c that belongs to three classes C_1, C_2 and C_3 with degrees $x_1 = 1$, $x_2 = 1$ and $x_3 = 0.3$ will not have the maximum degree of overlap since condition (3) of the previous definition is not satisfied. Even more, if the degrees are $x_1 = 1$, $x_2 = 1$ and $x_3 = 0$, from the second condition we will conclude that the n-dimensional degree of overlapping of this object into the classification system given by the classes C_1, C_2 and C_3 will be zero. This is the reason why this first extension of the original idea of overlap proposed in [BFMMO10] has been called n-*dimensional* overlap. Let us observe that this definition is closely related with the idea of intersection of n classes.

Example 6.43 It is easy to see that the following aggregation functions are n-dimensional overlap functions:

- The minimum powered by p. $G_O(x_1, \ldots, x_n) = \min_{1 \le i \le n} \{x_i^p\} = \left[\min_{1 \le i \le n} \{x_i\} \right]^p$ with $p > 0$. Note that for $p = 1$ we recover an averaging aggregation function.
- The product. $G_O(x_1, \ldots, x_n) = \prod_{i=1}^{n} x_i$.
- The Einstein product aggregation operator. $EP(x_1, \ldots, x_n) = \dfrac{\prod_{i=1}^{n} x_i}{1 + \prod_{i=1}^{n} (1 - x_i)}$.
- The sinus induced overlap $G_O(x_1, \ldots, x_n) = \sin \dfrac{\pi}{2} (\prod_{i=1}^{n} x_i)^p$ with $p > 0$.

The characterization results already introduced for bivariate overlap functions may be extended in a straight way for n-dimensional overlap functions. In particular we mention the following result.

Proposition 6.44 *Let A_n : $[0, 1]^n \longrightarrow [0, 1]$ be an aggregation function. If A_n is averaging, then A_n is an n-dimensional overlap function if and only if it is symmetric, continuous, has zero as absorbing element and satisfies $A_n(x, 1, \ldots, 1) \neq 1$ for any $x \neq 1$.*

We have the following theorem.

Theorem 6.45 *Let $G_1, \ldots G_m$ be n-dimensional overlap functions and let M : $[0, 1]^m \longrightarrow [0, 1]$ be a continuous and symmetric aggregation function such that if $M(x) = 0$ then $x_i = 0$ for some i and $M(x) = 1$ only if $x_i = 1$ for some i. Then the aggregation function G : $[0, 1]^n \longrightarrow [0, 1]$ defined as $G(x) = M(G_1(x), \ldots, G_m(x))$ is an n-dimensional overlap function.*

Note 6.46 Notice that, since any continuous and symmetric averaging aggregation function M satisfies the conditions of the previous Theorem, it is possible to conclude that any continuous symmetric averaging aggregation of n-dimensional overlap functions is also an n-dimensional overlap function.

As an illustration consider OWA functions.

Proposition 6.47 *Let $W = (w_1, \ldots, w_n) \in [0, 1]^n$ be a weighting vector. The following statements are equivalent:*

1. *The OWA function defined by the weighting vector W is an n-dimensional overlap function.*
2. *$w_n = 1$.*

Also for the case of quasi-aritmetic means the following holds.

Proposition 6.48 *Let g : $[0, 1] \to [-\infty, 0]$ be a continuous increasing bijection with $g(0) = -\infty$. Then the quasi-arithmetic mean M_g is an n-dimensional overlap function.*

6.3.6 Grouping Functions

The notion of grouping function was introduced in [BFMMO10, JBPPY13] by duality to overlap functions. Given n degrees of membership $x_i = \mu_{C_i}(c)$ for $i = 1, \ldots, n$ of an object c in the classes $C_1, \ldots, C_n$, a grouping function is supposed to yield the degree z up to which the combination (grouping) of the n classes $C_1, \ldots, C_n$ is supported.

Definition 6.49 (*Grouping function*) An n-dimensional function

$$G_G : [0, 1]^n \longrightarrow [0, 1]$$

is an n-dimensional grouping function if it satisfies the following conditions:

1. G_G is symmetric.
2. $G_G(x) = 0$ if and only if $x_i = 0$, for all $i = 1, \ldots, n$.
3. $G_G(x) = 1$ if and only if there exist $i \in \{1, \ldots, n\}$ with $x_i = 1$.
4. G_G is non-decreasing.
5. G_G is continuous.

Again, continuous t-conorms (their n-ary forms) and their convex combinations are prototypical examples of n-ary grouping functions.

Example 6.50 The maximum powered by p. $G_G(x_1, \ldots, x_n) = \max\limits_{1 \leq i \leq n} \{x_i^p\}$ with $p > 0$. In this case, note that for $p = 1$ we recover an averaging function.

Theorem 6.51 *Let G_O be an n-dimensional overlap function and let $N : [0, 1] \longrightarrow [0, 1]$ be a strict negation. Then the function $G : [0, 1]^n \longrightarrow [0, 1]$ defined as*

$$G(x_1, \ldots, x_n) = N(G_O(N(x_1), \ldots, N(x_n))$$

is an n-dimensional grouping function.

Theorem 6.52 *Let G_G be an n-dimensional grouping function and let $N : [0, 1] \longrightarrow [0, 1]$ be a negation. Then the function $G : [0, 1]^n \longrightarrow [0, 1]$ defined as*

$$G(x_1, \ldots, x_n) = N(G_G(N(x_1), \ldots, N(x_n))$$

is an n-dimensional overlap function.

Theorem 6.53 *Let $G_1, \ldots G_m$ be n-dimensional grouping functions and let $M : [0, 1]^m \longrightarrow [0, 1]$ be a continuous aggregation function such that $M(x) = 0$ if and only if $x_i = 0$ for some i and $M(x) = 1$ if and only if $x_i = 1$ for all i. Then the aggregation function $G : [0, 1]^n \longrightarrow [0, 1]$ defined as $G(x) = M(G_1(x), \ldots, G_m(x))$ is an n-dimensional grouping function.*

Corollary 6.54 *Let $G_1, \ldots, G_m$ be n-dimensional grouping functions and let $w_1, \ldots, w_m$ be nonnegative weights with $\sum\limits_{i=1}^{m} w_i = 1$. Then the convex sum $G(x) = \sum\limits_{i=1}^{m} w_i G_i(x)$ is also an n-dimensional grouping function.*

It is very important to note that for any t-norm T and any t-conorm S, the inequality $T(x, y) \leq S(x, y)$ holds for every $x, y \in [0, 1]$, and the same is true for n-dimensional t-norms and t-conorms. However, if G_O is an n-dimensional overlap function and G_G is an n-dimensional grouping function, it does not hold in general that $G_O(x_1, \ldots, x_n) \leq G_G(x_1, \ldots, x_n)$ for every $x_1, \ldots, x_n \in [0, 1]$, not even if G_G is the dual of G_O. This problem was considered in [MBBFI15] where the following result (for bivariate overlap and grouping functions) was introduced.

Proposition 6.55 *Let $f : [0, 1] \rightarrow [0, 1]$ be a continuous monotone function such that $f(x) = 0$ if and only if $x = 0$, and $f(x) = 1$ if and only if $x = 1$. Let O_f (G_f) denote the class of overlap functions G_O (grouping functions G_G) such that $G_O(x, 1) = f(x)$ ($G_G(x, 1) = f(x)$) for every $x \in [0, 1]$. Then:*

- *O_f and G_f are non-empty.*
- *Given any overlap function G_O (any grouping function G_G) there exists a grouping function $G_G \in G_f$ (overlap function $G_O \in O_f$) such that $G_O \leq G_G$.*

6.4 Generalized Bonferroni Mean

6.4.1 Main Definitions

In this section, we investigate modeling capabilities of the generalized Bonferroni means. Bonferroni's original function presented in [Bon50] was shown to be expressible as an aggregation system in [Yag09], composed of two arithmetic means and the product. This has an interesting interpretation involving the product of each argument with the average of the rest of the arguments. Yager [Yag09] suggested replacing the simple average with other mean type operators, in particular, the Ordered Weighted Averaging (OWA) function and the discrete Choquet integral. Generalizations of this aggregation system were also investigated in [MR09] (referred to as ABC-aggregation functions) and [BJMRY10, YBJ09] where the generalized Bonferroni mean was shown to be suitable for modeling various concepts, such as hard and soft partial conjunction and disjunction (see p. 58) [Duj07] and boundedness similar to k-intolerance [Mar07].

Introduced in the 1950s [Bon50], the Bonferroni mean remained forgotten until recently [Yag09].

Definition 6.56 (*Bonferroni mean*) Let $p, q \geq 0$ and $x_i \geq 0, i = 1, \ldots, n$. The Bonferroni mean is the function

$$B^{p,q}(\mathbf{x}) = \left(\frac{1}{n(n-1)} \sum_{i,j=1,i \neq j}^{n} x_i^p x_j^q \right)^{\frac{1}{p+q}}. \tag{6.9}$$

If $n = 2$ and $p = q$, the Bonferroni mean is equivalent to the geometric mean. If $q = 0$, the Bonferroni mean is the power mean (the power mean, in turn, has the geometric mean as a special case when $p = 0$). This function has a natural extension to the sum of triples $B^{p,q,r}$, or even to any $\mathbf{k}$-tuples $B^{\mathbf{k}}$.

We can easily see from Eq. (6.9) that $B^{p,q}$ is an averaging aggregation function. In the case of equal indices $p = q$ and $n > 2$, the Bonferroni mean models a soft partial conjunction. This means that we can have the case $B^{p,p}(\mathbf{x}) > 0$ even if some criteria are not satisfied, i.e. some of the x_i are equal to zero. However, since we are taking the sum of products, if there exists at least one pair $\{i, j\}$ such that $x_i, x_j > 0$ then it follows that $B^{p,p}(\mathbf{x}) > 0$. In other words, at least two criteria must be partially satisfied to avoid a zero score overall.

The interpretation of this characteristic could be similar to that of an OWA function with $w_1 = 0$, however there is a key difference. This type of OWA function excludes the greatest input from consideration in the score, and hence fails to satisfy some desirable properties, such as strict monotonicity in $]0, 1[^n$. The Bonferroni mean, on the other hand is strictly monotone on the domain $]0, 1]^n$. The parameters p, q make it reasonably flexible for modeling various degrees of conjunction/disjunction. Where the ratio $\frac{p}{q}$ approaches ∞ or 0, the Bonferroni mean behaves similar to the max operator (with the exception of near the boundary when one variable is 0).

Later in this section we will extend the Bonferroni mean to the sum of triples or **k**-tuples. In these cases, the minimum number of non-zero variables required to give an output greater than zero will be $|\mathbf{k}|$ whenever $n > |\mathbf{k}|$. Increasing $|\mathbf{k}|$ allows the users to specify an arbitrary number of criteria that must be met before the function will give a non-zero score.

By rearranging the terms, the Bonferroni mean is expressed as

$$B^{p,q}(\mathbf{x}) = \left(\frac{1}{n} \sum_{i=1}^{n} x_i^p \left(\frac{1}{n-1} \sum_{j=1, j\neq i}^{n} x_j^q \right) \right)^{\frac{1}{p+q}} . \tag{6.10}$$

We note here that the Bonferroni mean is the $(p+q)$-th root of the arithmetic mean, where each argument is the product of each x_i^p with the arithmetic mean of the remaining x_j^q.

Let us consider a special case $p = q = 1$, i.e.,

$$B^{1,1}(\mathbf{x}) = \left(\frac{1}{n} \sum_{i=1}^{n} x_i \left(\frac{1}{n-1} \sum_{j=1, j\neq i}^{n} x_j \right) \right)^{\frac{1}{2}} .$$

We see that each argument of the outer arithmetic mean is the product of the argument x_i with the average of all other $x_j, j \neq i$. So each term models a conjunction of the i-th criterion with the average satisfaction of the rest of the criteria,

$$x_i \text{ AND (the } average \text{ when } x_i \text{ is absent) .}$$

We will use the notation $\mathbf{x}_{j\neq i}$ to denote the vector in $\mathbb{I}^{n-1}$ that includes the arguments from $\mathbf{x} \in \mathbb{I}^n$ in each dimension except the i-th, $\mathbf{x}_{j\neq i} = (x_1, \ldots, x_{i-1}, x_{i+1}, \ldots, x_n)$.

Definition 6.57 (*Generalized Bonferroni mean*) [BJMRY10] Let $\mathbb{M} = \langle M_1, M_2, C \rangle$, with $M_1 : \mathbb{I}^n \to \mathbb{I}$, $M_2 : \mathbb{I}^{n-1} \to \mathbb{I}$ and $C : \mathbb{I}^2 \to \mathbb{I}$ aggregation functions,

C having the inverse diagonal d_C^{-1}. The generalized Bonferroni mean is given by

$$B_{\mathbb{M}}(\mathbf{x}) = d_C^{-1}\left(M_1(C(x_1, M_2(\mathbf{x}_{j\neq 1})), \ldots, C(x_n, M_2(\mathbf{x}_{j\neq n})))\right). \tag{6.11}$$

Example 6.58 Let $\mathbb{M} = \langle \frac{x_1^2 + 2x_2^2}{3}, x, x_1 x_2^2 \rangle$, which defines the parameters of a generalized Bonferroni mean. The diagonal will be $d_C(t) = t^3$, hence $d_C^{-1}(t) = \sqrt[3]{t}$ and

$$B_{\mathbb{M}}(\mathbf{x}) = \sqrt[3]{\frac{x_1^2 x_2^4 + 2x_1^4 x_2^2}{3}} = x_1 x_2 \sqrt[3]{\frac{2x_1^2 + x_2^2}{3x_1 x_2}}.$$

Example 6.59 We recover the original Bonferroni mean when $M_1 = M$ is the arithmetic mean, $M_2(\mathbf{x}) = (\frac{1}{n-1}\sum_{j\neq i}^{n} x_j^q)^{\frac{1}{q}}$ is the power mean, and $C(x, y) = x^p y^q$. It is easy to check that $d_C(t) = C(t, t) = t^{p+q}$ which gives $d_C^{-1}(t) = t^{\frac{1}{p+q}}$.

In the following subsections we will consider the interpretations and effects of replacing the two arithmetic means in (6.10) with other averaging functions following [BJMRY10, Yag09]. We will also investigate alternative conjunctive functions to the product, after presenting an overall general form and its properties. The proofs of the propositions are given in [BJMRY10].

6.4.2 Properties of the Generalized Bonferroni Mean

The following properties of the generalized Bonferroni mean were established in [BJMRY10].

Theorem 6.60 *The generalized Bonferroni defined in Eq. (6.11) is an aggregation function.*

Theorem 6.61 *For any $\mathbb{M} = \langle M_1, M_2, C \rangle$, with M_1, M_2 being averaging aggregation functions, $B_{\mathbb{M}}$ is an averaging aggregation function, independent of C.*

The reverse does not necessarily hold, i.e. if $B_{\mathbb{M}}$ is an averaging aggregation function it does not follow that both M_1 and M_2 are averaging, as the following example shows.

Example 6.62 Let $n = 5$, where $C, M_2 = T_P$ are conjunctive, and let $M_1 = \frac{1}{5}\sum_{i=1}^{5} x_i^{\frac{2}{3}}$ be disjunctive, the resulting generalized Bonferroni mean is

$$B_{\mathbb{M}}(\mathbf{x}) = \left(\frac{1}{5}\sum_{i=1}^{5}\left(x_i T_P(\mathbf{x}_{j\neq i})\right)^{\frac{2}{3}}\right)^{\frac{1}{2}} = \left(\prod_{i=1}^{5} x_i\right)^{\frac{1}{5}},$$

which is the geometric mean and hence idempotent.

Proposition 6.63 *If the aggregation functions M_1, M_2 are symmetric, then $B_{\mathbb{M}}$ is also symmetric, independently of C.*

In some applications it is necessary to model *absorbing elements* (also called *annihilators*), see Definition 1.30.

Proposition 6.64 *Let M_1 be an aggregation function with idempotent element a. If M_2 and C are aggregation functions with the same absorbing element a, then a is an absorbing element of $B_{\mathbb{M}}$, independently of M_1.*

Recall that the subscript d denotes the dual of an aggregation function.

Proposition 6.65 *The dual of a generalized Bonferroni mean $B_{\mathbb{M}}$, $\mathbb{M} = \langle M_1, M_2, C \rangle$ is given by $B_{\mathbb{M}_d}$ where $\mathbb{M}_d = \langle M_{1_d}, M_{2_d}, C_d \rangle$.*

6.4.3 Replacing the Outer Mean

Consider an averaging function M_1 (which can be symmetric or weighted) to replace the outer arithmetic mean. This gives the expression:

$$B_{M_1}(\mathbf{x}) = M_1(x_1 M(\mathbf{x}_{j \neq 1}), \ldots, x_n M(\mathbf{x}_{j \neq n}))^{\frac{1}{2}},$$

where M is the arithmetic mean of $n - 1$ arguments. When M_1 is a weighted mean, the weights corresponding to each x_i are naturally interpreted as the importance of each predicate x_i AND *the remaining.*

Example 6.66 Take M_1 as the projection to the first coordinate operator $M_1(\mathbf{x}) = x_1$, which can be seen as WAM with $\mathbf{w} = (1, 0, \ldots, 0)$. Then

$$B_{M_1}(\mathbf{x}) = \sqrt{x_1 M(\mathbf{x}_{j \neq 1})}.$$

This function takes the conjunction of this first criterion with the average of those remaining. This means that if x_1 is low, or the average of $\mathbf{x}_{j \neq 1}$ is low, the output will be low. The $w_1 = 1$ is then suggesting that x_1 is mandatory, but not sufficient. Similar interpretations follow for cases where $w_i = 0$ for all except a few i.

Example 6.67 Take M_1 as an OWA function with $\mathbf{w} = (1, 0, \ldots, 0)$. Then

$$B_{\max}(\mathbf{x}) = \sqrt{x_{(1)} M(x_{(2)}, \ldots, x_{(n)})},$$

i.e., the product of the highest input and the average of those remaining.

6.4.4 Replacing the Inner Mean

Now substitute an averaging function M_2 for the inner arithmetic mean. The Bonferroni mean can be expressed as:

$$B_{M_2}(\mathbf{x}) = \left(\frac{1}{n} \sum_{i=1}^{n} x_i M_2(\mathbf{x}_{j\neq i}) \right)^{\frac{1}{2}}. \tag{6.12}$$

The function B_{M_2} remains an averaging aggregation function [BJMRY10]. Let us look at some special cases.

Example 6.68 One interesting case is where M_2 is an OWA function $OWA_\mathbf{w}$. Then we have

$$B_{OWA_\mathbf{w}}(\mathbf{x}) = \left(\frac{1}{n} \sum_{i=1}^{n} x_i OWA_\mathbf{w}(\mathbf{x}_{j\neq i}) \right)^{\frac{1}{2}}.$$

Consider the case $w_1 = 1$, i.e. $OWA_\mathbf{w}(\mathbf{x}_{j\neq i}) = \max(\mathbf{x}_{j\neq i})$. This would then simplify to,

$$B_{OWA}(\mathbf{x}) = \left(x_{(1)} \left(\frac{x_{(2)}}{n} + \frac{1}{n} \sum_{i=2}^{n} x_{(i)} \right) \right)^{\frac{1}{2}}.$$

This is the product of the largest input and an arithmetic mean whose arguments are the remaining inputs with their highest duplicated.

Example 6.69 Now consider $OWA_\mathbf{w}(\mathbf{x}_{j\neq i}) = Median(\mathbf{x}_{j\neq i})$ for n even, i.e. $n = 2k$, $n - 1 = 2k - 1$. For a given input $\mathbf{x}$, we have two cases: we let $Median(\mathbf{x}_{j\neq i}) = x_{(k)}$ where $x_i \leq x_{(k+1)}$ and $Median(\mathbf{x}_{j\neq i}) = x_{(k+1)}$ where $x_i \geq x_{(k)}$. This results in

$$B_{Median}(\mathbf{x}) = \left(\frac{x_{(k+1)}}{n} \sum_{i=1}^{k} x_{(i)} + \frac{x_{(k)}}{n} \sum_{i=k+1}^{n} x_{(i)} \right)^{\frac{1}{2}}.$$

Let us now consider a weighted averaging function as M_2. Here we need to choose the weights appropriately, so that they are consistent with the application and inputs. Since M_2 is a mean of $n - 1$ arguments, and when it is multiplied with each x_i in (6.12), it averages a different set of arguments, it will be convenient to define its weighting vectors in some generic way, based on a weighting vector $\mathbf{w} \in [0, 1]^n$.

We define vectors $\mathbf{u}^i \in [0, 1]^{n-1}$, $i = 1, \ldots, n$ by

$$u_j^i = \frac{w_j}{\sum_{k\neq i} w_k} = \frac{w_j}{1 - w_i}, \quad w_i \neq 1. \tag{6.13}$$

Note that for every i, $\mathbf{u}^i$ sum to one. We can now use inner weighted means M_2, defined with respect to weighting vector $\mathbf{u}^i$ whenever M_2 is multiplied by x_i in (6.12) for all i.

6.4.5 Replacing the Product Operation

An important component of the Bonferroni mean is the product operation. Clearly, all three components, the two means and the product, will have an impact on the degree to which the function behaves conjunctively or disjunctively (the *andness* and *orness* values respectively). Suppose we have $M_1 = M_2 = M$, the arithmetic means, and we wish to see how replacing the product affects the function's behavior. We remind that Bonferroni mean models the operations

$$x_i \text{ AND (the } average \text{ when } x_i \text{ is absent) .}$$

We now substitute the product, which models AND, with other conjunctive functions C with invertible diagonals.

We express the generalized Bonferroni mean, with $p = q = 1$ as

$$B_C(\mathbf{x}) = d_C^{-1}(A(C(x_1, A(\mathbf{x}_{j\neq 1})), \ldots, C(x_n, A(\mathbf{x}_{j\neq n})))),$$

where d_C^{-1} is the inverse of the diagonal $d_C(t) = C(t, t)$.

The diagonal of a strict t-norm T with a generator g is continuous strictly increasing and is given by $d_T(t) = g^{-1}(2g(t))$, and hence invertible, with $d_T^{-1}(t) = g^{-1}(g(t)/2)$.

Example 6.70 Consider the Hamacher family of t-norms and conorms given by

$$T_\lambda^H(x, y) = \begin{cases} 0, & \text{if } \lambda = x = y = 0, \\ \frac{xy}{\lambda + (1-\lambda)(x+y-xy)} & \text{otherwise,} \end{cases}$$

$$S_\lambda^H(x, y) = \begin{cases} 1, & \text{if } \lambda = 0, x = y = 1, \\ \frac{x+y-xy-(1-\lambda)xy}{1-(1-\lambda)xy} & \text{otherwise.} \end{cases}$$

These functions have the special cases $T_1^H = T_P$ the product, and $S_1^H = S_P$ the probabilistic sum. The Hamacher family are convenient for the purpose of investigating the effect of C, since they provide a number of comparable functions. That is, given the minimum and maximum, (min and max), the geometric mean G, the arithmetic mean M and the quadratic mean Q, we have the inequality,

$$T_0^H \leq T_P \leq T_2^H \leq \min \leq G \leq M \leq Q \leq \max \leq S_2^H \leq S_P \leq S_0^H .$$

Using the formulas from Definition 2.2 and $n = 3$, the andness value can be calculated for each of the above functions. Table 6.2 shows the results.

Table 6.2 Influence of C on andness and orness degree (correct to 2 d.p.) of B_C

Choice for C	Andness (B_C)	Orness (B_C)
T_2^H	0.54	0.46
T_P (product)	0.57	0.43
T_0^H (Hamacher product)	0.63	0.37
Min	0.79	0.21
Geometric mean	0.58	0.42
Arithmetic mean	0.5	0.5
Power mean	0.44	0.56
Max	0.21	0.79
S_0^H	0.37	0.63
S_P (Probabilistic sum)	0.43	0.57
S_2^H (Einstein sum)	0.46	0.54

Clearly, when $C = M$, the arithmetic mean, the Bonferroni mean reduces to the arithmetic mean. It is interesting to note however, that the maximum andness is reached where C is the minimum, i.e. the strongest t-norm. As C approaches the drastic product, the andness degree approaches the neutral value of 0.5. These behaviors are due to the effect of the inverse diagonal d_C^{-1}. For $C = \min$, as for all averaging functions, the inverse diagonal is given by $d_C^{-1}(t) = t$, however for $C = T_P$, the inverse diagonal is $\sqrt{t}$, increasing the values to a greater degree than they are pulled down by the product itself.

6.4.6 Extensions to B_M^K

We have mentioned the concept of soft and hard partial conjunction and disjunction in Chap. 2 (p. 55). The ability of the Bonferroni mean to express these concepts requires extensions to B^k. The generalized model is also capable of expressing the Bonferroni mean of triples, $B^{p,q,r}(\mathbf{x})$. This extension of the Bonferroni mean is given by

$$B^{p,q,r}(\mathbf{x}) = \left(\frac{1}{n(n-1)(n-2)} \sum_{i,j,k=1,i\neq j\neq k}^{n} x_i^p x_j^q x_k^r \right)^{\frac{1}{p+q+r}},$$

which permits the alternative formulation,

$$B^{p,q,r}(\mathbf{x}) = \left(\frac{1}{n} \sum_{i,j=1,i\neq j}^{n} x_i^r \left(\frac{1}{(n-1)(n-2)} \sum_{i\neq j\neq k}^{n} x_j^p x_k^q \right) \right)^{\frac{1}{p+q+r}}.$$

We note here the similarity between the standard Bonferroni mean $B^{p,q}$ in Eq. (6.9) and the inner sum of this equation. The only difference here is that the $(p+q)$-th root is absent. By taking this into account with our choice of C, we have

$$B^{p,q,r}(\mathbf{x}) = \left(\frac{1}{n} \sum_{i=1}^{n} x_i^r (B^{p,q}(\mathbf{x}_{j\neq i}))^{p+q} \right)^{\frac{1}{p+q+r}}.$$

We can hence express this extension of the Bonferroni mean in terms of the generalized Bonferroni Mean, i.e. $B^{p,q,r} = B_{\mathbb{M}}$ with $\mathbb{M} = \langle A, B^{p,q}, x^r y^{p+q} \rangle$.

Let us now look at the special case $p = q = r = 1$ again. We can generalize the Bonferroni mean iteratively as follows

Definition 6.71 (*Iterative generalized Bonferroni mean*) The iterative generalized Bonferroni mean is

$$B_{\mathbb{M}}^{it}(\mathbf{x}) = d_{it}^{-1} \left(M_1(C(x_i, d_C(B_{\mathbb{M}}(\mathbf{x}_{j\neq i}))|_{i=1,\dots,n}) \right),$$

where $d_{it}(t) = C(t, d_C(t))$.

With the choices $M_1 = M_2 = M$ and C being the product, we recover $B^{1,1,1}$. It is immediate that this is an averaging aggregation function. Here we note that functions B^k allow us to model expressions like

$$x_i \text{ AND } x_j \text{ AND (the } average \text{ when } x_i, x_j \text{ are absent) },$$

so that satisfaction of both criteria i and j, and the average of the rest is required. As shown in the next section, this can be useful when modelling mandatory requirements.

6.4.7 Boundedness of the Generalized Bonferroni Mean

We now establish some bounds that may be useful for applications of the generalized Bonferroni mean.

Partial Conjunction and Disjunction

Proposition 6.72 *For a conjunctive C fixed, the strongest Bonferroni mean $\overline{B}_{\mathbb{M}}$ has $\mathbb{M} = \langle \max, \max, C \rangle$, and*

$$\overline{B}_{\mathbb{M}}(\mathbf{x}) = d_C^{-1}(\max_{i=1,\dots,n} \{C(x_i, \max_{j=1,\dots,n, j\neq i} x_j)\}).$$

Arrange the inputs in non-increasing order such that $x_{(1)} \geq x_{(2)} \geq \cdots \geq x_{(n)}$, i.e. $\max(\mathbf{x}) = x_{(1)}$. We then have

$$B_{\langle \max, \max, C \rangle}(\mathbf{x}) = d_C^{-1}(C(x_{(1)}, x_{(2)})) \leq d_C^{-1}(x_{(2)}).$$

We see from this that regardless of the choices of averaging functions M_1, M_2, the function $B_\mathbb{M}$ will be bounded by $d_C^{-1}(x_{(2)})$. The behavior of d_C^{-1} will depend on C, whose diagonal satisfies $d_C(t) \leq t$. We hence have $d_C^{-1}(t) \geq t$. In other words, the weaker (more conjunctive) C, the stronger the inverse diagonal and hence the stronger the $B_\mathbb{M}$. As C approaches the minimum, $B_\mathbb{M}$ becomes more conjunctive.

It follows from $B_\mathbb{M}(\mathbf{x}) \leq d_C^{-1}(x_{(2)})$ that,

$$B_\mathbb{M}(1, 0, \ldots, 0) = 0.$$

This has an interesting interpretation: at least two non-zero inputs are needed to make $B_\mathbb{M}$ positive. Other behavior can be gathered from the function C and the resulting d_C^{-1}. We have

$$x_{(2)} \leq B_{\langle \max, \max, C \rangle}(\mathbf{x}) \leq d_C^{-1}(C(x_{(1)}, x_{(2)})).$$

Consider now generalizations of $B^{1,1,1}$. The strongest iterative generalized Bonferroni mean will be $\overline{B}_\mathbb{M}^{it} = B_{\langle \max, \max, C \rangle}^{it} = B_{\langle \max, B_{\langle \max, \max, C \rangle}, C \rangle}$. Since $d_C(B_{\langle \max, \max, C \rangle}(\mathbf{x})) = C(x_{(1)}, x_{(2)})$, the resulting iterative Bonferroni mean takes the form,

$$\overline{B}_\mathbb{M}^{it}(\mathbf{x}) = d_{it}^{-1}(C(x_i, C(x_j, x_k))) \leq d_{it}^{-1}(x_{(3)}),$$

where x_i, x_j, x_k are a combination of the three highest inputs (in some particular order, which is irrelevant here). It follows that at least 3 non-zero inputs are required for $B_\mathbb{M}^{it} > 0$. Continuing this way, we get a mean which requires at least $4, 5, \ldots$, and so on inputs to be non-zero. Hence $B_\mathbb{M}^{it}$ is capable of modeling such averages, where a certain number of criteria must be satisfied.

By using duality, we get an averaging operator, in which several inputs are sufficient (if 2 criteria are fully satisfied, the result is 1), yet contribution of all the inputs is accounted for.

Let us now have a look at M_1 being a projection to the first coordinate operator $proj_1$. Then $B_\mathbb{M}(\mathbf{x}) \leq d_C^{-1}(x_1)$. So satisfaction of the first criterion ($x_1 \neq 0$) is necessary for the result to be non-zero. Therefore we model a mandatory requirement.

Unlike other weighted aggregation functions, however, $B_\mathbb{M}$ still takes into account not only x_1 but all the rest of the arguments. Here we model hard partial conjunction with respect to the first argument, but we insist on accounting for contribution of the other arguments.

Continuing this way, and using iterative generalized Bonferroni mean, $B_{\langle proj_1, B_{\langle proj_1, M_2, C \rangle}, C \rangle}$, we obtain hard partial conjunction with respect to the first two arguments, but maintain contribution of all other arguments. This way any number of arguments can be made mandatory.

Example 6.73 A university considers scholarship applications by aggregating 6 individual subject scores with the additional requirement that a minimum of 80 % should be achieved in both English (x_1) and Mathematics (x_2).

A Bonferroni mean $B_{\mathbb{M}}$ is used with $\mathbb{M} = \langle proj_1, B_{proj_1,A(\mathbf{x}),C_1}, C_2 \rangle$. In addition, x_1, x_2 are defined by threshold transformations based on the raw scores x_i' such that

$$x_i = \begin{cases} x_i', & x_i' \geq 0.8; \\ 0, & \text{otherwise.} \end{cases}$$

Thus $B_{\mathbb{M}}$ considers all 6 subjects in the aggregation process, but ensures that the minimum requirements are met.

Three students are compared below using $B_{\mathbb{M}}$, the standard arithmetic mean M and the geometric mean G.

Student	x_1'	x_2'	x_1	x_2	x_3	x_4	x_5	x_6	$B_{\mathbb{M}}$	M	G
s_1	0.8	0.9	0.8	0.9	0.7	0.8	0.3	0.8	0.710	0.717	0.678
s_2	0.9	0.8	0.9	0.8	0.8	0.9	1	0	0.728	0.733	0
s_3	0.7	1	0	1	0.8	1	0.8	0.9	0	0.75	0

The main advantages of the Bonferroni mean are highlighted by this example. Firstly, that a score is obtained for s_3 using the arithmetic mean, even though the minimum requirement for English is not met. Even if the subjects were weighted using a WAM so that only the average of English and mathematics scores were considered, the resulting function would still give an overall output of 0.5, and further could not distinguish between s_1 and s_2. On the other hand, the geometric mean ensures that the requirement for English is met, but still penalizes s_2 for a score of 0 in a non-mandatory subject. We see also that $B_{\mathbb{M}}$ can still be easily interpreted as an average of the inputs, which is its main advantage over the use of a conjunctive rule to model the mandatory criteria.

Example 6.74 Consider the following requirement: for the aggregated value to be positive, both x_1 and at least two other inputs must be positive. We can model this with the generalized Bonferroni mean $B_{\langle proj_1, B_{\langle max,A,T_P \rangle}, T_P \rangle}$, which results in the formula

$$B_{\langle proj_1, B_{\langle max,A,T_P \rangle}, T_P \rangle} = \left(x_1 x_{(1)} \frac{1}{n-2} \sum_{j \neq (1), j \neq 1} x_j \right)^{1/3}.$$

Clearly the requirement is satisfied, while at the same time the resulting function is idempotent and strictly monotone on $]0, \infty)^n$, and there are no other mandatory inputs except x_1.

6.4.8 k-intolerance Boundedness

The concept of k-intolerance was introduced by Marichal [Mar07] in the context of fuzzy measures (capacities) and the Choquet integral. The property, as well as its dual property—k-tolerance, places bounds on the function with respect to the k-th highest/lowest argument.

Definition 6.75 (k-intolerant) Let $k \in \{1, \ldots, n\}$, and $x_{(1)} \leq \cdots \leq x_{(n)}$ be a non-decreasing permutation of the input vector such that $x_{(k)}$ is the k-th lowest input. An aggregation function $f : \mathbb{I}^n \to \mathbb{I}$ is *at most k-intolerant* if $f(\mathbf{x}) \leq x_{(k)} \ \forall \mathbf{x} \in \mathbb{I}^n$. If, in addition $f \nleq x_{(k-1)}$, it is said to be k-intolerant.

Definition 6.76 (k-tolerant) Let $k \in \{1, \ldots, n\}$, and $x_{(1)} \geq \cdots \geq x_{(n)}$ be a non-increasing permutation of the input vector such that $x_{(k)}$ is the k-th highest input. An aggregation function $f : \mathbb{I}^n \to \mathbb{I}$ is *at most k-tolerant* if $f(\mathbf{x}) \geq x_{(k)} \ \forall \mathbf{x} \in \mathbb{I}^n$. If, in addition $f \ngeq x_{(k+1)}$, it is said to be k-tolerant.

The generalized Bonferroni mean is also capable of expressing this through its components [BJMRY10].

Proposition 6.77 *Let $C(x, y) = \min(x, y)$, $d_C^{-1}(t) = t$. If M_1 is at most a-intolerant and M_2 is at most b-intolerant, $B_{\mathbb{M}}$ will be at most $\min(a, b)$-intolerant.*

Where C is conjunctive, the boundedness will be associated with the inverse diagonal d_C^{-1}.

Proposition 6.78 *Let C be a conjunctive aggregation function, with $d_C(t) = C(t, t)$. If M_1 is at most a-intolerant and M_2 is at most b-intolerant, then the function $B_{\mathbb{M}}$ is at most h-intolerant, where $h = \max(a, b + 1)$.*

Example 6.79 Let $\mathbb{M} = \langle \min, \min, xy \rangle$. M_1 and M_2 are both 1-intolerant (i.e. conjunctive), given $x_{(1)} \leq \cdots \leq x_{(n)}$,

$$B_{\mathbb{M}}(\mathbf{x}) = \sqrt{\min(x_{(1)}x_{(2)}, x_{(2)}x_{(1)}, x_{(3)}x_{(1)}, \ldots, x_{(n)}x_{(1)})}$$
$$= \sqrt{x_{(1)}x_{(2)}}.$$

For $x_{(1)} = x_{(2)}$, $B_{\mathbb{M}}(\mathbf{x}) = x_{(1)}$, however if $x_{(1)} < x_{(2)}$ then it follows $x_{(1)} < B_{\mathbb{M}} < x_{(2)}$. Hence $B_{\mathbb{M}}$ is not bounded by $x_{(\min(a,b))}$ but rather by $\sqrt{x_{(\min(a,b))}}$.

Corollary 6.80 *Let C be a disjunctive aggregation function, with $d_C(t) = C(t, t)$. If M_1 is at most a-tolerant, M_2 is at most b-tolerant ($a, b \in \{1, \ldots, n\}$), then function $B_{\mathbb{M}}$ is at most h-tolerant, where $h = \max(a, b + 1)$.*

6.4.9 Generated t-norm and Generated Quasi-arithmetic Means as Components of $B_{\mathbb{M}}$

Consider a system where a generator $g : [0, 1] \to [0, \infty]$ with $g(1) = 0$ is used to generate an Archimedean t-norm C, as well as to generate the quasi-arithmetic means M_1 and M_2. We allow the possibility that the means will have two different weighting vectors, $\mathbf{w}$ and $\mathbf{u}$.

Proposition 6.81 [BJMRY10] *A generalized Bonferroni mean with weighted quasi-arithmetic means M_1, M_2 and t-norm C generated by the same function g, will be a weighted quasi-arithmetic mean generated by g.*

In some situations the weights will simplify to a general expression. Given the n-dimensional vector $\mathbf{u}'$ such that $\mathbf{u}$ is the $n - 1$-dimensional weighting vector such that $u_j = \frac{u_j'}{1-u_i'}$, the weighting vector $\mathbf{v}$ is given by,

$$v_i = \frac{1}{2}(w_i + u_i' \sum_{j=1,j\neq i}^{n} \frac{w_j}{1 - u_j'}).$$

In the case where $\mathbf{w} = \mathbf{u}'$, i.e. where the same importance is allocated to each variable, this simplifies to

$$v_i = \frac{w_i}{2}(1 + \sum_{j=1,j\neq i}^{n} \frac{w_j}{1 - w_j}).$$

This last equation has somewhat the effect of drawing the weights closer to $\frac{1}{n}$, i.e. less dispersion.

If generalized OWA are used for the means, the result will draw upon a similar argument. The weights for the OWA can be determined from the weighting vectors as:

$$v_i = \frac{1}{2}(w_i + u_{i-1} \sum_{j=1}^{i-1} w_j + u_i \sum_{j=i+1}^{n} w_j).$$

Example 6.82 Suppose $n = 4$, $\mathbf{w} = (0.5, 0.2, 0.3, 0)$ and $\mathbf{u} = (0.1, 0.2, 0.7)$,

$$v_1 = 0.5(0.5 + 0.1(0.2 + 0.3 + 0)) = 0.275$$
$$v_2 = 0.5(0.2 + 0.1(0.5) + 0.2(0.3 + 0)) = 0.155$$
$$v_3 = 0.5(0.3 + 0.2(0.5 + 0.2) + 0.7(0)) = 0.22$$
$$v_4 = 0.5(0 + 0.7(0.5 + 0.2 + 0.3)) = 0.35$$

Clearly, if $w_i = \frac{1}{n}$ and $u_i = \frac{1}{n-1}$, we will have $v_i = \frac{1}{n}$ and the function will be symmetric.

6.5 Consistency and Stability

6.5.1 Motivation

One problem that arises in decision making and information fusion is how to deal with data of varying dimension. If comparing two items based on multiple criteria or a group of experts' opinions, it may be that some evaluations are missing. Similarly, when fusing the readings of sensors, it could be that not all of the readings are available at all times and we need a global evaluation based on some subset. Thus we are looking for families of aggregation functions that produce consistent outputs regardless input cardinality.

Some aggregation functions, such as t-norms, t-conorms and uninorms, are associative, and therefore have a natural way of defining n-variable instances. Quasi-arithmetic means (with equal weights) is another example where the whole family of functions is defined consistently. In contrast, defining consistently weighted means and ordered weighted means does represent a significant challenge.

Some recent studies have analyzed the notion of stability from the viewpoint of consistency between members of a family of aggregation functions defined for varying dimension [CMS15, GMRR12, RGRM11]. Whilst stability and robustness of aggregation is usually thought of in terms of concepts like Lipschitz continuity [BCJ10, CM01] (see Sect. 1.3.5), where a small increase to one of the inputs should not result in a drastic increase to the output, it also makes sense that the inclusion of an additional input should not drastically alter the aggregated value if it is representative of the rest. From Yager's self-identity property [YR97] (Definition 1.49), the authors of [GMRR12, RGRM11] consider the consistency of various classes of aggregation functions (the concept they refer to as stability). The idea is that given a set of inputs, if we add the aggregated value of these inputs as a new input, the overall output should not change. A function is considered to be stable if the new input can be aggregated both from the right and the left (i.e., as either the last or the first argument respectively).

In this section, we adopt the definitions from [BJ13, GMRR12, GRMRB14, RGMR13, RGRM11], and present the results for quasi-arithmetic means, quasi-medians and other averaging functions that can be defined as penalty-based functions [Bel09]. We then consider in detail weighted aggregation functions. We shall use the terms *stability* and *consistency* interchangeably in this section.

For weighted functions, the i-th input is usually representative of the source, particular criterion, expert etc., so requiring a function to have the same output regardless of whether a particular input corresponds with the first or n-th index could be counterintuitive. We focus on the notion of R-strict stability and draw upon the notion of penalty-based aggregation functions to explore the stability of different aggregation function families.

6.5.2 *Strictly Stable Families*

Following from Yager's self-identity property [YR97] (see also [Cal+00]), Rojas et al. propose the following conditions for stability of a family of aggregation functions [RGRM11].

Definition 6.83 *(L- and R-stability)* Let $\{f_n : \mathbb{I}^n \to \mathbb{I}, n \in N\}$ be a family of aggregation functions. Then it is said that:

1. $\{f_n\}_n$ is R-strictly stable if

$$f_n(x_1, x_2, \ldots, x_{n-1}, f_{n-1}(x_1, x_2, \ldots, x_{n-1})) = f_{n-1}(x_1, x_2, \ldots, x_{n-1}).$$

2. $\{f_n\}_n$ is L-strictly stable if

$$f_n(f_{n-1}(x_1, x_2, \ldots, x_{n-1}), x_1, x_2, \ldots, x_{n-1}) = f_{n-1}(x_1, x_2, \ldots, x_{n-1}).$$

3. $\{f_n\}_n$ is LR-strictly stable if both properties hold simultaneously.

The geometric means and arithmetic means with respect to a weighting vector with equal weights, the maximum, minimum, and median are LR-strictly stable, while the weighted counterparts of these means and the OWA, in general, are unstable.

These strict stability conditions can be considered recursively for the definitions of aggregation functions of n and $n - 1$ dimensions. In [CM94] some rules for defining such functions using a sequence of 2-variate weighted means were considered from the viewpoint of consistency and computability. The following proposition was established for weighted geometric, arithmetic and harmonic means in [Cal+00] and then in [GMRR12].

Proposition 6.84 *Let* $\mathbf{w}^n = (w_1^n, \ldots, w_n^n) \in [0, 1]^n, n \in \mathbb{N}$ *be a sequence of weighting vectors such that* $\sum\limits_{i=1}^{n} w_i^n = 1$ *holds* $\forall n \geq 2$. *Then, the family of weighted means defined by these weights is R-strictly stable if and only if the following holds.*

$$w_i^n = (1 - w_n^n) \cdot w_i^{n-1}, \ \forall n \in \mathbb{N},$$

and L-strictly stable if and only if

$$w_{i+1}^n = (1 - w_1^n) \cdot w_i^{n-1}, \ \forall n \in \mathbb{N}.$$

From Proposition 6.84, we can obtain the following corollary.

Corollary 6.85 *For weighted means, a sequence of weighting vectors* $\mathbf{w}^n$, *with* $n = 2, 3, \ldots, N$ *is LR-strictly stable if and only if there exists a* $\lambda \geq 0$ *such that* $w_i^n = \lambda w_{i-1}^n$ *for* $i = 2, 3, \ldots, n$.

Proof We express the 2-dimensional weighting vector in terms of λ, with $\mathbf{w}^2 = (w, \lambda w)$. For L-strict stability we require the ratio between the 2nd and 3rd inputs to be the same,

$$w_2^3 : w_3^3 = w : \lambda w,$$

while for R-strict stability we require

$$w_1^3 : w_2^3 = w : \lambda w.$$

From this it follows that the 3-dimensional weighting vector must have the ratio

$$w_1^3 : w_2^3 : w_3^3 = w : \lambda w : \lambda^2 w.$$

Since we require $w(1 + \lambda + \lambda^2) = 1$, the value of λ follows from the solution to

$$\lambda = -1 \pm \sqrt{1 - 4\left(1 - \frac{1}{w}\right)},$$

which has a unique feasible solution for all $0 < w \le 1$.

The same reasoning applies as n increases to 4, 5, etc.   □

This means we can determine all $\mathbf{w}^n$ from the ratio between the weights for $\mathbf{w}^2$. The value of $\lambda = 1$ leads to the stable family of weighting vectors with equal weights. If $\lambda = 0$ we have the family of weighting vectors with $w_1 = 1$, $w_i = 0$ otherwise, while the limiting case of $\lambda = \infty$ will have $w_n = 1$, $w_i = 0$ otherwise. If we have $\lambda = 2$, for example, the weighting vectors for $n = 2$ and $n = 3$ respectively will be $\mathbf{w}^2 = (1/3, 2/3)$, $\mathbf{w}^3 = (1/7, 2/7, 4/7)$.

Given that the weights assigned to inputs in weighted aggregation are often indicative of the importance of the source, it could usually be assumed that a new input would be aggregated in the n-th position and that R-strict stability might be more appropriate to consider for applications. We will consider penalty-based aggregation functions to determine properties on the weights for weighted functions.

6.5.3 R-strict Stability

From Proposition 6.84 it follows that a family of weighted means cannot be consistently defined for all $n \ge 2$ such that it is both L- and R-strictly stable unless every pair of sequential weights satisfies $w_i = \lambda w_{i-1}$. On the other hand, the usual interpretation of weighting vectors is that the weight w_i reflects the importance of the input x_i, so it may not necessarily make sense in applications to shift the indices of the inputs the way we do when f_{n-1} is aggregated in the first position.

We will hence restrict the following considerations to the notion of R-strict stability. We will do this in terms of penalty-based aggregation operators of the

form given in Eq. (5.2). Expressing the functions in this way allows us to generalize the results for a number of important aggregation families, including weighted
quasi-arithmetic means and weighted quasi-medians.

We use the notation $\mathbf{x}$ and $\mathbf{x}_{i \neq n}$ to denote the respective input vectors $(x_1, x_2, \ldots,$
$x_{n-1}, x_n)$ and $(x_1, x_2, \ldots, x_{n-1})$. For aggregation functions expressed in terms of
their penalties, we have the following proposition proved in [BJ13],

Proposition 6.86 *Given a family $\{f_n\}_n$ of faithful penalty-based aggregation operators with Eq. (5.2)*

$$P(\mathbf{x}, y) = \sum_{i=1}^{n} w_i p(x_i, y),$$

*if the weighting vectors $\mathbf{w}^n$ and $\mathbf{w}^{n-1}$ associated with the penalty expressions for
each $f_n(\mathbf{x})$ and $f_{n-1}(\mathbf{x}_{i \neq n})$ satisfy:*

$$w_i^n = \lambda_n w_i^{n-1}, i = 1, \ldots, n-1,$$

each $\lambda_n \geq 0$ a constant,[4] then the family is R-strictly stable.

As a corollary, we obtain that all unweighted quasi-arithmetic means and quasi-
medians are R-stable (take $\lambda_n = 1$), which extends the results of [RGRM11] for
arithmetic, geometric means and power means and the median.

The next proposition shows that in the case of $p(x, y)$ being *differentiable*, this
relationship between the weighting vectors is also the necessary condition for stability
[BJ13].

Proposition 6.87 *For a family $\{f_n\}_n$ of faithful penalty-based aggregation operators
where the penalty function $p(x, y)$ is differentiable in y, R-strict stability holds if and
only if the weighting convention described in Proposition 6.86 is satisfied.*

This relationship between the weighting vectors of the weighted arithmetic means,
harmonic means, quadratic means and power means was established in [GMRR12],
and then in [BJ13] for the R-strict stability of all penalty-based functions with $p(x, y)$
differentiable. To show that differentiability of p is essential, consider the cases like
$p(x, y) = |x - y|$ where the weighting convention is still sufficient, but is not necessary
to guarantee R-strict stability. Example 6.88 illustrates this last point.

Example 6.88 Consider the lower weighted median (Definition 3.46) resulting from
the penalty expression $p(x, y) = |x - y|$ and the 2-dimensional weighting vector,
$\mathbf{w}^2 = (0.4, 0.6)$. We have the following two situations:

$x_1 \geq x_2$, from which we obtain $\mathbf{u} = (0.4, 0.6)$ and $x_{(k)} = x_2$;
or
$x_2 \geq x_1$, which gives $\mathbf{u} = (0.6, 0.4)$ and $x_{(k)} = x_2$ again.

---

[4]Here we allow the relationship $w_i^n = 0, \forall i \neq n, w_n^n > 0$ that results for $\lambda_n = 0$.

In fact, for any 2-dimensional weighting vector with $w_1^2 < w_2^2$, it will follow that $Med_{\mathbf{w}}(\mathbf{x}) = x_2$.

For 3 inputs, we then have $x_3 = x_2$ and any weighting vector with $w_1^3 < 0.5$ will result in a weighted median that is R-strictly stable with respect to the 2-dimensional case. For instance, the relationship between the weighting vectors $\mathbf{w}^2 = (0.4, 0.6)$ and $\mathbf{w}^3 = (0.45, 0.3, 0.25)$ is R-strictly stable for weighted medians, even though the ratio $w_1 : w_2$ is not preserved.

This merely shows that R-strict stability may not be the best indicator of consistency for penalty-based aggregation functions defined with respect to a non-differentiable penalty. When $p(x, y)$ is differentiable, however, and $w_n^n \neq 1$, R-strict stability is equivalent to the preservation of the ratios between each of the weights. We have the following useful corollaries that follow from Propositions 6.86 and 6.87.

Corollary 6.89 *All weighted quasi-arithmetic means are R-strictly stable if and only if for any sequence of weights* $\mathbf{w}^n$, $\mathbf{w}^{n-1}$ *it holds that*

$$w_i^n = (1 - w_n^n)w_i^{n-1}, i = 1, 2, \ldots, n - 1. \tag{6.14}$$

Proof Direct from Proposition 6.87 with $\lambda_n = (1 - w_n^n)$ and the ability to model these functions in the form of Eq. (5.2). $\quad\square$

For quasi-weighted medians in the next proposition, we have only *if* part.

Corollary 6.90 *All weighted quasi-medians are R-strictly stable if for any sequence of weights* $\mathbf{w}^n$, $\mathbf{w}^{n-1}$ *it holds that*

$$w_i^n = (1 - w_n^n)w_i^{n-1}, i = 1, 2, \ldots, n - 1.$$

Proof Direct from Proposition 6.86 with $\lambda_n = (1 - w_n^n)$. $\quad\square$

Corollary 6.91 *A family of α-quantile operators is R-strictly stable with $\alpha = \frac{c}{1+c}$ provided c is fixed for all n.*

Proof Direct from Proposition 6.86 with $p(x, y)$ defined as it is in Sect. 5.3.1, item 9, and equal weights for all i. $\quad\square$

The α-quantile operator includes special cases of the median (with $c = 1$) the maximum ($c = \infty$) and the minimum ($c = 0$), all of which can be defined with respect to its penalty expression with equal weights. This result does not extend to k-order statistics, however, since c depends on n and the penalty expressions would differ for f_n and f_{n-1}.

The following example illustrates the application of the weighting convention in Corollary 6.89.

Example 6.92 Consider the weighting vector $\mathbf{w}^4 = (\frac{3}{20}, \frac{5}{20}, \frac{8}{20}, \frac{4}{20})$. The R-strictly stable weighting sequence for $n = 2, 3, 4$ would be,

$$\mathbf{w}^2 = (w_1^2, w_2^2)$$
$$= \left(\frac{3}{8}, \frac{5}{8}\right)$$
$$\mathbf{w}^3 = ((1 - w_3^3)w_1^2, (1 - w_3^3)w_2^2, w_3^3)$$
$$= \left(\frac{3}{16}, \frac{5}{16}, \frac{8}{16}\right)$$
$$\mathbf{w}^4 = ((1 - w_4^4)w_1^3, (1 - w_4^4)w_2^3, (1 - w_4^4)w_3^3, w_4^4)$$
$$= \left(\frac{3}{20}, \frac{5}{20}, \frac{8}{20}, \frac{4}{20}\right).$$

Example 6.93 (Stable student evaluation) Students competing for a scholarship are evaluated against 4 criteria: exam marks (40%), interview (30%), application letter (10%) and 1 written reference (20%). Due to unforseen circumstances, however, the decision needs to be made earlier than anticipated and the reference for many students is yet to arrive. In order to be as fair as possible, students with all data available have their scores aggregated with respect to the full weighting vector $\mathbf{w}^4 = (0.4, 0.3, 0.1, 0.2)$, while for those students with a missing reference, a stable weighting vector is defined such that $w_i^3 = w_i^4/(1 - 0.2)$. This gives $\mathbf{w}^3 = (0.5, 0.375, 0.125)$.

Although Proposition 6.86 extends to a number of important averages including weighted means and quantile operators, it cannot be used to establish R-strict stability for more general aggregation functions that require a reordering step in their calculation such as the OWA function. The OWA function in general is neither L- nor R-strictly stable. Some conditions that make OWA functions L- or R- stable are presented in [BJ13].

Next we turn to the problem of learning weights of L- or R-stable averages, when we only have the observed input/output data and don't know the relative importance of each input. The next Example follows from Example 6.93.

Example 6.94 (Learning consistent weights) The scholarship assessment panel is unconvinced that the proportional importance allocated to the criteria properly reflects the students' potential. After the first year, they have performance data available for all the candidates, along with the data used to award the scholarships (since the late references were not used, their score data is still unavailable). An example of such data is given in Table 6.3.

The problem is now to learn both a 3- and 4-dimensional weighting vector (stable with respect to one another) from the data set in order to approximate the importance of each criteria in assessing the academic potential of each student.

Table 6.3 Scholarship and performance scores with data missing for some students

Student	s_1	s_2	s_3	s_4	s_5
Exam (x_1)	0.5	0.6	0.2	0.5	0.7
Interview (x_2)	0.8	0.3	0.7	0.9	0.8
App. Letter (x_3)	0.4	0.6	0.1	0.3	0.7
Reference (x_4)	–	0.6	0.4	–	–
Performance (y)	0.51	0.63	0.62	0.71	0.78

6.5.4 Learning Consistent Weights

We present a method recently proposed in [BJGRM15]. Suppose we wish to learn a weighted quasi-arithmetic mean that best fits the data of the form in Table 6.3. We note that we are essentially required to learn 7 parameters, i.e. the aggregation weights that best model the output y as a function of the respective 3- and 4-dimensional input vectors. If we learn $\mathbf{w}^3$ and $\mathbf{w}^4$ separately, the weights may not satisfy the consistency relation (6.14) and therefore could not be considered stable—in this case, they would give conflicting approximations of the importance for each criterion. It would also result in fewer data with respect to the number of variables we need to learn. Lastly, the relationship in Eq. (6.14) is not linear with respect to the weights and therefore could not be incorporated into a simple optimization algorithm.

Since a consistent weighting vector leads to aggregation behavior such that

$$A_n(x_1, \ldots, x_{n-1}, A_{n-1}(x_1, \ldots, x_{n-1}))$$

$$= A_{n-1}(x_1, \ldots, x_{n-1}),$$

one approach to dealing with the missing values is to assign $x_4 = y$, i.e. replace the missing datum with the observed y value and then simply apply the program given by (2.11) in Sect. 2.3.7, which is converted to a linear program as on p. 37. Namely we solve

$$\text{Minimize}_{\mathbf{w}} \quad \sum_{k=1}^{K} r_k^+ + r_k^-, \tag{6.15}$$

$$\text{s.t.} \quad f_{\mathbf{w}}(\mathbf{x}_k) - r_k^+ + r_k^- = y_k, k = 1, \ldots, K,$$
$$w_1 + w_2 + \cdots + w_n = 1,$$
$$r_k^+, r_k^- \geq 0,$$
$$w_i \geq 0, \forall i.$$

The problem with this, however is that if the data does not exactly correspond with data that would be generated by a stable aggregation function (i.e. if there is noise or natural variation) then the weight allocation will be exaggerated for w_4, especially if the number of 3-dimensional inputs is high since in these cases $|x_4 - y_k| = 0$.

We therefore consider the problem as the following bilevel optimization problem

$$\underset{\alpha, \mathbf{w}^{n-1}}{\text{Minimize}} \sum_{j=1}^{J} |f_\mathbf{w}(\mathbf{x}_j) - y_j| + \sum_{k=1}^{K} |f_\mathbf{w}(\mathbf{x}_k) - y_k|,$$

where $\alpha = (1 - w_n^n)$, J and K represent the number of observed data of $n - 1$ and n dimensions respectively, and $\mathbf{x}_j \in [0, 1]^{n-1}$ and $\mathbf{x}_k \in [0, 1]^n$.

In the case of fitting quasi-arithmetic means, for fixed α, we can still minimize with respect to the same objective as in (6.15) and sum the residuals, however we impose the following constraints.

For $\mathbf{x}_j \in [0, 1]^{n-1}$, we have

$$\left(\sum_{i=1}^{n-1} w_i g(x_{ji})\right) - r_j^+ + r_j^- = g(y_j), \tag{6.16}$$

then for $\mathbf{x}_k \in [0, 1]^n$, we can use Eq. (6.14) and our α, giving us,

$$\alpha \left(\sum_{i=1}^{n-1} w_i g(x_{ki})\right) + w_n g(x_{kn}) - r_k^+ + r_k^- = g(y_k). \tag{6.17}$$

Since α is a scalar, these constraints remain linear with respect to $\mathbf{w}^{n-1}$. We remind that w_n is obtained directly from α and hence is also a fixed constant in this step of the minimization process. We then only require the constraints such that the weights in $\mathbf{w}^{n-1}$ sum to 1 and the residuals are nonnegative.

6.5.5 Consistency and Global Monotonicity

In [CMS15, MC97] the authors proposed the concept of global monotonicity of (extended) aggregation functions. This concept is closely related to stability.

Definition 6.95 (*Calvo-Mayor order relation*) Let $\mathbf{x} \in \mathbb{I}^n$ and $\mathbf{y} \in \mathbb{I}^m$. The Calvo-Mayor order relation $\mathbf{x} \leq_{CM} \mathbf{y}$ is defined by:

- If $n = m$, $x_i \leq y_i$ for all $i = 1, \ldots, n$;
- If $n < m$, $x_i \leq y_i$ for $i = 1, \ldots, n$, $\max(x_1, \ldots, x_n) \leq \min(y_{n+1}, \ldots, y_m)$;
- If $n > m$, $x_i \leq y_i$ for $i = 1, \ldots, m$, $\max(x_{m+1}, \ldots, x_n) \leq \min(y_1, \ldots, y_m)$.

Definition 6.96 (*Global monotonicity*) Let $F = \{f_n : \mathbb{I}^n \to \mathbb{I}, n \in N\}$ be a family of aggregation functions (i.e., an extended aggregation function). Then F is globally monotone if $F(\mathbf{x}) \leq F(\mathbf{y})$ whenever $\mathbf{x} \leq_{CM} \mathbf{y}$.

Proposition 6.97 *The family of aggregation functions* $\{f_n : \mathbb{I}^n \to \mathbb{I}, n \in N\}$ *is globally monotone if and only if*

$$f_{n+1}(x_1, \ldots, x_n, \wedge x_i) \leq f_n(_1, \ldots, x_n) \leq f_{n+1}(x_1, \ldots, x_n, \vee x_i) \qquad (6.18)$$

for all $\mathbf{x} \in \mathbb{I}^n$ *and* $n \geq 1$, *where* $\wedge x_i$ *and* $\vee x_i$ *are the minimum and maximum of the components of* $\mathbf{x}$.

Then Calvo et al. [CMS15] describe four compatible ways to introduce consistency into the aggregation process: (1) associativity, (2) decomposability, (3) self-identity and (4) duplication ($F(x_1, \ldots, x_n, x_i) = F(x_1, \ldots, x_n)$). Note that self-identity and duplication imply idempotency, and continuity and decomposability together also imply idempotency.

The main result of [CMS15] is that aggregation functions that either

- are idempotent and decomposable;
- satisfy self-identity property; or
- satisfy duplication property

are globally monotone.

References

[AGMB01] A. Amo, D. Gómez, J. Montero, G. Biging, Relevance and redundancy in fuzzy classification systems. Mathw. Soft Comput. **8**, 203–216 (2001)

[AE91] L.-E. Andersson, T. Elfving, Interpolation and approximation by monotone cubic splines. J. Approx. Theory **66**, 302–333 (1991)

[Bel00] G. Beliakov, Shape preserving approximation using least squares splines. Approximate Theory Appl. **16**, 80–98 (2000)

[Bel02] G. Beliakov, Monotone approximation of aggregation operators using least squares splines. Int. J. Uncertainty Fuzziness Knowl Based Syst. **10**, 659–676 (2002)

[Bel03] G. Beliakov, How to build aggregation operators from data? Int. J. Intell. Syst. **18**, 903–923 (2003)

[Bel05] G. Beliakov, Monotonicity preserving approximation of multivariate scattered data. BIT **45**, 653–677 (2005)

[Bel07] G. Beliakov, Construction of aggregation operators for automated decision making via optimal interpolation and global optimization. J. Ind. Manage. Optim. **3**, 193–208 (2007)

[Bel09] G. Beliakov, Construction of aggregation functions from data using linear programming. Fuzzy Sets Syst. **160**, 65–75 (2009)

[BBF11] G. Beliakov, H. Bustince, J. Fernandez, The median and its extensions. Fuzzy Sets Syst. **175**, 36–47 (2011)

[BC07] G. Beliakov, T. Calvo, Construction of aggregation operators with noble reinforcement. IEEE Trans. Fuzzy Syst. **15**, 1209–1218 (2007)

[BCJ10]   G. Beliakov, T. Calvo, S. James, On Lipschitz properties of generated aggregation functions. Fuzzy Sets Syst. **161**, 1437–1447 (2010)

[BCL07]   G. Beliakov, T. Calvo, J. Lázaro, Pointwise construction of Lipschitz aggregation operators with specific properties. Int. J. Uncertainty Fuzziness Knowl Based Syst. **15**, 193–223 (2007)

[BJ13]   G. Beliakov, S. James, Stability of weighted penalty-based aggregation functions. Fuzzy sets Syst. **226**, 1–18 (2013)

[BJGRM15]   G. Beliakov, S. James, D. Gómez, J. T. Rodríguez, J. Montero, Learning stable weights for data of varying dimension, in *Proceedings of the AGOP Conference*, Katowice, Poland (2015)

[BJMRY10]   G. Beliakov, S. James, J. Mordelová, T. Rückschlossová, R.R. Yager, Generalized Bonferroni mean operators in multicriteria aggregation. Fuzzy Sets Syst. **161**, 2227–2242 (2010)

[BPC07]   G. Beliakov, A. Pradera, T. Calvo, *Aggregation Functions: A Guide for Practitioners Studies in Fuzziness and Soft Computing*, vol. 221 (Springer, Berlin, 2007)

[Bon50]   C. Bonferroni, Sulle medie multiple di potenze. Boll. Matematica Ital. **5**, 267–270 (1950)

[BBPS06]   H. Bustince, E. Barrenechea, M. Pagola, F. Soria, Weak fuzzy S-subsethood measures. Overlap index. Int. J. Uncertainty Fuzziness Knowl. Based Syst. **14**, 537–560 (2006)

[BFMMO10]   H. Bustince, J. Fernandez, R. Mesiar, J. Montero, R. Orduna, Overlap functions. Nonlinear Anal. Theory Methods Appl. **72**, 1488–1499 (2010)

[BMM09]   H. Bustince, J. Montero, R. Mesiar, Migrativity of aggregation operators. Fuzzy Sets Syst. **160**, 766–777 (2009)

[Bus+12]   H. Bustince et al., Grouping, Overlap, and generalized bientropic functions for fuzzy modeling of pairwise comparisons. IEEE Trans. Fuzzy Syst. **20**, 405–415 (2012)

[CKKM02]   T. Calvo, A. Kolesárová, M. Komorníková, R. Mesiar, Aggregation operators: properties, classes and construction methods, in *Aggregation Operators. New Trends and Applications*, ed. by T. Calvo, G. Mayor, R. Mesiar. (Physica-Verlag, Heidelberg, 2002), pp. 3–104

[CMS15]   T. Calvo, G. Mayor, J. Suñer, Globally monotone extended aggregation functions, in *Enric Trillas: A Passion for Fuzzy Sets*, ed. by L. Magdalena, J.L. Verdegay, F. Esteva. (Springer, Berlin, 2015), pp. 49–66

[CM01]   T. Calvo, R. Mesiar, Stability of aggregation operators, in: *2nd Conference of the Europe Society for Fuzzy Logic and Technology*, Leicester (2001), pp. 475–478

[CP04]   T. Calvo, A. Pradera, Double aggregation operators. Fuzzy Sets Syst. **142**, 15–33 (2004)

[Cal+00]   T. Calvo et al., Generation on weighting triangles associated with aggregation functions. Int. J. Uncertainty Fuzziness Knowl. Based Syst. **8**, 417–451 (2000)

[CM94]   V. Cutello, J. Montero, Recursive families of OWA operators, in *Proceedings of FUZZIEEE*. Piscataway, NJ, (1994), pp. 1137–1141

[DMDB05]   H. De Meyer, B. De Baets, Fitting piecewise linear copulas to data, in *4th Conference of the European Society for Fuzzy Logic and Technology*. Barcelona, Spain (2005)

[DVV93]   M. Delgado, J.L. Verdegay, M.A. Vila, On aggregation operations of linguistic labels. Int. J. Intell. Syst. **8**, 351–370 (1993)

[Duj07]   J.J. Dujmovic, Continuous preference logic for system evaluation. IEEE Trans. Fuzzy Syst. **15**, 1082–1099 (2007)

[DS08]   F. Durante, P. Sarkoci, A note on the convex combinations of triangular norms. Fuzzy Sets Syst. **159**, 77–80 (2008)

[FR07]   J. Fodor, I.J. Rudas, On continuous triangular norms that are migrative. Fuzzy Sets Syst. **158**, 1692–1697 (2007)

[Gin58]   C. Gini, *Le Medie*. Milan (Russian translation, Srednie Velichiny, Statistica, Moscow, 1970): Unione Tipografico-Editorial Torinese (1958)

[GMRR12]   D. Gómez, J. Montero, J.T. Rodríguez, K. Rojas, Stability in aggregation operators, in *Proceedings of IPMU*. Catania, Italy (2012)

[GRMBB15]   D. Gómez, J.T. Rodríguez, J. Montero, H. Bustince, E. Barrenechea, n-dimensional overlap functions, in *Fuzzy Sets and Systems in press*. doi:10.1016/j.fss.2014.11.023 (2015)

[GRMRB14]   D. Gómez, K. Rojas, J. Montero, J.T. Rodríguez, G. Beliakov, Consistency and stability in aggregation operators : an application to missing data problems. Int. J. Comput. Intell. Syst. **7**, 595–604 (2014)

[Gra04]   M. Grabisch, The Choquet integral as a linear interpolator, in *10th International Conference on Information Processing and Management of Uncertainty*. Perugia, Italy (2004), pp. 373–378

[GL05a]   M. Grabisch, C. Labreuche, Bi-capacities—Part I: definition, Möbius transform and interaction. Fuzzy Sets Syst. **151**, 211–236 (2005)

[GL05b]   M. Grabisch, C. Labreuche, Bi-capacities—Part II: the Choquet integral. Fuzzy Sets Syst **151**, 237–259 (2005)

[HHV97]   F. Herrera, E. Herrera-Viedma, Aggregation operators for linguistic weighted information. IEEE Trans. Syst Man Cybern **27**, 646–656 (1997)

[HHV00]   F. Herrera, E. Herrera-Viedma, Linguistic decision analysis: steps for solving decision problems under linguistic information. Fuzzy Sets Syst. **115**, 67–82 (2000)

[HHV03]   F. Herrera, E. Herrera-Viedma, A study of the origin and uses of the ordered weighted geometric operator in multicriteria decision making. Int. J. Intell. Syst. **18**, 689–707 (2003)

[HHVV96]   F. Herrera, E. Herrera-Viedma, J.L. Verdegay, Direct approach processes in group decision making using linguistic OWA operators. Fuzzy Sets Syst. **79**, 175–190 (1996)

[HHVV98]   F. Herrera, E. Herrera-Viedma, J.L. Verdegay, Choice processes for non-homogeneous group decision making in linguistic setting. Fuzzy Sets Syst. **94**, 287–308 (1998)

[JBPPY13]   A. Jurio, H. Bustince, M. Pagola, A. Pradera, R.R. Yager, Some properties of overlap and grouping functions and their application to image thresholding. Fuzzy Sets Syst. **229**, 69–90 (2013)

[KK05]   E.P. Klement, A. Kolesárová, Extension to copulas and quasi-copulas as special 1-Lipschitz aggregation operators. Kybernetika **41**, 329–348 (2005)

[Koj07]   I. Kojadinovic, A weight-based approach to the measurement of the interaction among criteria in the framework of aggregation by the bipolar Choquet integral. Europ. J. Oper. Res. **179**, 498–517 (2007)

[KM07]   I. Kojadinovic, J.-L. Marichal, Entropy of bi-capacities. Eur. J. Oper. Res. **178**, 168–184 (2007)

[KMM02]   A. Kolesárová, E. Muel, J. Mordelová, Construction of kernel aggregation operators from marginal values. Int. J. Uncertainty Fuzziness Knowl. Based Syst. **10**, 37–49 (2002)

[MBBFI15]   N. Madrid, A. Burusco, H. Bustince, J. Fernandez, I. Perfilieva, Upper bounding overlaps by groupings. Fuzzy Sets Syst. **264**, 76–99 (2015)

[Mar07]   J.-L. Marichal, k-intolerant capacities and Choquet integrals. Eur. J. Oper. Res. **177**, 1453–1468 (2007)

[MPR03]   R.A. Marques Pereira, R.A. Ribeiro, Aggregation with generalized mixture operators using weighting functions. Fuzzy Sets Syst. **137**, 43–58 (2003)

[MC97]   G. Mayor, T. Calvo, Extended aggregation functions, in *IFSA'97*. vol. 1. Prague, (1997), pp. 281–285

[MN96]   R. Mesiar, V. Novák, Open problems from the 2nd International conference of fuzzy sets theory and its applications. Fuzzy Sets Syst. **81**, 185–190 (1996)

[MV99]   R. Mesiar, D. Vivona, Two-step integral with respect to fuzzy measure. Tatra Mount. Math. Publ. **16**, 359–368 (1999)

[MZA15]   A. Mesiarová-Zemánková, K. Ahmad, Averaging operators in fuzzy classification systems. Fuzzy Sets Syst. **270**, 53–73 (2015)

[MR09]   J. Mordelová, T. Rückschlossová, ABC-aggregation functions, in *Proceedings of the 5th International Summer School on Aggregation Operators*. Palma de Mallorca, Spain (2009)

[NT05]   Y. Narukawa, V. Torra, Graphical interpretation of the twofold integral and its generalization. Int. J. Uncertainty Fuzziness Knowl. Based Syst. **13**, 415–424 (2005)

[RGMR13] K. Rojas, D. Gómez, J. Montero, J.T. Rodríguez, Strictly stable families of aggregation operators. Fuzzy Sets Syst. **228**, 44–63 (2013)

[RGRM11] K. Rojas, D. Gómez, J. T. Rodríguez, J. Montero, Some properties of consistency in the families of aggregation operators, in *Eurofuse 2011*, ed. by B. De Baets et al. (Springer, 2011), pp. 169–176

[Sam04]   S. Saminger, Aggregation of bi-capacities, in *10th International Conference on Information Processing and Management of Uncertainty*, Perugia, Italy (2004), pp. 335–342

[SM03]   S. Saminger, R. Mesiar, A general approach to decomposable bi-capacities. Kybernetika **39**, 631–642 (2003)

[SK88]   N.S. Sapidis, P.D. Kaklis, An algorithm for constructing convexity and monotonicity preserving splines in tension. Comp. Aided Geom. Des. **5**, 127–137 (1988)

[Sch66]   D.G. Schweikert, An interpolation curve using a spline in tension. J. Math. Phys. **45**, 312–317 (1966)

[TN07]   V. Torra, Y. Narukawa, *Modeling Decisions. Information Fusion and Aggregation Operators* (Springer, Berlin, 2007)

[TW80]   J.F. Traub, H. Wozniakowski, *A General Theory of Optimal Algorithms* (Academic Press, New York, 1980)

[XG07]   L. Xie, M. Grabisch, The core of bicapacities and bipolar games. Fuzzy Sets Syst. **158**, 1000–1012 (2007)

[Yag09]   R.R. Yager, On generalized Bonferroni mean operators for multi-criteria aggregation. Int. J. Approximate Reasoning **50**, 1279–1286 (2009)

[YBJ09]   R.R. Yager, G. Beliakov, S. James, On generalized Bonferroni means, in *Proceedings Eurofuse, Conference* (Pamplona, Spain, 2009)

[YF94]   R.R. Yager, D. Filev, *Essentials of Fuzzy Modelling and Control* (Wiley, New York, 1994)

[YR97]   R.R. Yager, A. Rybalov, Noncommutative self-identity aggregation. Fuzzy Sets Syst. **85**, 73–82 (1997)

[Zad75a] L. Zadeh, The concept of a linguistic variable and its application to approximate reasoning. Part I. Inf. Sci. **8**, 199–249 (1975)

[Zad75b] L. Zadeh, The concept of a linguistic variable and its application to approximate reasoning. Part II. Inf. Sci. **8**, 301–357 (1975)

[Zad75c] L. Zadeh, The concept of a linguistic variable and its application to approximate reasoning. Part III. Inf. Sci. **9**, 43–80 (1975)

Chapter 7
Non-monotone Averages

Abstract Monotonicity is a fundamental property of aggregation. However not all means are monotone increasing. This chapter presents a weaker notion of directional monotonicity of averages. The robust estimators of location, mixture functions, Gini and Lehmer means, as well as density based means and medians are treated in the framework of weak monotonicity. Cone monotonicity and monotonicity with respect to coalitions are introduced and discussed. The notion of directional monotonicity is presented and pre-aggregation functions are formalized.

7.1 Motivation

As we already have seen, not all the averages are monotone increasing in all arguments. The means of Bajraktarevic and their special cases like Gini and Lehmer means and also mixture functions are not always monotone. The most notable non-monotone average is the mode, being the most frequent input, which is routinely used in statistics.[1] The mode is not monotone as the following example shows. Taking the vectors $\mathbf{x} = (1, 1, 2, 2, 3, 3, 3)$, $\mathbf{y} = (1, 1, 0, 0, 0, 0, 0)$, and $\mathbf{z} = (1, 1, 1, 1, 1, 1, 1)$, then $Mode(\mathbf{x}) = 3$, $Mode(\mathbf{x} + \mathbf{y}) = 2$ and $Mode(\mathbf{x} + \mathbf{z}) = 4$, which means that a positive increment to the argument can result in an increased or decreased output. Characterising the averages Gini [Gin58], p. 64, writes that an average of several quantities is a value obtained as a result of a certain procedure, which equals to either one of the input quantities, or a new value that lies in between the smallest and the largest input. No monotonicity requirement is postulated.

On the other hand, standard monotonicity restricts the robustness of the average against gross errors in the inputs, often called outliers. The value of the average is supposed to be a representative of the inputs. Yet an outlier can influence the output of a monotone averaging function pulling it in its own direction away from the main group of inputs, thus making the output less representative. Averages employed as robust estimators of location [RL03] should therefore be not monotone.

---

[1] In general the mode is multivalued, so in order to make it a single-valued function, a convention is needed to select one of the multiple outputs, e.g. the smallest.

© Springer International Publishing Switzerland 2016   251
G. Beliakov et al., *A Practical Guide to Averaging Functions*,
Studies in Fuzziness and Soft Computing 329,
DOI 10.1007/978-3-319-24753-3_7

An important question arises: should we just abandon monotonicity altogether, or are there other kinds of monotonicity, which would be less restrictive than the standard monotonicity yet meaningful in terms of the applications?

It seems that some sort of monotonicity is necessary for the averages to ensure their behaviour is consistent with the role they play. For instance, robust estimators of location are required to be shift-invariant, so that the change of the origin does not affect the relative position of the output.

Consider an application from image processing mentioned in Chap. 1, image reduction (p. 8). Image reduction refers to reducing the resolution of an image so that it can fit a particular display, or to speed up its analysis [BBP12, PFBMB15]. Local averaging is a computationally efficient way of image reduction. One natural condition is that uniform changes in intensity of the input image (image lightening or darkening) should not lead to the opposite changes in intensity of the output image. On the other hand, the averaging must be robust to the noise in the image, so that even large increases/decreases in the intensity of the individual pixels (due to corruption) do not lead to the undesired changes in the intensity of the output. In fact discarding such outliers may lead to changes in the intensity in the opposite direction [BLVW14, BLVW15, Wil13], as to minimize the dissimilarity between the non-corrupt inputs and the output. So we see that the averaging functions employed for image reduction might be non-monotone, yet this application requires some weaker type of monotonicity.

In this chapter we will discuss several types of monotonicity that extend the range of averaging aggregation functions, and we also provide interpretations for these mathematical concepts from the applications point of view. This chapter is based on the recent results in [BW14, CBW14, WB14, WB15, WBC14], from which we take the terminology and the relevant definitions.

7.2 Weakly Monotone Functions

The definition of weak monotonicity provided herein is prompted by applications and intuition, which suggests that it is reasonable to expect that a representative value of the inputs does not decrease if all the inputs are increased by the same amount (or shifted uniformly). The relative positions of the inputs are not changed in this case, and if our goal is to get a value representative of the inputs, it seems logical to request that such a value is governed by the relative rather than absolute positions of the inputs.

One example here is averaging in image processing (p. 8). A group of pixels of an image is replaced by their average, in particular to remove the noise in the image. Even if a non-monotone average is applied, such as the mode, we still expect that a uniform lightening or darkening of the original image does not change the output in the direction *opposite* to those uniform changes. We expect that if a group of pixels was to increase in luminosity uniformly, their average (a representative value) should not decrease.

Wilkin and Beliakov [WB15, WBC14] introduced the notion of weak monotonicity, which implies that the same increase in *all* the inputs should not lead to a decrease in the output. Weakly monotone averaging functions include two groups—monotone functions and shift-invariant functions.

A formal definition that conveys this property is as follows.

Definition 7.1 (*Weakly monotone function*) A function f is called **weakly monotone** increasing (or directionally monotone in the direction **1**) if $f(\mathbf{x} + a\mathbf{1}) \geq f(\mathbf{x})$ for any $a > 0$, $\mathbf{1} = \underbrace{(1, 1, \ldots, 1)}_{n \text{ times}}$, such that $\mathbf{x}, \mathbf{x} + a\mathbf{1} \in \mathbb{I}^n$.

Note 7.2 If f is directionally differentiable in its domain then weak monotonicity is equivalent to non-negativity of the directional derivative $D_{\mathbf{1}}(f)(\mathbf{x}) \geq 0$.

Note 7.3 Evidently monotonicity implies weak monotonicity, hence all aggregation functions are weakly monotone. We shall describe weakly monotone functions that are not monotone as *proper weakly monotone* functions. By Definition 1.46 all shift-invariant functions are also weakly monotone.

Note 7.4 The set of weakly monotone increasing functions forms a cone in the linear vector space of weakly monotone (increasing or decreasing) functions.

Many of the examples presented in the sequel involve minimization of some function P which unlike the penalty functions in the Definition 5.1 in Chap. 5 are not quasi-convex. Therefore we need to drop the condition that P is quasi-convex and replace Definition 5.1 with the following.

Definition 7.5 (*Quasi-penalty function*) The function $P : \mathbb{I}^{n+1} \to \mathbb{R}$ is a quasi-penalty function if it satisfies:

1. $P(\mathbf{x}, y) \geq c \quad \forall \mathbf{x} \in \mathbb{I}^n, y \in \mathbb{I}$;
2. $P(\mathbf{x}, y) = c$ if and only if all $x_i = y$; and,
3. $P(\mathbf{x}, y)$ is lower semi-continuous in y for any $\mathbf{x}$,

for some constant $c \in \mathbb{R}$ and any closed, non-empty interval $\mathbb{I}$.

Note that the third condition ensures the existence of the minimum and a non-empty set of minimisers. In the case where the set of minimisers of P is not an interval, we need to adopt a reasonable rule for selecting the value of the penalty-based function f. We suggest stating in advance that in such cases we choose the infimum of the set of minimisers of P. From now on P will refer to quasi-penalty functions.

7.2.1 Basic Properties of Weakly Monotone Functions

Let us establish some useful properties of weakly monotone averages.[2] Consider the function $f : \mathbb{I}^n \to \mathbb{I}$ formed by the composition $f(\mathbf{x}) = A(B_1(\mathbf{x}), B_2(\mathbf{x}))$, where A, B_1 and B_2 are means.

Proposition 7.6 *If A is monotone and B_1, B_2 are weakly monotone, then f is weakly monotone.*

Proof By weak monotonicity $B_i(\mathbf{x} + a\mathbf{1}) \geq B_i(\mathbf{x})$ implies that $\exists \delta_i \geq 0$ such that $B_i(\mathbf{x} + a\mathbf{1}) = B_i(\mathbf{x}) + \delta_i$, with $\mathbf{x}, \mathbf{x} + a\mathbf{1} \in \mathbb{I}^n$. Thus $f(\mathbf{x} + a\mathbf{1}) = A(b_1 + \delta_1, b_2 + \delta_2)$, where $b_i = B_i(\mathbf{x})$. The monotonicity of A ensures that $A(b_1 + \delta_1, b_2 + \delta_2) \not< A(b_1, b_2)$ and hence $f(\mathbf{x} + a\mathbf{1}) \geq f(\mathbf{x})$ and f is weakly monotone. $\qquad\square$

Proposition 7.7 *If A is weakly monotone and B_1, B_2 are shift-invariant, then f is weakly monotone.*

Proof Shift invariance implies that $\forall a : B_i(\mathbf{x} + a\mathbf{1}) = B_i(\mathbf{x}) + a$, with $\mathbf{x}, \mathbf{x} + a\mathbf{1} \in \mathbb{I}^n$. Thus $f(\mathbf{x} + a\mathbf{1}) = A(b_1 + a, b_2 + a)$, where $b_i = B_i(\mathbf{x})$. The weak monotonicity of A ensures that $A(b_1 + a, b_2 + a) \not< A(b_1, b_2)$ and hence $f(\mathbf{x} + a\mathbf{1}) \geq f(\mathbf{x})$ and f is weakly monotone. $\qquad\square$

Consider functions of the form $\varphi(\mathbf{x}) = (\varphi(x_1), \varphi(x_2), \ldots, \varphi(x_n))$.

Proposition 7.8 *If f is weakly monotone and φ is an affine function then the φ-transform $f_\varphi(\mathbf{x}) = g(\mathbf{x}) = \varphi^{-1}(f(\varphi(\mathbf{x})))$ is weakly monotone.*

Proof $\varphi(x) = \alpha x + \beta, \alpha \neq 0$ and hence $\varphi(x + a) = \alpha(x + a) + \beta = \alpha x + \beta + \alpha a = \varphi(x) + c$. Hence, for $\alpha > 0$

$$
\begin{aligned}
g(\mathbf{x} + a\mathbf{1}) &= \varphi^{-1}(f(\varphi(x_1 + a), \ldots, \varphi(x_n + a))) \\
&= \varphi^{-1}(f(\varphi(\mathbf{x}) + c\mathbf{1})) \\
&= \frac{f(\varphi(\mathbf{x}) + c\mathbf{1}) - \beta}{\alpha} \\
&\geq \frac{f(\varphi(\mathbf{x})) - \beta}{\alpha} = \varphi^{-1}(f(\varphi(\mathbf{x})))
\end{aligned}
$$

by weak monotonicity of f. For $\alpha < 0$ the result is analogous, as in this case $c \leq 0$ for $a \geq 0$, and the direction of the inequality remains. Hence $g(\mathbf{x} + a\mathbf{1}) \geq g(\mathbf{x})$ and g is weakly monotone. $\qquad\square$

Note that unlike in the case of standard monotonicity, a nonlinear φ-transform does not always preserve weak monotonicity, see Sect. 7.9.

---

[2]Many of the results here apply not only to averaging functions but the focus of this book is on averages.

Corollary 7.9 *The dual f_d of a weakly monotone function f is weakly monotone under the standard negation.*

The following result is relevant to an application of weakly monotone averages in image processing, discussed in Sect. 7.8.

Theorem 7.10 *Let $h : \mathbb{I}^n \to \mathbb{I}$ be a shift-invariant function, and g be a function. Let f be a penalty based averaging function with the quasi-penalty P depending on the terms $g(x_i - h(\mathbf{x}))(x_i - y)^2$. Then f is shift-invariant and hence weakly monotone.*

Proof Let

$$\mu = \arg\min_y P\left(g(x_1 - h(\mathbf{x}))(x_1 - y)^2, \ldots, g(x_n - h(\mathbf{x}))(x_n - y)^2\right).$$

Then

$$\arg\min_y P(\mathbf{x} + a\mathbf{1}, y) = \arg\min_y P\left(g(x_1 + a - h(\mathbf{x} + a\mathbf{1}))(x_1 + a - y)^2, \ldots\right.$$

$$\left. \ldots, g(x_n + a - h(\mathbf{x} + a\mathbf{1}))(x_n + a - y)^2\right) =$$

(by shift invariance)

$$= \arg\min_y P\left(g(x_1 - h(\mathbf{x}))(x_1 + a - y)^2, \ldots, g(x_n - h(\mathbf{x}))(x_n + a - y)^2\right)$$

$$= \mu + a.$$

$\square$

Note 7.11 Indeed we need not restrict ourselves to quasi-penalty functions with the terms depending on the squares of differences $(x_i - y)^2$. Functions D that depend on the differences $x_i - y$ with the minimum $D(0)$ will satisfy the above proof and satisfy the conditions on P with regards to the existence of solutions to (5.1). That is, we can use quasi-penalties which depend on the terms $g(x_i - f(\mathbf{x}))D(x_i - y)$. In particular, Huber type functions from robust regression can be used as functions D.

Locally internal means are not necessarily weakly monotone, as illustrated by the following example.

Example 7.12 Take $\mathbf{x} = (x_1, x_2) \in [0, 1]^2$ and

$$f(\mathbf{x}) = \begin{cases} \min(\mathbf{x}), & \text{if } x_1 + x_2 \geq 1 \\ \max(\mathbf{x}) & \text{otherwise} \end{cases}$$

which is internal with values in the set $\{\min(\mathbf{x}), \max(\mathbf{x})\}$. Consider the point $\mathbf{x} = (\frac{1}{4}, 0)$. If $t \leq \frac{3}{4}$ then the function

$$g(t) = f(\mathbf{x} + t\mathbf{1}) = f(\frac{1}{4} + t, t) = \begin{cases} \frac{1}{4} + t, & \text{if } t \leq \frac{3}{8} \\ t & \text{otherwise} \end{cases}$$

is neither increasing or decreasing. Hence this f is not weakly monotone for all $\mathbf{x} \in \mathbb{I}^2$.

7.3 Robust Estimators of Location

The methods of robust statistics [RL03] aim at outperforming the classical statistical methods in the presence of outliers, or, generally, when the assumptions about the underlying distributions of data are not justified. The outliers could be due to gross errors in recording the data (for example, when values like 999 are used to denote missing quantities, or when there was equipment failure). Such methods become more and more relevant with the ever growing quantities of data recorded, as manual identification of erroneous values is no longer an option, and the chances of having gross errors increase with the size of the data set.

On the other hand, the values far away from the main group of data may not be due to errors but to some unknown phenomenon, and as such these values themselves are of the main interest. In either case the outlying inputs need to be identified and then either removed or examined more closely.

Robust estimators of location deliver values representative of the majority of the data (in one or several dimensions) similar to any average, such as the arithmetic mean. However, the arithmetic mean can be easily distorted by even one single outlier (for example consider the mean of $(1, 2, 2, 3, a)$ when $a \to \infty$). As a consequence, the arithmetic mean is no longer a good representative of the main cluster of inputs, and furthermore, it does not allow one to identify the outlier by its distance from the average. In our example all the inputs will have similar distances from the average and no one will stand out.

The use of the median instead of the arithmetic mean alleviates this problem of robustness against the outliers. Indeed the value of the median is not affected even if up to half of the inputs are outliers. However, the median unnecessarily removes all but one central input, and therefore may not be as representative of the inputs as one desires.

The trimmed mean (see Sect. 3.2) takes into account a number of inputs while removing a specified proportion of data seen as potential outliers. It is more robust than the arithmetic mean and is used when an estimate of the proportion of inputs that are corrupted can be made. We recall that the trimmed means are examples of central OWA functions. A problem with the trimmed means is that one has to be sure that no more than $\alpha\%$ inputs are outliers. There are more powerful methods that can automatically identify and remove up to 50% of outliers while taking into account all the rest of the inputs. We review some of these estimators of locations in this section.

7.3.1 Mode

Perhaps the most widely used estimator of location is the mode, being the most frequent input. The mode is the minimiser of the quasi-penalty function

$$P(\mathbf{x}, y) = \sum_{i=1}^{n} p(x_i, y), \quad \text{where} \quad p(x_i, y) = \begin{cases} 0, & \text{if } x_i = y \\ 1 & \text{otherwise} \end{cases}.$$

It follows that $fF(\mathbf{x} + a\mathbf{1}) = \arg\min_y P(\mathbf{x} + a\mathbf{1}, y) = \arg\min_y \sum_{i=1}^{n} p(x_i + a, y)$, which is minimised for the value $y = f(\mathbf{x}) + a$. Hence, $f(\mathbf{x} + a\mathbf{1}) = f(\mathbf{x}) + a$ and thus the mode is shift-invariant. Hence the mode is weakly monotone.

Note 7.13 Note that the mode is a multi-valued function. For example, the mode of $(1, 1, 2, 2, 3, 4, 5)$ is not defined uniquely. In this case we use a suitable convention. The quasi-penalty P associated with the mode is not quasi-convex, and as such it may have several minimisers that do not form a convex set. A convention is needed as to which minimiser is selected, e.g., the smallest or the largest. Other examples of non-monotone means that follow also involve quasi-penalties, and the same convention as for the mode is adopted. The discrete scales can be also considered, cf. [KMM07].

7.3.2 Shorth

The remaining estimators of location presented here compute their value using the shortest contiguous sub-sample of $\mathbf{x}$ containing at least half of the values. The candidate sub-samples are the sets $X_k = \{x_j : j \in \{k, k+1, \ldots, k + \lfloor \frac{n}{2} \rfloor\}$, $k = 1, \ldots, \lfloor \frac{n+1}{2} \rfloor$ (Fig. 7.1). The length of each set is taken as $\|X_k\| = \left| x_{k+\lfloor \frac{n}{2} \rfloor} - x_k \right|$ and thus the index of the shortest sub-sample is

$$k^* = \arg\min_i \|X_i\|, \quad i = 1, \ldots, \left\lfloor \frac{n+1}{2} \right\rfloor.$$

Under the translation $\bar{\mathbf{x}} = \mathbf{x} + a\mathbf{1}$ the length of each sub-sample is unaltered since $\|\bar{X}_k\| = \left| \bar{x}_{k+\lfloor \frac{n}{2} \rfloor} - \bar{x}_k \right| = \left| (x_{k+\lfloor \frac{n}{2} \rfloor} + a) - (x_k + a) \right| = \left| x_{k+\lfloor \frac{n}{2} \rfloor} - x_k \right| = \|X_k\|$ and thus k^* remains the same.

Fig. 7.1 The contiguous
sub-samples of the data X_k
and the shortest contiguous
sub-sample $X^* = X_{k*}$

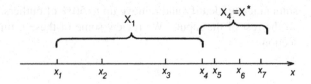

The shorth [And+72] is the arithmetic mean of X_{k*}. The shorth is given by

$$Shorth(\mathbf{x}) = \frac{1}{h} \sum_{i=1}^{h} x_i, \ x_i \in X_{k*}, \ h = \left\lfloor \frac{n}{2} \right\rfloor + 1.$$

Since the set X_{k*} is unaltered under translation and the arithmetic mean is shift-invariant, then the shorth is shift-invariant and hence weakly monotone.

7.3.3 Least Median of Squares (LMS)

Consider now the Least Median of Squares estimator [Rou84], which is the midpoint of X_{k*}. The LMS can be computed by minimisation of the quasi-penalty function

$$P(\mathbf{x}, y) = median \left\{ (x_i - y)^2 \, | y \in \mathbb{I}, \ x_i \in X_{k*} \right\}.$$

The value y minimises the quasi-penalty $P(\mathbf{x} + a\mathbf{1}, y)$, given by

$$\min_y P(\mathbf{x} + a\mathbf{1}, y) = \min_y median \left\{ (x_j + a - y)^2 \, | y \in \mathbb{I}, x_j \in X_{k*} \right\} = P(\mathbf{x}, \mu),$$

is clearly $y = \mu + a$. Hence, $LMS(\mathbf{x} + a\mathbf{1}) = LMS(\mathbf{x}) + a$ and the LMS is shift-invariant and weakly monotone.

7.3.4 Least Trimmed Squares (LTS)

The Least Trimmed Squares estimator [RL03] rejects up to 50 % of the data values as outliers and minimises the squared residuals using the remaining data. The LTS uses the quasi-penalty function

$$P(\mathbf{x}, y) = \sum_{i=1}^{h} r_{(i)}^2,$$

where $r_{(i)} = S_i(\mathbf{r})$ is the ith order statistic of $\mathbf{r}$, $r_k = x_k - y$ and $h = \lfloor \frac{n}{2} \rfloor + 1$. If σ is the order permutation of $\{1, \ldots, n\}$ such that $\mathbf{r}_{\infty} = \mathbf{r}_{\nearrow}$, then the minima of P occur when $P_y = -2 \sum_{i=1}^{h} (x_{\sigma(i)} - y) = 0$, which implies that the minimum value is

$\mu = \frac{1}{h} \sum_{i=1}^{h} x_{\sigma(i)}$. Since $S_k(\mathbf{x})$ is shift-invariant then $S_i(\mathbf{r} + a\mathbf{1}) = r_{\sigma(i)} + a$ and thus

$$P(\mathbf{x} + a\mathbf{1}, y) = \sum_{i=1}^{h} v_{\sigma(i)}^2,$$

where $v_k = ((x_k + a) - y)$. It follows that the value y that minimises $P(\mathbf{x} + a\mathbf{1}, y)$ is $y = \mu + a$, hence the LTS is shift-invariant and thus weakly monotone.

7.3.5 Least Trimmed Absolute Deviations (LTA)

Similar to the LTS, the Least Trimmed Absolute deviations estimator [HO99] rejects up to 50 % of the data values as outliers and minimises the absolute residuals using the remaining data. The LTA uses the quasi-penalty function

$$P(\mathbf{x}, y) = \sum_{i=1}^{h} |r_{(i)}|,$$

where $r_{(i)} = S_i(\mathbf{r})$ is the ith order statistic of $\mathbf{r}$, $r_k = x_k - y$ and $h = \lfloor \frac{n}{2} \rfloor + 1$. If σ is the order permutation of $\{1, \ldots, n\}$ such that $\mathbf{r}_{\infty} = \mathbf{r}_{\nearrow}$, then the minima of P occur when $P_y = -2 \sum_{i=1}^{h} (x_{\sigma(i)} - y) = 0$, which implies that the minimum value is

$\mu = \frac{1}{h} \sum_{i=1}^{h} x_{\sigma(i)}$. Since $S_k(\mathbf{x})$ is shift-invariant then $S_i(\mathbf{r} + a\mathbf{1}) = r_{\sigma(i)} + a$ and thus

$$P(\mathbf{x} + a\mathbf{1}, y) = \sum_{i=1}^{h} v_{\sigma(i)}^2,$$

where $v_k = ((x_k + a) - y)$. It follows that the value y that minimises $P(\mathbf{x} + a\mathbf{1}, y)$ is $y = \mu + a$, hence the LTS is shift-invariant and thus weakly monotone.

Maximum trimmed likelihood methods are obtained by using functions other than the squares and absolute deviations on the right-and side of the trimmed sum in the penalty formula [BC93, HLn97].

7.3.6 The Least Winsorized Squares Estimator

The least Winsorized squares estimator is given by

$$LWS(\mathbf{x}) = \arg\min_y \sum_{i=1}^{h} r_{(i)}^2 + (n-h)r_{(h)}^2.$$

It is shift-invariant and weakly monotone.

7.3.7 OWA Penalty Functions

Quasi-penalty functions having the form

$$P(\mathbf{x}, y) = \sum_{i=1}^{n} w_i S_i \left((\mathbf{x} - y\mathbf{1})^2 \right)$$

define regression operators f, see [YB10]. Consider the following results dependent on the weight vector $\Delta = (w_1, \ldots, w_n)$.

1. $\Delta = \mathbf{1}$ generates Least Squares regression and f is monotone and hence weakly monotone;
2. $\Delta = (0, \ldots, 0, 1)$ generates Chebyshev regression and f is monotone and hence weakly monotone;
3. Since all the terms $S_i \left((\mathbf{x} - y\mathbf{1})^2 \right)$ are constant under transformation $(\mathbf{x}, y) \rightarrow (\mathbf{x} + a\mathbf{1}, y + a)$ (cf Theorem 7.10), the OWA regression operators are shift-invariant for any choice of the weight vector Δ.
4. For $\Delta = \begin{cases} (0, \ldots, 0_{k-1}, \frac{1}{2}, \frac{1}{2}, 0, \ldots 0) & n = 2k \text{ is even} \\ (0, \ldots, 0_{k-1}, 1, 0, \ldots, 0) & n = 2k - 1 \text{ is odd} \end{cases}$ then f is the Least Median of Squares operator and hence shift invariant and weakly monotone; and
5. For $\Delta = (1, \ldots, 1_h, 0, \ldots, 0)$, $h = \lfloor \frac{n}{2} \rfloor + 1$ then f is the Least Trimmed Squares operator and hence is shift-invariant.

In the cases 3–5 the OWA regression operators are not monotone.

7.4 Lehmer and Gini Means

7.4.1 Lehmer Means

The Lehmer means are a subclass of mixture functions (Sect. 2.4.7), where the weighting function w is the power function $w(t) = t^q$. We recall their definition (note that we assumed equal weights w_i when compared to Definition 2.40).

Definition 7.14 (*Lehmer mean*) The mapping $L^q : [0, \infty)^n \to [0, \infty), q \in \mathbb{R}$ given by

$$L^q(x_1, \ldots, x_n) = \frac{\sum_{i=1}^{n} x_i^{q+1}}{\sum_{i=1}^{n} x_i^{q}} \tag{7.1}$$

is called the Lehmer mean.

Note that $L^q(\mathbf{0}) = 0$ is defined by continuity. Some representative plots are presented in Sect. 2.4.1. Note that the Lehmer mean is homogeneous (of degree one), and hence it is sufficient to analyse its monotonicity properties on some finite domain $[0, b]^n$. It is known that the Lehmer mean is monotone for $q \in [-1, 0]$. Another useful property of L^q (which is also valid for Gini means in a more general form) is the ordering $L^p \leq L^q$ when $p \leq q$ [Bul03].

Let us now establish sufficient conditions for weak monotonicity of the Lehmer means. We start with the case for which weak monotonicity does not hold.

Proposition 7.15 *The Lehmer mean L^q is not weakly monotone for $q \in {]}0, 1[$.*

Proof It is sufficient co consider the limits of the partial derivative when $\mathbf{x} \to (1, 0, \ldots, 0)$. For example, when $n = 2$,

$$\lim_{x \to 0^+} \frac{\partial L^q(x, 1)}{\partial x} = \frac{x^{2q} - (q+1)x^q - qx^{q-1}}{(x^q + 1)^2} = \lim_{x \to 0^+} \frac{-qx^{q-1}}{1} = -\infty.$$

$\square$

Proposition 7.16 *The Lehmer mean of n arguments is weakly monotone on $[0, \infty)^n$ for $q \leq 0$.*

Proof Consider the directional derivative

$$(\mathbf{D_1}L^q)(\mathbf{x}) = \frac{1}{\sqrt{n}} \frac{1}{\left(\sum_{j=1}^{n} x_j^q\right)^2} \sum_{i=1}^{n} \left((q+1)x_i^q \sum_{j=1}^{n} x_j^q - qx_i^{q-1} \sum_{j=1}^{n} x_j^{q+1} \right) \quad (7.2)$$

$$= \frac{1}{\sqrt{n}} \left((q+1) - q \left(\sum_{i=1}^{n} x_i^{q-1} \right) \left(\sum_{i=1}^{n} x_i^{q+1} \right) \Big/ \left(\sum_{j=1}^{n} x_j^q \right)^2 \right),$$

which needs to be non-negative. Omitting the positive factor, we need

$$q + 1 - q \frac{L^q(\mathbf{x})}{L^{q-1}(\mathbf{x})} \geq 0. \quad (7.3)$$

Since the Lehmer means are ordered by q it means that the ratio $\frac{L^q(\mathbf{x})}{L^{q-1}(\mathbf{x})} \geq 1$. There-fore for weak monotonicity we need

$$\frac{q+1}{q} \leq 1 \leq \frac{L^q(\mathbf{x})}{L^{q-1}(\mathbf{x})},$$

which holds for all $q < 0$, which proves the result.   □

Unfortunately, this proof does not work for positive q, as the inequality charac-terising q takes the opposite direction. The following result from [BS15] establishes that weak monotonicity of Lehmer means depends on the dimension of the input vector.

Theorem 7.17 *Let $L^q : [0, \infty)^n \to [0, \infty)$ be the Lehmer mean (7.1), $q > 1$. Then L^q is weakly monotone increasing for*

$$n \leq 1 + \frac{(\sqrt{q} + 1)^{2(q-1)}}{(q-1)^{q-1}}. \quad (7.4)$$

The proof presented in [BS15] is based on the analysis of power sums following a technique from [Rez83].

Let us analyse the weak monotonicity condition (7.4). Note that weak monotonic-ity depends on the number of arguments n and that the right hand side of (7.4) is strictly increasing with q and unbounded, and therefore for larger q the function L^q is weakly monotone for a larger number of arguments.

The special case of $q = 1$ is obtained directly from (7.3) by noticing that the maximum of $\frac{L^1}{L^0}$ is reached when $\mathbf{x} = (1, 0, \ldots, 0)$, which gives the condition $2 \geq \frac{L^1}{L^0}(\mathbf{x}) = n$, so we obtain the condition for weak monotonicity $n \leq 2$, and hence the following corollary.

Corollary 7.18 *The contra-harmonic mean (L^q with $q = 1$) is weakly monotone only for two arguments.*

7.4.2 Gini Means

The Gini means are an interesting class of the means of Bajraktarevic M_w^g where the weighting functions w and the scaling function g are power functions. We recall their definition (note that we assumed equal weights w_i when compared to Definition 2.41). Plots of some Gini means are presented in Sect. 2.4.1.

Definition 7.19 (*Gini mean*) Let $p, q \in \mathbb{R}$. The Gini mean is the mapping

$$
G^{p,q}(\mathbf{x}) = \begin{cases} \left(\dfrac{\sum\limits_{i=1}^{n} x_i^p}{\sum\limits_{i=1}^{n} x_i^q} \right)^{1/p-q} , & \text{if } p \neq q, \\[2em] \left(\prod\limits_{i=1}^{n} x_i^{x_i^p} \right)^{1/\sum\limits_{i=1}^{n} x_i^p} , & \text{if } p = q. \end{cases} \tag{7.5}
$$

Of course the Lehmer means are a special case of Gini means. In this section we present the weak monotonicity results obtained in [BCW15, BS15].

Theorem 7.20 *The Gini mean* (7.5) *for* $p, q \in \mathbb{R}$ *is weakly monotone on* $[0, \infty)^n$ *if either*

1. $q \leq 0$ and $p \geq q$, or
2. $p \leq 0$ and $p \leq q$ or
3. $(n-1) \left(\left(\frac{q}{p-1} \right)^{p-1} \left(\frac{q-1}{p} \right)^{q-1} \right)^{\frac{1}{p-q}} \leq 1$ and $((p \geq q \geq 1)$ or $(1 \leq p \leq q1))$.

Proof For $p = q$ the Gini mean becomes [Bul03]

$$
G^{p,p}(\mathbf{x}) = \left(\prod x_i^{x_i^p} \right)^{1/\sum x_i^p} ,
$$

which is monotone. The cases $q = 0$ or $p = 0$ result in (monotone) power means.

Since $G^{p,q} = G^{q,p}$ we only consider the case $p > q$. Let $m = p - q > 0$. To ensure weak monotonicity of the Gini mean we need to show that the sum of its partial derivatives is non-negative. In the case $q \in (0, 1)$ the partial derivative at $\mathbf{x} = (a, b)$ when $a \to 0^+$ tends to $-\infty$, so the Gini mean is not weakly monotone for such q.

Consider the case (a). By adding the partial derivatives of the m-th power of $G^{p,q}$ we get the condition

$$
p \sum_{i=1}^{n} x_i^{p-1} \sum_{i=1}^{n} x_i^q - q \sum_{i=1}^{n} x_i^{q-1} \sum_{i=1}^{n} x_i^p \geq 0,
$$

and then (dropping the summation indices)

$$p - q \frac{\sum x_i^{q-1} \sum x_i^p}{\sum x_i^{p-1} \sum x_i^q} = p - q \frac{G^{p,q}(\mathbf{x})}{G^{p-1,q-1}(\mathbf{x})} \geq 0,$$

Since $G^{p,q} \leq G^{s,t}$ as long as $p \leq s, q \leq t$ [Bul03], the ratio in the above formula is greater than or equal to 1, hence for $q < 0$ we get

$$\frac{p}{q} \leq 1 \leq \frac{G^{p,q}}{G^{p-1,q-1}},$$

and therefore $p \geq q$ as required. The case (b) is obtained by symmetry with respect to p and q.

The proof of the case (c) is more technical and is presented in [BCW15]. □

Two other sufficient conditions for weak monotonicity are presented in [BS15].

Theorem 7.21 *Let $G^{p,q} : [0, \infty)^n \to [0, \infty)$ be the Gini mean (7.5) with $1 \leq q < p$. Then $G^{p,q}$ is weakly monotone increasing for $1 \leq q \leq p - 1$, $2q > p$ and*

$$n \leq 1 + \left(\frac{p + 2\sqrt{q(p-q)}}{2q - p} \right)^{\frac{2q-p}{p-q}}.$$

Note 7.22 For $p = q + 1$ and $q = m > 1$ we obtain condition (7.4) for Lehmer means.

Theorem 7.23 *Let $G^{p,q} : [0, \infty)^n \to [0, \infty)$ be the Gini mean (7.5) with $1 < q < p$. Then $G^{p,q}$ is weakly monotone increasing for*

$$n \leq 1 + \frac{(r^*)^{1+\frac{t}{q}} (tr^* - (2q + t))}{(2q + t)r^* - t},$$

where $r^* = A + \sqrt{A(A - 2)} - 1$, $A = \frac{p}{q} \frac{2(q+t)^2}{(2q+t)t}$ and $t \geq 0$ is a number satisfying $t \geq 2q - p$ and $t \leq q - 1$.

We note that numerically the choice $t = q - 1$ delivers larger (i.e., less restrictive) bounds on n. Also with this choice and with $p = q + 1$ we obtain after some simplifications condition (7.4) for Lehmer means.

It is interesting to examine the behaviour of the quantities r and A, and hence n as functions of p and q (and t) when these parameters are increasing. If both $p, q \to \infty$ and $t = q - 1$ we get an interesting limit $\lim_{q \to \infty} n(2q, q) = 17$ (note that n is an integer). However, if $p \to \infty$ for a fixed q, then $n \to \infty$. This is also the case when p is increasing faster than q, for example if $p = q^k$, $k > 1$.

Also note that because $G^{p,q}$ are symmetric in p and q, the two theorems also work when p and q exchange roles.

7.5 Mixture Functions

Mixture functions are a subclass of Bajraktarevic means, which generalize quasi-arithmetic means [Baj58, Baj63]. In mixture functions the inputs are averaged as in a weighted mean, but the weights depend on the inputs. Weights can thus be chosen so as to alternatively emphasize or de-emphasize the small or large inputs. When applied to averaging functions measuring the distance between the inputs, such means allow for de-emphasizing contributions from outliers. An equivalent class of functions was also presented in [Ped09] under the name of statistically grounded aggregation operators. It was shown in [MS06, MSV08] that mixture operators can be represented as penalty based functions.

Definition 7.24 (*Mixture function*) Let $w : \mathbb{I} \to [0, \infty)$. The mixture function is the mapping

$$M_w(\mathbf{x}) = \frac{\sum\limits_{i=1}^{n} w(x_i) x_i}{\sum\limits_{i=1}^{n} w(x_i)}. \tag{7.6}$$

Let us consider three cases for ensuring mixture functions are monotone increasing. In each case consider the weights $w : \mathbb{I} \to [0, \infty)$, and let $\mathbb{I} = [0, 1]$. Then sufficient conditions for monotonicity are:

1. $w \geq w'$ for any increasing, piecewise differentiable weighting function [MPR03];
2. $w(x) \geq w'(x)(1 - x)$ for all $x \in [0, 1]$: [MS06, MSV08];
   or, if we fix the dimension n of the domain,
3. $\frac{w^2(x)}{(n-1)w(1)} + w(x) \geq w'(x)(1 - x), x \in [0, 1], n > 1$: [MS06].

Analogous results have been obtained for decreasing weighting functions using duality (with respect to the standard negation). Taking the dual weighting function $w^d(x) = w(1 - x)$, the resulting mixture function is the dual to M_w; that is, $M_{w^d} = 1 - M_w$. Duality preserves both weak and standard monotonicity. Additionally, M_w is invariant to scaling of the weight functions (i.e., $M_{\alpha w} = M_w \ \forall \alpha \in \mathbb{R} \setminus \{0\}$).

A sufficient condition for weak monotonicity is obtained by adding the partial derivatives of M_w:

$$(\sum_{j=1}^{n} w(x_j))^2 + \sum_{j=1}^{n} w(x_j) \sum_{j=1}^{n} w'(x_j) x_j - \sum_{j=1}^{n} w'(x_j) \sum_{j=1}^{n} w(x_j) x_j \geq 0. \tag{7.7}$$

7.5.1 Some Special Cases of Weighting Functions

Unlike the conditions from [MS06, MSV08], it appears there is no simple way to characterize weighting functions w which satisfy (7.7). We shall now look at some special but prototypical cases of the weighting functions, see [CBW14, WBC14].

Proposition 7.25 *Let M_w be a mixture function defined by (7.6) and $w : \mathbb{I} \to [0, \infty)$. For the following generators, the functions M_w are:*

1. *$w(x) = e^{ax+b}$, $a \in [-1, 1]$: monotone;*
2. *$w(x) = e^{ax+b}$, $a \in \mathbb{R}$: shift-invariant and hence weakly monotone;*
3. *$w(x) = \ln(1 + x)$: weakly monotone for $n = 2$ and $x \geq 0.1117$, $n = 3$ and $x \geq 0.2647$, and $x \geq 0.4547$ as $n \to \infty$; and,*

for all $\mathbf{x} \in [0, 1]^n$.

Proof The first generator trivially satisfies the first condition above for $a \in [0, 1]$, $w'(x) \leq w(x)$, and for negative a the result is obtained using duality.

In the second case this generator fails conditions 1 and 2 for $a > 1$. However

$$M_w(\mathbf{x} + t\mathbf{1}) = \frac{\sum e^{ax_i+b+at}(x_i + t)}{\sum e^{ax_i+b+at}} = \frac{\sum e^{ax_i+b}(x_i + t)}{\sum e^{ax_i+b}} = M_w(\mathbf{x}) + t.$$

For the third generator we check the three conditions of monotonicity stated above. For the first condition $w(x) = \ln(1+x) \geq w'(x) = \frac{1}{1+x}$ when $x \geq LW(1) \approx 0.7632$ (LW is the Lambert W function, defined by the relation $x = LW(x)e^{LW(x)}$ for $x \geq -e^{-1}$). Thus the mixture function generated by $w(x) = \ln(1+x)$ is monotone in $[0.7632, \infty)$. For the second sufficient condition we obtain that the mixture operator is monotone in $[0.4547, \infty)$ and from the third condition we find that the mixture operator generated by $w(x) = \ln(1 + x)$ is monotone in $[0.3708, \infty)$ for $n = 2$.

Now, if we consider weak monotonicity, it turns out that the mixture function with such a weight is weakly monotone for all such x that satisfy

$$\ln(2) + (n - 1)\left(\ln(x + 1) + \frac{x - 1}{x + 1}\right) \geq 0.$$

This inequality follows from the fact that the directional derivative of M_w is proportional to the sum of partial derivatives of $F(\mathbf{x}, y) = \sum_{i=1}^{n} \ln(1 + x_i)(x_i - y)$, that $F_{x_{(1)}} \geq \ln(1 + x_{(1)})$ and that all other partial derivatives achieve their minimum with respect to y when $y = x_{(1)}$. The directional derivative is the smallest when $x_{(1)}$ is the largest ($x_{(1)} = 1$), from which the above inequality is derived. By solving this inequality for x numerically for fixed n, we obtain that for $n = 2$, x must be larger than or equal to 0.1117, for $n = 3$, $x \in [0.2647, \infty)$, and so on. When $n \to \infty$ the smallest x approaches $0.4547\ldots$, which is consistent with the standard monotonicity. $\square$

The next result proved in [BCW15] is particularly interesting. It shows that if the weighting function is Gaussian, $w(x) = \exp(-(x-a)^2/b^2)$, weak monotonicity holds irrespective of the values a, b provided $b^2 \geq 1$.

Proposition 7.26 *Let M_w be a mixture function defined by* (7.6) *with the generator* $w(x) = e^{-(\frac{x-a}{b})^2}$, *then M_w is weakly monotone for all $a, b \in \mathbb{R}$, $b^2 \geq 1$ and* $\mathbf{x} \in [0,1]^n$.

Thus Gaussian weighting functions become particularly important in the construction of mixture functions. By varying the parameters a and b we obtain variously monotone increasing, monotone decreasing (both convex and concave) and unimodal quasi-concave weighting functions, which will all be proper weakly monotone functions. Mixture functions with Gaussian weighting functions feature prominently in signal and image processing applications.

7.5.2 Affine Weighting Functions

A linear weighting function is appealing due to its simplicity, but as we have seen, the Lehmer mean L^1 is weakly monotone only for two arguments, which significantly limits its application. In this section we discuss a modification of the weighting function in L^1 which would ensure the result is weakly monotone.

Theorem 7.27 [BS15] *Let $M_w : [0,1]^n \to [0,1]$ be a mixture function defined by* (7.6) *with an affine weighting function $w(x) = x + l$, $l \geq 0$. Then M_w is weakly monotone increasing for*

$$l \geq \frac{\sqrt{2}-1}{2} \approx 0.207107. \tag{7.8}$$

Proof On the basis of sufficient condition (6.12), we get

$$n^2 l^2 + 2nl \sum_{i=1}^{n} x_i + 2\left(\sum_{i=1}^{n} x_i\right)^2 - n \sum_{i=1}^{n} x_i^2 \geq 0, \tag{7.9}$$

whence

$$l^2 + \frac{2l \sum_{i=1}^{n} x_i}{n} + \frac{2\left(\sum_{i=1}^{n} x_i\right)^2}{n^2} - \frac{\sum_{i=1}^{n} x_i^2}{n} \geq 0. \tag{7.10}$$

If we substitute well-known descriptive statistics—the arithmetic mean $\bar{x} = \dfrac{1}{n}\sum\limits_{i=1}^{n} x_i$ and the variance $s_x^2 = \overline{x^2} - \bar{x}^2$ into the previous inequality, we obtain the condition $l^2 + 2l\bar{x} + (\bar{x}^2 - s_x^2) \geq 0$. From the solution of the last inequality we have

$$l \geq s_x - \bar{x}. \qquad (7.11)$$

We are interested in the largest value of the right hand side of (7.11) for a fixed n and also for any n.

Unlike the Lehmer mean, the mixture function with an affine weighting function w is not homogeneous, and we cannot apply the same argument as in Theorem 7.17 to identify the vector $\mathbf{y}$ which delivers the minimum of the left hand side of (7.10). In fact the minimisers of that expression without constraints are unbounded. However, since we consider function M_w on the unit cube, we can formulate minimisation of the left hand side of (7.9) (or (7.10)) as a quadratic programming problem with box constraints.

The expression in (7.9) can be written as follows.

$$H(\mathbf{x}) = \mathbf{x}^T Q \mathbf{x} + \mathbf{c}^T \mathbf{x} + D \geq 0,$$

where $c_i = 2nl$, $i = 1, \ldots, n$, $D = n^2 l^2$ and matrix Q given by

$$Q = 2J_n - nI_n,$$

where I_n is the identity matrix of size n and J_n the square matrix of size n with all elements equal to 1.

To find the minimum of H we solve

$$\text{Minimize } H(\mathbf{x})$$
$$\text{subject to } 0 \leq x_i \leq 1, i = 1, \ldots, n. \qquad (7.12)$$

This is an indefinite quadratic programming problem with box constraints, because the matrix Q has positive and negative eigenvalues, namely the eigenvalues are $n, -n, \ldots, -n$ (the negative eigenvalue has multiplicity $n - 1$). These eigenvalues follow from the following considerations: J_n eigenvalues $\lambda_1 = n$ and $\lambda_2 = 0$ (multiplicity $n - 1$) (consider one eigenvector $(1, \ldots, 1)$ and the $n - 1$ vectors from the nullspace of J_n). Matrix cJ_n has eigenvalues cn and 0. Finally matrix $cJ_n - aI_n$ has eigenvalues $\tilde{\lambda}_1 = cn - a$ and $\tilde{\lambda}_2 = -a$ (multiplicity $n-1$). In our case $Q = 2J_n - nI_n$ and hence the eigenvalues of Q are $n, -n, \ldots, -n$.

By Proposition 2.3 in [DAPT97], at least $n - 1$ constraints are active at a local optimum of H. As such, the minima and maxima of H happen at the edges of the unit cube.

Next, let us show that at the edges of the unit cube the objective H is concave. Fix all but one component of $\mathbf{x}$ at either 0 or 1. The quadratic term in H becomes

$(2 - n)x_i^2$, which is a concave function for $n \geq 2$. Therefore the local (and hence global) minima of H happen at the vertices of the unit cube.

Therefore, without loss of generality let $(\underbrace{1, 1, \ldots, 1}_{k\text{-times}}, \underbrace{0, 0, \ldots, 0}_{(n-k)\text{-times}})$ be the input

vector, at which H attains the smallest value. Then $\bar{x} = \dfrac{k}{n}$ and $s_x^2 = \dfrac{k}{n} - \dfrac{k^2}{n^2}$. Using

substitution $d = \bar{x} = \dfrac{k}{n}$ we can rewrite condition (7.11) as follows

$$l \geq \sqrt{d - d^2} - d. \tag{7.13}$$

We need to find the largest value of $\sqrt{d - d^2} - d$, which is positive, hence $k \leq n - k$. That means l is non-negative for input vectors with lower numbers of ones than zeros.

Without loss of generality, we can assume that the expression on the right side of inequality (7.13) is a continuous differentiable function. This function has the maximum at the point $d_{max} = \dfrac{2 - \sqrt{2}}{4}$, hence we immediately obtain condition (7.8) for any n. $\qquad\qquad\square$

Note 7.28 Of course, for a specific value of n d_{max} is not necessarily reached, as d takes a discrete set of values. Therefore we will obtain less restrictive bounds on l by selecting the largest (over k) value of $\frac{\sqrt{k(n-k)}-k}{n}$, while d_{max} remains the (conservative) global bound. For example, for $n = 2$ we get $l \geq 0$, which is consistent with our result in Theorem 7.17.

7.5.3 Linear Combinations of Weighting Functions

Proposition 7.29 Let $w : \mathbb{I} \to [0, \infty)$ be given by $w(x) = u(x) + v(x)$ where $u, v : \mathbb{I} \to [0, \infty)$. Then the mixture operator M_w defined by (7.6) is weakly monotone if the mixture operators with generators u and v are also weakly monotone.

Proof Consider the function $F^{(w)}$ related to the penalty function defining the mixture operator M_w. It is a matter of a simple calculation to note that any partial derivative of $F^{(w)}(\mathbf{x}, z) = \sum_{i=1}^{n} g(x_i)(x_i - y)$ is

$$
\begin{aligned}
F_{x_i}^{(w)} &= w'(x_i)(x - y) + w(x_i) \\
&= (u'(x_i) + v'(x_i))(x - y) + u(x_i) + v(x_i) \\
&= (u'(x_i)(x - y) + u(x_i)) + (v'(x_i)(x - y) + v(x_i)) \\
&= F_{x_i}^{(u)} + F_{x_i}^{(v)}.
\end{aligned}
$$

Hence, $F_{x_i}^{(w)} \geq 0$ if $F_{x_i}^{(u)} \geq 0$ and $F_{x_i}^{(v)} \geq 0$. Therefore, the directional derivative of M_w will be non-negative if the directional derivatives of M_u and M_v are non-negative. □

Observe that the last result can be extended for any number of weighting functions, so that, if the weighting function of a mixture function M_w is in the form $w = \sum\limits_{i=1}^{r} u_i$ the mixture function M_w is weakly monotone if the mixture functions M_{u_i} are also weakly monotone.

7.5.4 The Duals of Lehmer Mean and Other Mixture Functions

Duality preserves monotonicity and boundary conditions of aggregation functions, idempotency, symmetry, as well as the associativity and continuity. Linear transformations (such as standard negations) also preserve weak monotonicity [BCW15]. Concerning the mixture functions, we have the following results.

Proposition 7.30 [š08] *Let $M_w : [0, b]^n \to [0, b]$ be a mixture function* (7.6). *Then*

$$(M_w)_d = M_{w^*},$$

where $w^* : [0, b] \to [0, \infty)$ is given by $w^*(x) = w(b - x)$ and $(M_w)_d = M_w$ if and only if, $w(x) = w(b - x)$ for all $x \in [0, b]$, i.e., if $w = w^*$.

Proof For any $(x_1, \ldots, x_n) \in [0, b]^n$, we have

$$(M_w)_d(x_1, \ldots, x_n) = b - M_w(b - x_1, \ldots, b - x_n)$$

$$= b - \frac{\sum\limits_{i=1}^{n} w(b - x_i) \cdot (b - x_i)}{\sum\limits_{i=1}^{n} w(b - x_i)} = \frac{\sum\limits_{i=1}^{n} w^*(x_i) \cdot x_i}{\sum\limits_{i=1}^{n} w^*(x_i)} = M_{w^*}(x_1, \ldots, x_n).$$

□

Corollary 7.31 *Let $L^m : [0, b]^n \to [0, b]$ be the Lehmer mean* (7.1) *as a mixture function with weighting function $w(x_i) = x_i^m$ and $L_*^m : [0, b]^n \to [0, b]$ with weighting function $w(x_i^*) = (b - x_i)^m$.*
Then

$$L_d^m(x_1, \ldots, x_n) = b - \frac{\sum\limits_{i=1}^{n} (b - x_i)^m (b - x_i)}{\sum\limits_{i=1}^{n} (b - x_i)^m} = \frac{\sum\limits_{i=1}^{n} (b - x_i)^m x_i}{\sum\limits_{i=1}^{n} (b - x_i)^m} = L_*^m(x_1, \ldots, x_n).$$

Corollary 7.32 *The dual Lehmer mean L_d^m is weakly monotone under the same conditions on n and m as the Lehmer mean in Theorem 7.17.*

7.6 Density Based Means and Medians

7.6.1 Density Based Means

The main motivation behind the density based means in [AY13] was to define ana-logues of the weighted arithmetic mean to filter outliers, by using the weights that depend on the density of the data, so that inputs closer to the main group of data have higher weights than those far away. The Angelov and Yager [AY13] develop their means in the context of stream processing, in which only recent data are avail-able and older data are represented by a few statistical quantities. Notably, by using a Cauchy kernel they develop recursive update formulas, suitable for on-line, real-time stream processing applications. To simplify our presentation, we state the formulas from [AY13] without referring to the current time instance, assuming it is fixed. Once the mathematical properties are established they are easily ported to the stream processing context as in [AY13].

Let d_{ij} denote the distance between inputs x_i and x_j. In [AY13] the authors used Euclidean and cosine distances (in case the inputs x_i are vectors themselves), although here we restrict our consideration to Euclidean or Minkowski distances. The *density based mean* is defined as follows.

Definition 7.33 (*Density based mean*) The density based mean is the mapping

$$DBM(\mathbf{x}) = \sum_{i=1}^{n} w_i(\mathbf{x})x_i, \qquad (7.14)$$

where

$$w_i(\mathbf{x}) = \frac{u_i(\mathbf{x})}{\sum_{j=1}^{n} u_j(\mathbf{x})} = \frac{K_C(\frac{1}{n}\sum_{j=1}^{n} d_{ij}^2)}{\sum_{k=1}^{n} K_C(\frac{1}{n}\sum_{j=1}^{n} d_{kj}^2)}, \qquad (7.15)$$

$d_{ij} = |x_i - x_j|$ and where K_C is the Cauchy kernel given by

$$K_C(t) = (1+t)^{-1}. \qquad (7.16)$$

Note that the formula (7.14) is more general than the mixture operators because w_i depends on all input data, rather than merely on x_i. Also note that the penalty associated to (7.14) is

$$P(\mathbf{x}, y) = \sum w_i(\mathbf{x})(x_i - y)^2. \qquad (7.17)$$

This penalty also differs from (5.2) in that now the weights w_i depend on all the components of the input vector. Consequently the sufficient conditions for monotonicity of mixture functions from [MPR03, MS06, MSV08, RMP03], see p. 265, where w_i depends only on x_i, are not applicable.

The density based means are not monotone in the usual sense, which can be seen from the following example: take $n = 20$, $\mathbf{x} = (1, 0, 0, \ldots, 0)$ and $\mathbf{y} = (3, 0, 0, \ldots, 0)$, and hence $\mathbf{x} \leq \mathbf{y}$. Then $\mathbf{w}(\mathbf{x}) = (\frac{7}{254}, \frac{13}{254}, \ldots, \frac{13}{254})$ and $\mathbf{w}(\mathbf{y}) = (\frac{29}{3658}, \frac{191}{3658}, \ldots, \frac{191}{3658})$, which can be computed easily by noticing that for vector $\mathbf{x}$: $\sum_j d_{1,j}^2 = n - 1 = 19$ and $\sum_j d_{i,j}^2 = 1$ for $i \neq 1$, and for $\mathbf{y}$: $\sum_j d_{1,j}^2 = 9(n - 1) = 171$ and $\sum_j d_{i,j}^2 = 9$ for $i \neq 1$. Then applying formula (7.14) we get $F_{\mathbf{w}(\mathbf{x})}(\mathbf{x}) = \frac{7}{254} > \frac{87}{3658} = F_{\mathbf{w}(\mathbf{y})}(\mathbf{y})$, hence monotonicity does not hold.

We shall now explore the issues of bounds preservation and weak monotonicity of (7.14) by applying Propositions 2.52 and 5.11, following the approach in [BW14]. Furthermore, we present several extensions of the formulas (7.15), (7.16).

Theorem 7.34 *The density based mean (7.14) is bound preserving (and hence a mean), and weakly monotone.*

Proof The first assertion comes from Proposition 2.52 and the fact that (7.14) is based on the associated weighted arithmetic mean. The second assertion is supported by the fact that the weights depend only on the pairwise distances between the inputs, which remain constant under translation, and then follows from penalty representation (7.17) and Proposition 5.11 with $G_i = w_i$, since all shift-invariant functions are weakly monotone.[3]   □

Our next goal is to generalize the density based means in the following ways.

7.6.2 Density Based Medians and ML Estimators

The weighted arithmetic mean is an example of the maximum likelihood (ML) estimator, which is based on the assumption that the inputs follow Gaussian distribution around some value, estimated by the mean. If the distribution of the inputs is different, then other estimators, such as the median (when the distribution is Laplace), are more appropriate. By modifying the penalty (7.17) we define density based analogues of the median and other ML estimators.

Our first generalization arises by changing the penalty from

$$P(\mathbf{x}, y) = \sum w_i(\mathbf{x})(x_i - y)^2 \tag{7.18}$$

---

[3] We refer here only to the standard distances in linear vector spaces. The cosine distance mentioned in [AY13] is not a proper metric and the distances are affected by translation.

to

$$P(\mathbf{x}, y) = \sum w_i(\mathbf{x})|x_i - y|, \tag{7.19}$$

from which we obtain a density based weighted median. This approach was taken in [Ped09] and further explored in [MSV08], in the case of w_i being a function of one argument x_i. In our case the weights depend on the density, which is a function of all inputs $\mathbf{x}$, hence we get a more general construction we call the density based median.

By using other functions instead of the squared or absolute differences we will obtain other maximum likelihood type estimators. All these estimators will be bounds preserving and weakly monotone (by applying Propositions 2.52 and 5.11), and if all w_i are the same functions, the estimators will also be symmetric. These functions will reduce to the corresponding standard averaging aggregation functions (with constant weights) in the limiting case of equal distances between all data.

One interesting example is based on Huber's loss function, where we minimise

$$P(\mathbf{x}, y) = \sum w_i(\mathbf{x})\rho(|x_i - y|), \tag{7.20}$$

where

$$\rho(t) = \begin{cases} \frac{1}{2}t^2, & \text{if } t \le \delta, \\ \delta(t - \frac{\delta}{2}) & \text{otherwise.} \end{cases}$$

Another example is the loss function $\rho(t) = \log(\cosh(t))$, which behaves similarly to Huber's function. These two estimators arose in robust statistics [RL03], where one aims at limiting the contribution of the outliers. Huber's estimator behaves like the least squares estimator for the data close to the average (the parameter δ quantifies the notion of closeness), yet it behaves like the (more robust) median for data further away, when ρ changes to a linear function and hence limits the contribution of potential outliers.

7.6.3 Modified Weighting Functions

The second generalization of the density based means is produced by modifying the weights. We note that any set of (non-negative) functions G_i which depend only on the distances will ensure the result is a shift-invariant and hence a weakly monotone function. Instead of adding all pairwise distances as in [AY13], we use a more localised estimator of density, such as the variable kernel density estimators [TS92] and the nearest neighbour estimator. These density estimators are more accurate than the ones based on all pairwise distances, and can hence capture the notion of the density better. Specifically, such estimators limit the influences of the inputs far away from the point where the density is measured.

The nearest neighbour density estimator can be expressed using a simple modification to the formula (7.15). Let us fix i and k and order the distances (squared or otherwise) $d_{i,(1)}(\mathbf{x}) \leq d_{i,(2)}(\mathbf{x}) \leq \cdots \leq d_{i,(n)}(\mathbf{x})$. Now the kth nearest neighbour density estimator is

$$u_i^k(\mathbf{x}) = \frac{k}{2nd_{i,(k)}(\mathbf{x})}, \tag{7.21}$$

with typical $k \approx \sqrt{n}$.

A generalization of the above formula is obtained by using a kernel K:

$$u_i^{K,k}(\mathbf{x}) = \frac{1}{nd_{i,(k)}} \sum_{j=1}^{n} K\left(\frac{|x_i - x_j|}{d_{i,(k)}}\right). \tag{7.22}$$

A detailed analysis of various other non-parametric density estimators is presented in [TS92].

Finally, one can also use a trimmed mean of the distances instead of the mean in (7.15), i.e.,

$$u_i^T(\mathbf{x}) = K\left(\frac{1}{k}\sum_{j=1}^{k}(d_{i,(j)}(\mathbf{x}))^2\right), \tag{7.23}$$

or the median of the distances $d_{ij}(\mathbf{x}), j = 1, \ldots, n$, for a fixed i,

$$u_i^{med}(\mathbf{x}) = K\left(median((d_{ij}(\mathbf{x}))^2)\right) = K\left((d_{i,(\lceil \frac{n}{2} \rceil)}(\mathbf{x}))^2\right), \tag{7.24}$$

in order to obtain a robust estimator of the density [Rou84, RL03]. Here only the contribution of the k nearest data (or at most half of the data in the case of median distance) is accounted for. This way the outliers are discarded in density calculations, and hence the weights become more robust.

The weighting functions w_i are still computed by

$$w_i(\mathbf{x}) = \frac{u_i(\mathbf{x})}{\sum_{j=1}^{n} u_j(\mathbf{x})}. \tag{7.25}$$

In all cases mentioned above the weighting functions will remain constant under translations and therefore the resulting density based mean will be shift-invariant and bound-preserving.

7.7 Mode-Like Averages

The mode, discussed in Sect. 7.3.1, is based on minimising the quasi-penalty function

$$P(\mathbf{x}, y) = \sum_{i=1}^{n} p(x_i, y), \quad \text{where} \quad p(x_i, y) = \begin{cases} 0, & \text{if } x_i = y \\ 1 & \text{otherwise} \end{cases}. \quad (7.26)$$

For a finite number of inputs the mode is rather inconvenient as, for instance, if all the inputs are distinct then any input can be taken as the value of the mode. That is, the quasi-penalty P will have n distinct minimizers. In this section we discuss modifications to the mode function which will make it better suited as a representative value of the input vector. We will follow the approach in [BLVW15, Wil13].

Our interest is in identifying a sufficiently compact and large cluster of inputs whose representative value (such as the center) can be taken as a representative of the whole input vector. Firstly, we aim at removing the outliers from consideration, and therefore the representative value is calculated not over the whole input but over a cluster. Secondly, we want to identify a range where the inputs are frequent, so that the output is representative of this range. Lastly, we also require the output to be representative of a sufficiently large number of inputs, which may or may not be a majority. These requirements are not dissimilar to those in the density based means approach in Sect. 7.6, but here we use a different construction method.

We should note that the two requirements: having a large number of inputs in the cluster, and having a compact cluster, can be contradictory. Consider the inputs presented in Fig. 7.2. Here we can identify a very compact group of 3 inputs on the right and a larger but less compact group on the left. Which of those clusters is more representative? On the one hand the cluster on the left represents the majority, but the cluster on the right is more compact and consistent (which can be thought of as a vocal minority). The distribution from which these inputs were generated is in fact bimodal (also shown in the figure), and its (main) mode is on the right. The averaging methods we considered so far will not identify the mode as it is not representative of the majority of the inputs, and, of course, using the mode function is pointless for inputs that are all distinct.

Fig. 7.2 The density of a bimodal distribution and the data generated from it

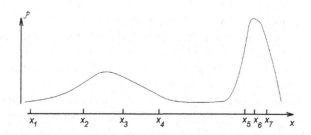

We shall model the mentioned requirements of having a large and compact cluster by using a penalty function based on a weighted sum of partial penalties given by

$$P(\mathbf{x}, y) = \sum_{i=1}^{n} w_i(y) p(x_i, y) \tag{7.27}$$

where

$$p(x_i, y) = \begin{cases} r_i, & \text{if } r_i \leq \tau, \\ \beta\tau, & \text{if } r_i > \tau, \end{cases} \tag{7.28}$$

and where

$$\tau = \alpha \max(\varepsilon, r_{(k)}) \text{ and } \alpha > 0, 0 \leq \beta \leq 1, 2 \leq k \leq n.$$

The parameters of this model are described below.

Let us denote the residuals by $r_i = |x_i - y|$. Then $r_{(k)}$ denotes the kth smallest element of the set of ordered (ascending) values of r_i, given the aggregate value y. Now, let us fix the value y for the moment. The penalty for the deviation of y from x_i is based on a threshold τ which is computed from an estimate of the density of the inputs near y, namely using the distance to the kth closest datum to y multiplied by a constant α (the small parameter ε is technical, to avoid a zero threshold when k inputs coincide).

Hence if the input x_i is within α times distance to the kth input closest to y, the penalty is the distance r_i, whereas in the other case is it β times the threshold. Importantly, the increase in r_i does not lead to further increase of the penalty, as x_i is thought of as an outlier. The threshold τ controls the spread of the cluster, and β imposes a penalty for the data excluded from that cluster. In LTS and LTA type averages the outliers are simply excluded, provided that the cluster has at least half of the data, whereas here the cluster can have fewer inputs yet we apply a penalty for clusters of smaller size. The larger β, the more emphasis is on the cluster size versus compactness. The parameter k controls how many inputs (closest to y) are used in estimating the local density (and define τ). The parameter α tunes the cluster boundary as a function of the minimum non-zero difference between the inputs. In the limiting case of $\beta = 0, \alpha = 1, \varepsilon = 0$ and $k = \lfloor \frac{n+1}{2} \rfloor$ we obtain the LTA function. When $\alpha = 0$ we obtain the mode function.

By minimizing the total penalty P we select the output y representative of a compact (as the differences r_i will be smaller for compact clusters) yet sufficiently large cluster of inputs (as the terms $\beta\tau$ will penalise small clusters). The minimization over y can be performed by taking y from the set $\{x_1, \ldots, x_n\}$ to make numerical computations faster, in which case we obtain a locally internal average.

Also notice that we included the dependence of the weighting function w_i on y. This done to accommodate further information about the relations between the inputs, not reflected in their values. For instance, in the works [BLVW15, Wil13] the presented mode-like function was used to average the intensities of the pixels in the image reduction task. Since the pixels in an image are also related geometrically, it makes sense to account for such relations through the weights w_i. The inputs close to each other both spatially and in the intensity space were favored. If no such extra relation is specified, no weights w_i are necessary.

The penalty based function produced by minimizing P is neither monotone, nor continuous (nor it is intended to be), but it is certainly weakly monotone, as the terms p depend on the distances between the output and the inputs. It is also clearly an average, for one being a locally internal function, but also because the values $y < x_{(1)}$ and $y > x_{(n)}$ result in higher penalty than those at $y = x_{(1)}$ and $y = x_{(n)}$ respectively.

The penalties in (7.28) can be further modified by using other ways of estimating the local density. For example, we can define the threshold τ as α times the mean of k smallest differences $r_{(i)}$. We can also use the squared partial penalties p^2 in (7.27). The values y at which P is computed can be selected as the midpoints/medians of every contiguous subset of the inputs, instead of the inputs themselves.

7.8 Spatial-Tonal Filters

In image processing, the well known class of spatial-tonal filters includes the mode filter [WB01], bilateral filter [TM98] and anisotropic diffusion [PM90] among others. This is an important class of filters developed to preserve edges within images when performing tasks such as filtering or smoothing. While these filters are commonly expressed in integral notation over a continuous space, they are implemented in discrete form over a finite set of pixels that take on finite values in a closed interval. It can be shown that this class of functions is given (in discrete form) by the averaging function

$$f_{\mathbf{w}}^g(\mathbf{x}; x_1) = \frac{\sum_{i=1}^n w_i g(|x_i - x_1|) x_i}{\sum_{i=1}^n w_i g(|x_i - x_1|)}, \tag{7.29}$$

where the weights w_i are nonlinear and non-convex functions of the locations of the pixels, which have intensity x_i. In all practical problems the locations are constant and hence can be pre-computed to produce the constant weight vector $\mathbf{w} = (w_1, w_2, \ldots, w_n)$. The pixel x_1 is the pixel to be filtered/smoothed such that its new value is $\bar{x}_1 = f_{\mathbf{w}}^g(\mathbf{x}; x_1)$.

The function $f_{\mathbf{w}}^g$ is nonlinear and not monotone. It is easily shown to be expressed as a penalty-based function with penalty

$$P(\mathbf{x}, y) = \sum_{i=1}^{n} w_i g(|x_i - x_1|)(x_i - y)^2.$$

In image filtering applications it is known that this penalty minimizes the mean squared error between the filtered image and the noisy source image [Ela02]. By Theorem 7.10 it follows directly that the filter $f_{\mathbf{w}}^g$ is shift-invariant and hence weakly monotone. Furthermore, Theorem 7.10 permits us to generalize this class of filters to be those penalty based averaging functions having penalty function

$$P(\mathbf{x}, y) = \sum_{i=1}^{n} w_i g(|x_i - f(\mathbf{x})|)(x_i - y)^2, \qquad (7.30)$$

or even further using other bivariate function $D : \mathbb{I}^2 \to \mathbb{R}$ (as discussed in Note 7.11)

$$P(\mathbf{x}, y) = \sum_{i=1}^{n} w_i g(|x_i - f(\mathbf{x})|)D(x_i, y), \qquad (7.31)$$

for some averaging function f.

The implication of replacing x_1 with $f(\mathbf{x})$ in the scaling function g is that we may use any shift-invariant aggregation of $\mathbf{x}$, which allows us to account for the possibility that x_1 is itself an outlier within the local region of the image. For example, we could use the median, the mode or the shorth for $f(\mathbf{x})$. This provides an interesting result and invites further research in the application of weakly monotone means to spatial-tonal filtering and smoothing problems.

7.9   Transforms

One important question related to weak monotonicity is its preservation under a φ-transform. We already know that weak monotonicity is preserved under linear transforms, and in particular under standard negation. We also know that standard monotonicity is preserved under the φ-transform when φ is a monotone bijection. In this section we discuss nonlinear φ-transforms of weakly monotone functions. Our results are based on [BCW15, WB15, WBC14].

For the φ-transform, if φ is nonlinear then f may or may not be weakly monotone for all $\mathbf{x}$, which can be observed by example.

Example 7.35 Take $\mathbf{x} = (1, 8, 16, 35, 47.9)$ and $\varphi(t) = \sqrt{t}$, then $\varphi(\mathbf{x}) = (1, 2\sqrt{2}, 4, \sqrt{35}, \sqrt{47.9})$ and $\varphi(\mathbf{x}+1) = (\sqrt{2}, 3, \sqrt{17}, 6, \sqrt{48.9})$. If f is the shorth (we proved the weak monotonicity of the shorth in Sect. 7.3) then $f(\varphi(\mathbf{x})) = 5.61$ and $f(\varphi(\mathbf{x}+1)) = 2.84$. As $\varphi^{-1}(t) = t^2$ clearly $5.61^2 > 2.84^2$ and f_φ is not weakly monotone.

It turns out that weak monotonicity is not generally preserved, although of course for specific weakly monotone functions this property will be preserved for some specific choices of φ.

Proposition 7.36 *The only functions φ which preserve weak monotonicity of all weakly monotone functions are linear functions.*

Proof Let us find an example of a shift-invariant (and hence weakly monotone) function whose φ-transform is not weakly monotone in general. Take $D(t) = \sin(kt)$, $k > 0$ and let $f(x, y) = D(y - x) + \frac{x+y}{2}$, which is obviously shift-invariant. Taking derivatives of f we obtain

$$f_x = -k \cos(y - x) + \frac{1}{2}$$

and

$$f_y = k \cos(y - x) + \frac{1}{2}.$$

It follows that

$$f_x + f_y = 1 \geq 0.$$

Now consider a continuous piecewise differentiable φ-transform of f given by $f_\varphi(x, y) = \varphi^{-1}(f(\varphi(x), \varphi(y)))$ and the sum of its partial derivatives with respect to the inputs. We need only consider the function $f(\varphi(x), \varphi(y))$ to observe the result, in which case we find that

$$(f(\varphi(x), \varphi(y)))_x + (f(\varphi(x), \varphi(y)))_y$$
$$= D'(\varphi(y) - \varphi(x))(\varphi'(y) - \varphi'(x)) + \frac{\varphi'(x) + \varphi'(y)}{2}$$
$$= k \cos(k(\varphi(y) - \varphi(x)))(\varphi'(y) - \varphi'(x)) + \frac{\varphi'(x) + \varphi'(y)}{2}.$$

The first term is zero when $\varphi'(x) = \varphi'(y)$ for all x, y. That is, φ' is constant and hence φ is a linear function. In the other cases if the first term is positive for (x, y) then it is negative for (y, x). The derivative φ' is bounded at least for some choices of x and y, so the second term is bounded, but the first term can be made arbitrarily large in absolute value by selecting a sufficiently large k, and negative by exchanging x and y if necessary. Hence the sum of the partial derivatives can be negative for some pairs (x, y) and thus $f_\varphi(x, y)$ is not always weakly monotone. $\square$

Note that in Proposition 7.56 we could take the product $f(x, y) = D(y-x) \cdot (x+y)$ (which is weakly monotone but not shift-invariant) and obtain an analogous result.

It remains to be seen whether there are subclasses of weakly monotone functions which preserve weak monotonicity under some non-linear φ-transforms. The next result shows that for some shift-invariant functions, such as the shorth and LMS estimators, weak monotonicity is preserved under φ-transforms with increasing bijections φ such that $\ln(\varphi')$ is concave.

Consider the robust estimators of location that are calculated using the shortest contiguous sub-sample of $\mathbf{x}$ containing at least half of the values, see Sect. 7.3. These estimators are shift-invariant. The candidate sub-samples are the sets $X_k = \{x_j : j \in \{k, k+1, \ldots, k + \lfloor \frac{n}{2} \rfloor\}\}$, $k = 1, \ldots, \lfloor \frac{n+1}{2} \rfloor$. The length of each contiguous set is taken as $\|X_k\| = \left| x_{k+\lfloor \frac{n}{2} \rfloor} - x_k \right|$ and thus the index of the shortest sub-sample is

$$ k^* = \arg \min_i \|X_i\|, \quad i = 1, \ldots, \left\lfloor \frac{n+1}{2} \right\rfloor. $$

Under the translation $\bar{\mathbf{x}} = \mathbf{x} + a\mathbf{1}$ the length of each sub-sample is unaltered since $\|\bar{X}_k\| = \left| \bar{x}_{k+\lfloor \frac{n}{2} \rfloor} - \bar{x}_k \right| = \left| (x_{k+\lfloor \frac{n}{2} \rfloor} + a) - (x_k + a) \right| = \left| x_{k+\lfloor \frac{n}{2} \rfloor} - x_k \right| = \|X_k\|$ and thus k^* remains the same.

We shall need the following two technical lemmas, whose proofs are given in [BCW15].

Lemma 7.37 *Let φ be any twice differentiable bijection, with $\varphi(\mathbf{x}) = (\varphi(x_1), \ldots, \varphi(x_n))$ and $\varphi^{-1}(\mathbf{x}) = (\varphi^{-1}(x_1), \ldots, \varphi^{-1}(x_n))$. Then the function $\varphi\left(\varphi^{-1}(\mathbf{x}) + c\mathbf{1}\right)$ is concave if and only if: $\ln \varphi'$ is concave when $\varphi' > 0$; or, $\ln |\varphi'|$ is convex for $\varphi' < 0$, for every $c \in \mathbb{R}, c \geq 0$ and $\mathbf{x} \in \mathbb{I}^n$.*

Lemma 7.38 *Let $\mathbf{x}$ be ordered such that $x_i \leq x_j$ for $i < j$ and let X_i denote the subset $\{x_i, \ldots, x_{i+k}\}$ for some fixed k. Let $\Delta X_i = \|X_i\| = |x_{i+k} - x_i|$ denote the length of the interval containing X_i. If φ is a concave increasing function then for $i < j$, $\Delta X_j \leq \Delta X_i$ implies that $\Delta \varphi(X_i) \geq \Delta \varphi(X_j)$, where $\Delta \varphi(X_i) = |\varphi(x_{i+k}) - \varphi(x_i)|$.*

Proposition 7.39 *Consider a robust estimator of location f based on the shortest contiguous half of the data. Let $\varphi(\mathbf{x}) = (\varphi(x_1), \varphi(x_2), \ldots, \varphi(x_n))$, with φ any twice differentiable, strictly increasing invertible function such that $\ln \varphi'$ is concave. Then $f(\mathbf{x} + a\mathbf{1}) \geq f(\mathbf{x})$ implies $\varphi^{-1}(f(\varphi(\mathbf{x} + a\mathbf{1}))) \geq (\varphi^{-1} f(\varphi(\mathbf{x})))$.*

Proof Denote by $\mathbf{y} = \varphi(\mathbf{x})$ and $\mathbf{x} = \varphi^{-1}(\mathbf{y})$ (functions are applied component wise) and we have $\varphi' > 0$. We need to show that

$$ f(\psi_c(\mathbf{y})) = f(\varphi(\varphi^{-1}(\mathbf{y}) + c\mathbf{1})) \geq f(\mathbf{y}). $$

By Lemma 7.37 function ψ_c is concave. By Lemma 7.38 we have $\Delta X_i \leq \Delta X_j \Rightarrow \Delta \psi_c(X_i) \leq \Delta \psi_c(X_j)$ for $i \geq j$. Hence the starting index of the shortest half cannot decrease after the transformation ψ_c and the result follows directly. $\square$

Corollary 7.40 *Let f be a robust estimator of location based on the shortest contiguous half of the data. Let $\varphi(\mathbf{x}) = (\varphi(x_1), \varphi(x_2), \ldots, \varphi(x_n))$, with φ any twice*

Table 7.1 Selected φ-transforms that preserve weak monotonicity of robust estimators of location

$\varphi(x)$	Increasing for	$y = \ln \varphi'(x)$	y''	Concave
e^{ax}	$a > 0$	$\ln a + ax$	0	$x \in \mathbb{R}$
e^{ax^2}	$a > 0$	$\ln 2a + \ln x + ax^2, x > 0$	$2\vert a\vert - \frac{1}{x^2}$	$x \in \mathbb{R}^+$
$e^{\frac{(x-a)^2}{b^2}}$	all $a, b \in \mathbb{R}, x \le a$	$\ln\left(-\frac{2}{b^2}(x-a)\right) - \frac{(x-a)^2}{b^2}$	$\frac{-1}{x-a} - \frac{2}{b^2}$	$x \in (0, \frac{1}{\sqrt{2a}}]$
x^p	$x \ge 0, p > 0$	$\ln p + (p-1)\ln x$	$\frac{1-p}{x^2}$	$p > 1, x \ge 0$
$\frac{e^x + e^{-x}}{2}$	$x \ge 0$	$\ln \frac{1}{2} + \ln\left(e^x - e^{-x}\right)$	$1 - \frac{(e^x + e^{-x})}{(e^x - e^{-x})}$	$x \ge 0$

Table 7.2 Selected φ-transforms that do not preserve weak monotonicity of robust estimators of location, because $y'' > 0$

$\varphi(x)$	Increasing for	$y = \ln \varphi'(x)$	y''
$\ln x$	$x > 0$	$-\ln x$	$1/x^2$
$\frac{e^x - e^{-x}}{2}$	$x \in \mathbb{R}$	$\ln \frac{1}{2} + \ln\left(e^x + e^{-x}\right)$	$1 - \frac{(e^x - e^{-x})}{(e^x + e^{-x})}$
$\tan x$	$x \in \mathbb{R}/\{(2n+1)\frac{\pi}{2} : n \in \mathbb{N}\}$	$\ln\left(1 + \tan^2 x\right)$	$2(1 + \tan^2 x)$

differentiable, strictly decreasing invertible function such that $\ln \vert\varphi'\vert$ *is convex. Then the φ-dual of f is weakly monotone.*

Proposition 7.39 serves as a simple test to verify that a given φ-transform preserves weak monotonicity of averages such as the shorth and LMS estimator. For example $\varphi(x) = e^x$ preserves weak monotonicity while $\varphi(x) = \ln(x)$ does not.

Example 7.41 Table 7.1 shows some cases of functions φ that preserve weak monotonicity of robust estimators under the φ-transform, which follows from Proposition 7.39. Table 7.2 shows examples of φ that do not preserve weak monotonicity of the robust estimators of location under the φ-transform.

We remark that while the robust estimators of location are shift-invariant, their nonlinear φ-transforms are not, hence we obtain a much broader class of weakly monotone functions than shift-invariant functions.

7.10 Cone Monotone Functions

7.10.1 Formal Definitions and Properties

In this section we examine the spectrum of averaging functions which lie in between standard and weak monotonicity. For example, one kind of monotonicity would be monotonicity with respect the *majority* of the inputs, or perhaps, monotonicity with respect to a certain *coalition* of inputs. An important question is how we represent and model such types of monotonicity. Here we aim at developing a mathematical formalism to systematically deal with various kinds of monotonicity, interpret them and examine their properties. Specifically we discuss the relation between the directional, cone monotonicity and monotonicity with respect to groups of variables following [BCW14].

Let us examine a spectrum of functions between monotone and weakly monotone functions. Consider a cone C lying in the positive octant $\mathbb{R}_+^n = \{\mathbf{x} \in \mathbb{R}^n | x_i \geq 0, i = 1, \ldots, n\}$.

Definition 7.42 (*Directional monotonicity*) Let $\mathbf{u} \in R_+^n$ be a non-zero vector. A function $f : \mathbb{I}^n \to \mathbb{I}$ is called **directionally monotone** in the direction $\mathbf{u}$ if

$$f(\mathbf{x} + a\mathbf{u}) \geq f(\mathbf{x})$$

for any $a > 0$, such that $\mathbf{x}, \mathbf{x} + a\mathbf{u} \in \mathbb{I}^n$.

Definition 7.43 (*Cone monotonicity*) Let $C \subseteq \mathbb{R}_+^n$ be a nonempty cone. A function $f : \mathbb{I}^n \to \mathbb{I}$ is called **cone monotone** with respect to C if f is directionally monotone in any direction $\mathbf{u} \in C$. The set of functions monotone with respect to C will be denoted by Mon_C.

Remark 7.44 Evidently weakly monotone functions are directionally monotone in the direction $\mathbf{u} = (1, 1, \ldots, 1)$. Standard monotone functions are cone monotone for $C = \mathbb{R}_+^n$.

Let us visualise the notion of cone monotonicity. The key here is a suitable representation of a cone. Rather than dealing with (usually infinite) sets of vectors representing a cone C, we can use a set $\mathcal{C}$ representing a cone. One way of doing it is to use the intersection of C with the unit sphere, as illustrated in Fig. 7.3. The cone C is defined by the set of unit vectors $\{\mathbf{u} | \mathbf{u} \in \mathcal{C}\}$. However a more convenient way is to use the intersection of C with the hyperplane $P = \{\mathbf{x} \in \mathbb{R}^n | x_1 + x_2 + \cdots, x_n = 1\}$. In this case we get a flat set $\mathcal{C} = C \cap P$, illustrated in Fig. 7.4.

Consider a few examples.

Fig. 7.3 Intersection of a cone C with the unit sphere

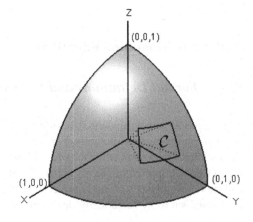

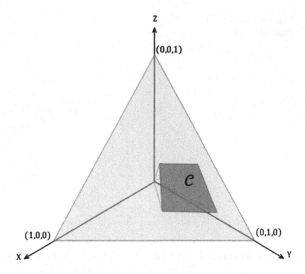

Fig. 7.4 Intersection of a cone C with the plane P

Example 7.45 Let $\mathcal{U}$ be the convex hull of $\{e_1, \ldots, e_n\}$. Then functions $f \in Mon_U$ are the usual monotone increasing functions.

Example 7.46 Let $C = \{(\frac{1}{n}, \ldots, \frac{1}{n})\}$. Then functions $f \in Mon_C$ are weakly monotone functions.

The next proposition is quite simple, but very helpful for the analysis of cone monotonicity.

Proposition 7.47 *Let $\mathcal{A}$ be a set of vectors in P. If a function f is cone monotone with respect to $\mathcal{A}$, then it is also cone monotone with respect to the convex hull $CH(\mathcal{A})$.*

Consequently it makes sense to discuss monotonicity only with respect to convex subsets of $\mathcal{U}$. In the sequel CH will denote a convex hull. The subsequent propositions are also helpful.

Proposition 7.48 *If $f \in Mon_C$ then any positive multiple $af \in Mon_C$, $a \geq 0$. Conversely, if $f \notin Mon_C$ then $af \notin Mon_C$, $a \geq 0$.*

Proposition 7.49 *If $f_1, f_2 \in Mon_C$ then a positive combination $af_1 + bf_2 \in Mon_C$, $a, b \geq 0$.*

Proposition 7.50 *If $f_1 \in Mon_C$ and $f_2 \in Mon_D$, then a positive combination $af_1 + bf_2 \in Mon_{C \cap D}$, $a, b \geq 0$.*

The above results naturally extend to positive combinations of more than two functions.

Proposition 7.51 *If $C \subset D$ then $f \in Mon_D \Rightarrow f \in Mon_C$, i.e. $Mon_D \subset Mon_C$.*

Fig. 7.5 The set $\mathcal{C}$ in
Example 7.54

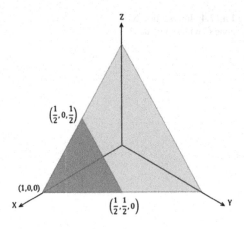

As the convex hull of two sets $\mathcal{A}$ and $\mathcal{B}$ contains their union, we have

Proposition 7.52 *If* $f \in Mon_A$ *and* $f \in Mon_B$, *then* $f \in Mon_{A \cup B}$.

As a consequence of the properties of the directional derivative we have

Proposition 7.53 *Let* f *be a differentiable function. Then* $f \in Mon_C$ *if and only if* $\mathbf{u}^T \nabla f(\mathbf{x}) \geq 0$ *for all* $\mathbf{x} \in \mathbb{I}^n$ *and* $\mathbf{u} \in C$.

Example 7.54 Let $n = 3$ and $C = CH(\{(1, 0, 0), (\frac{1}{2}, \frac{1}{2}, 0), (\frac{1}{2}, 0, \frac{1}{2})\})$, see Fig. 7.5. The function $f(\mathbf{x}) = x_1 - x_2 - x_3$ belongs to Mon_C. Furthermore, linear functions of the form $f(\mathbf{x}) = ax_1 + bx_2 + cx_3$ belong to Mon_C iff a, b, c satisfy $a \geq 0, a + b \geq 0$ and $a + c \geq 0$, which follows from Proposition 7.53.

The next proposition justifies the name *weak monotonicity* we discussed earlier.

Proposition 7.55 *If* f *is a symmetric function which is cone monotone (for any C), then it is weakly monotone.*

Proof Take any $\mathbf{u} \in C$. By symmetry, f is also directionally monotone with respect to $\mathbf{v} = \mathbf{u}_\sigma$, for any permutation σ of $(1, 2, \ldots, n)$. But the sum over all permutations $\sum_\sigma \mathbf{u}_\sigma = K(1, 1, \ldots, 1)$ for some constant $K > 0$. By Propositions 7.47 and 7.48 f is monotone with respect to $(1, 1, \ldots, 1)$. □

The cone containing just one vector $\mathbf{1}$ is the smallest nonempty cone that a symmetric function can be monotone with respect to. Thus we have the weakest type of cone monotonicity for symmetric functions.

The next results relate to preservation of directional monotonicity under φ-transforms.

Proposition 7.56 *The only functions* φ *which preserve directional monotonicity of all weakly monotone functions from* Mon_C, $C = \{\mathbf{u}\}$, *are linear functions.*

Corollary 7.57 *If f is directionally monotone with respect to $\mathbf{u}$, its dual f_d with respect to the standard negation is also directionally monotone with respect to $\mathbf{u}$.*

Corollary 7.58 *Function $f \in Mon_C$ if and only if its dual f_d (with respect to the standard negation) $f_d \in Mon_C$.*

7.10.2 Verification of Cone Monotonicity

For differentiable functions verification of cone monotonicity can be performed by calculating the gradient of f. Indeed to check cone monotonicity of a given f we need to ensure that $D_u(f) = \mathbf{u}^T \nabla f \geq 0$ for all vectors $\mathbf{u} \in C$. Note that if C is a polytope, then verification is needed only for its vertices, as the inequality will follow automatically for all linear convex combinations of the vertices.

Example 7.59 Consider Lehmer means $f = L^q$ given by (7.1). It was established in [WB15] that for $q > 1$ the partial derivatives are bounded as follows

$$L^q_{x_{\sigma(1)}}(\mathbf{x}) \geq \frac{x^q_{\sigma(1)}}{\sum x^q_i}, \quad L^q_{x_i}(\mathbf{x}) \geq -\frac{x^q_{\sigma(1)}}{\sum x^q_i}\left(\frac{q-1}{q+1}\right)^{q-1}, \quad i = 1, \ldots, n, \ i \neq \sigma(1),$$

where $x_{\sigma(1)} \geq \cdots \geq x_{\sigma(n)}$. Then for a vector $\mathbf{u}$ such that $\sum u_i = 1$, $u_i \geq 0$ the directional derivative is

$$D_u(L^q)(\mathbf{x}) = c x^q_{\sigma(1)}\left(u_{\sigma(1)} - (1 - u_{\sigma(1)})\left(\frac{q-1}{q+1}\right)^{q-1}\right),$$

for some $c = \dfrac{1}{\sqrt{\sum u_i^2 \sum x_i^q}} > 0$, and $u_{\sigma(1)}$ corresponding to the largest x_i. By resolving $D_u(f)(\mathbf{x}) \geq 0$ we get that

$$u_{\sigma(1)} \geq \frac{1}{1 + \left(\frac{q+1}{q-1}\right)^{q-1}} > 0.$$

Since the Lehmer mean is symmetric, it follows that any component u_i of $\mathbf{u}$ must satisfy the above inequality. Hence the Lehmer mean is cone monotone with respect to the set C defined by

$$u_i \geq \frac{1}{1 + \left(\frac{q+1}{q-1}\right)^{q-1}} > 0, \ i = 1, \ldots, n.$$

For example, for $q = 3$, it must be that all $u_i \geq \frac{1}{5}$. Since vectors with zero coordinates are not part of the set C, Lehmer mean is not majority monotone.

Non differentiable functions are more difficult to treat.

Example 7.60 Consider the robust estimators of location that are calculated using the shortest contiguous sub-sample of **x** containing at least half of the data, as described in Sect. 7.3. In particular, let us focus on the shorth. The shorth is shift-invariant and hence weakly monotone.

Let us show that the shorth is *proper weakly monotone*, i.e., it is not monotone with respect to any α-majority for which $\lfloor \alpha n \rfloor + 1 < n$.

Take these inputs $\mathbf{x} = (1, 5, 7, 9, 12)$. Then $Shorth(\mathbf{x}) = 7$. Now take $\mathbf{y} = \mathbf{x} + (3, 3, 0, 3, 3) = (4, 8, 7, 12, 15)$. Then $Shorth(\mathbf{y}) = 6 < 7$, i.e., an increase in four out of five inputs leads to a decrease of the value of the shorth function.

7.10.3 Construction of Cone Monotone Lipschitz Functions

We shall now establish tight pointwise bounds on functions from Mon_C based on the values of these functions at specified points. We will consider the set of Lipschitz functions with a fixed Lipschitz constant M, $Lip(M)$. Lipschitz properties of functions ensure that small changes in the arguments do not produce drastic changes in function values. The bounds we are about to establish will help us construct cone monotone functions based on a set of predefined values, using pointwise construction method presented in [Bel05, Bel06, Bel07, BCL07], see *Lipschitz approximation* section on p. 40 and Sect. 6.1.5.

Suppose that we have a number of data, the pairs $(\mathbf{x}^k, y^k), k = 1, \dots, K$ of inputs and desired outputs. Our goal is to construct an averaging cone monotone Lipschitz function f which interpolates the data. The Lipschitz condition guarantees that changes to the outputs are limited to a multiple of changes in the inputs, which is essential for numerical stability of averaging in applications.

Probelm 7.61 (*Pointwise construction of a cone monotone function*) Find a function $f \in Lip(M) \cap Mon_C$, such that $f(\mathbf{x}^k) = y^k$ for all $k = 1, \dots, K$ and $f(a, \dots, a) = a$, $f(b, \dots, b) = b$.

Of course we assume that such a function exists, i.e., the data set allows interpolation with a function from $Lip(M) \cap Mon_C$. If this is not the case (for instance the chosen Lipschitz constant M is too small), there are ways to smoothen the data by using quadratic or linear programming techniques, see [Bel05, Bel07].

The method developed in [Bel05, Bel06, BCL07] is based on the concept of optimal interpolation. Any interpolation algorithm A produces an error E_A no smaller than the intrinsic error of the problem $E_P = \inf_A E_A$. The intrinsic error is the radius of the set of possible solutions to Problem 7.61. The central algorithm delivers an interpolant whose error is E_p, and it consists in identifying the tight upper and lower bounds on the values of f at any $\mathbf{x}$, see [TW80]. Formally we obtain the solution to the problem

$$\min_{g \in \mathcal{F}} \max_{h \in \mathcal{F}} \max_{\mathbf{x} \in \mathbb{I}^n} |h(\mathbf{x}) - g(\mathbf{x})|$$

$$\text{subject to } g(\mathbf{x}^k) = h(\mathbf{x}^k) = y^k, k = 1, \ldots, K,$$

where $\mathcal{F} = \{f \in Lip(M) \cap Mon_C | f(a, \ldots, a) = a, f(b, \ldots, b) = b\}$. Among all functions from our set $\mathcal{F}$ that interpolate the data, we choose the one that minimises the largest possible difference to a function from $\mathcal{F}$ in the L_∞ norm. In other words, we select the function which delivers the smallest interpolation error in the worst case scenario.

Of course the set $\mathcal{F}$ can be further restricted by using other conditions, and specifically the averaging behavior. That results in the standard bounds $\min \le f \le \max$, which we will account for in the final formula.

In the case of standard monotonicity, the solution f to Problem 7.61 by the central algorithm is given as [Bel05, Bel06, Bel07] (see Sect. 6.1.5),

$$f(\mathbf{x}) = \frac{1}{2}(\underline{A}(\mathbf{x}) + \overline{A}(\mathbf{x})),$$

with

$$\underline{A}(\mathbf{x}) = \max\{\sigma_l(\mathbf{x}), B_l(\mathbf{x})\}, \quad \overline{A}(\mathbf{x}) = \min\{\sigma_u(\mathbf{x}), B_u(\mathbf{x})\}, \tag{7.32}$$

$$B_l(\mathbf{x}) = \max\{a, b - M||(b, \ldots, b) - \mathbf{x}||\}, \quad B_u(\mathbf{x}) = \min\{b, M||\mathbf{x}||\}, \tag{7.33}$$

$$\sigma_u(\mathbf{x}) = \min_k \{y^k + M||(\mathbf{x} - \mathbf{x}^k)_+||\},$$

$$\sigma_l(\mathbf{x}) = \max_k \{y^k - M||(\mathbf{x}^k - \mathbf{x})_+||\}, \tag{7.34}$$

where z_+ denotes the positive part of vector $\mathbf{z}$: $\mathbf{z}_+ = (\bar{z}_1, \ldots, \bar{z}_n)$, with

$$\bar{z}_i = \max\{z_i, 0\}.$$

Note that the solution is the mean of the tight upper and lower bounds, which are composed from the bounds σ_u, σ_l which result from the interpolation conditions $f(\mathbf{x}^k) = y^k$, and the generic bounds B_u, B_l which are the consequences of the boundary conditions in the Definition 1.5.

We now elaborate on the method from [Bel05, Bel06, Bel07] by changing from standard to cone monotonicity. The bounds σ_u, σ_l will change to the following

$$\sigma_u^C(\mathbf{x}) = \min_k \{y^k + M||(\mathbf{x} - C_-(\mathbf{x}^k))||\},$$

$$\sigma_l^C(\mathbf{x}) = \max_k \{y^k - M||(C_+(\mathbf{x}^k) - \mathbf{x})||\}, \tag{7.35}$$

where $C_-(\mathbf{x}^k)$ and $C_+(\mathbf{x}_k)$ denote the following cones centered at $\mathbf{x}^k$

$$C_-(\mathbf{a}) = \{\mathbf{x} \in \mathbb{R} | \mathbf{a} - \mathbf{x} \in C\}$$

and

$$C_+(\mathbf{a}) = \{\mathbf{x} \in \mathbb{R} | \mathbf{x} - \mathbf{a} \in C\}.$$

The constraints due to averaging behaviour are two extra bounds min and max. There could be other constraints imposed by the symmetry or values of f on the opposite diagonal, marginals, etc. specified in [Bel07, BC07, BCL07].

While the formal replacement of bounds σ_u, σ_l by σ_u^C, σ_l^C seems to be straightforward, it has important implications at the algorithmic level. The bounds σ_u, σ_l in the case of standard monotonicity are easily computed by simply taking the norm of the positive components of a vector. To calculate the distances to the cones C_-, C_+ is technically more challenging. Depending on the norm used, the following techniques are suitable.

First, note that the required distance to a cone C can be replaced by a distance to the polytope $C \cup \mathbf{0}$, in case the intersection of C with the plane P, C, is a polytope, e.g., when C is given as a convex hull of a finite number of points. For the Euclidean norm, there is Wolfe's algorithm [Wol76] which uses the vertex representation of a polytope, and therefore is quite convenient. In case of l_1 norm, the distance can be also calculated using linear programming methods.

Thus, pointwise construction of cone monotone functions using a set of data is certainly feasible using existing algorithms, although we should note that the algorithms needed are more sophisticated and computationally expensive compared to the case of standard monotonicity.

7.11 Monotonicity with Respect to Coalitions

7.11.1 Simple Majority

Next we tackle another type of monotonicity. Here we rely mostly on semantic interpretations of monotonicity. The technical aspects are best handled by using cone monotonicity, therefore it will be useful to determine the relation between the two concepts and define the methods for passing from one framework to another.

In the case of standard monotonicity the output cannot decrease if any of the inputs is increased (and none of the others decreases). In contrast, weak monotonicity is interpreted in this way: only if all the inputs are increased (by the same amount) the output cannot decrease.

It is helpful to consider temporarily strictly monotone functions, where an increase in one (or a group of) input(s) leads to an increase in the output (if the rest of the inputs are not decreased). Then standard monotonicity corresponds to dictatorship: each input is a "dictator", as an increase in any input produces the corresponding increase in the output. Weak monotonicity corresponds to unanimity: all inputs must increase. In between these two extremes lies the majority vote. The output increases only if a majority of inputs are increased by the same amount.

Definition 7.62 (*Monotonicity with respect to majority*) A function f is called monotone with respect to the (simple) majority of the inputs, if $f(\mathbf{x} + a\mathbf{u}) \geq f(\mathbf{x})$ for any $\mathbf{u}$ having at least $\lfloor \frac{n}{2} \rfloor + 1$ positive equal components, and $\mathbf{x}, \mathbf{x} + a\mathbf{u} \in \mathbb{I}^n$.

Let us formally represent the concept of monotonicity with respect to the majority of inputs using cone monotonicity. Let us define m vectors $\mathbf{u}_i$ of the form $\frac{1}{q}(\underbrace{1, \ldots, 1}_{q \text{ times}}, \underbrace{0, \ldots, 0}_{n-q \text{ times}})$ with exactly q ones and $n - q$ zeros, where $q = \lfloor \frac{n}{2} \rfloor + 1$.

There are $m = \binom{n}{q}$ such vectors. Then define further $\binom{n}{q+1}$ vectors with $q + 1$ ones and $n - q - 1$ zeros. Continue this way until we get the vector with n ones, $\mathbf{1}$. In other words, we define sets of vectors $U_j, j = 1, \ldots, \lceil \frac{n}{2} \rceil$ so that each set has all (properly scaled) vectors with $q = \lfloor \frac{n}{2} \rfloor + j$ ones and the rest zeros. Next define $\mathcal{C} = CH(\bigcup_j U_j)$. The functions monotone with respect to the majority are precisely cone monotone functions from $Mon_{\mathcal{C}}$.

Note that $\frac{1}{n}(1, 1, \ldots, 1) \in \mathcal{C}$, therefore the set $Mon_{\mathcal{C}}$ is a subset of weakly monotone functions. Also note that the set $\mathcal{C}$ is composed form vectors of the form $c_1\mathbf{u}_1 + c_2\mathbf{u}_2 + \cdots c_p\mathbf{u}_p$, with $\sum c_i = 1, c_i \geq 0$ and $p = \binom{n}{q} + \binom{n}{q+1} + \cdots + 1$, i.e. all convex combinations of the vertices of the polytope $\mathcal{C}$. Hence polytope $\mathcal{C}$ includes vectors with more than q positive components, which could be either all equal, as in the definition, or not equal due to using the convex hull.

On the other hand, if exactly q inputs increase, they must increase by the same amount. We can easily see this from the fact that C contains just one point in which the first q components are positive and the remaining $n - q$ are zeros. The positive components are all the same, hence the increase of the first q inputs must be by the same amount.

Notice that if the restrictions on the sizes of positive components are not specified, we fall back to the standard monotonicity in the limit, as one would be able to select the directional vectors $\mathbf{u}$ such that one component approaches 1, $q - 1$ components approach 0, an the remaining are 0, which would technically be a vector with q positive components, but infinitely close to the vector $\mathbf{e}_i$. The closure of the set $\mathcal{C}$ would be $\mathcal{U}$.

Example 7.63 Let $n = 3$. Take $\mathbf{u}_1 = \frac{1}{2}(1, 1, 0)$, $\mathbf{u}_2 = \frac{1}{2}(1, 0, 1)$, $\mathbf{u}_3 = \frac{1}{2}(0, 1, 1)$, and construct the convex hull of these vectors. The illustration is provided in Fig. 7.6. One function which is monotone with respect to this cone is

$$f(\mathbf{x}) = x_{(1)} + x_{(2)} - x_{(3)},$$

where the components of $\mathbf{x}_{()}$ are arranged in decreasing order. Function f is clearly symmetric, idempotent and is bounded by min from below. Its partial derivatives are

Fig. 7.6 The set $\mathcal{C}$ in
Example 7.63

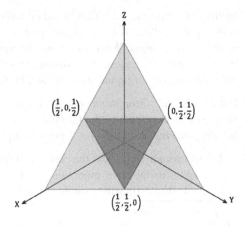

(1, 1, −1) in the domain of differentiability, and the directional derivative is non-negative with respect to the vectors $\mathbf{u}_i$, and hence this function is cone-monotone with respect to the convex hull of these vectors.

The majority can be quantified as half of the inputs, or any given proportion, e.g., 80 % majority. Another possibility is to have at least $\alpha\%$ of the inputs increasing to ensure f does not decrease. We can provide the following.

Definition 7.64 (*Monotonicity with respect to quantile*) A function f is called monotone with respect to the α-quantile of the inputs, if $f(\mathbf{x} + a\mathbf{u}) \geq f(\mathbf{x})$ for any $\mathbf{u}$ having at least $\lfloor \alpha n \rfloor + 1$ positive equal components and $\mathbf{x}, \mathbf{x} + a\mathbf{u} \in \mathbb{I}^n$.

Similarly to the case of the simple majority, we represent this type of monotonicity, which we abbreviate as α-monotonicity, using cone monotonicity. Let us define a collection of vectors $\{\mathbf{u}_i\}$ in which every vector has exactly $q = \lfloor \alpha n \rfloor + 1$ ones and $n - q$ zeros. There are $m = \binom{n}{q}$ such vectors. Add vectors $\mathbf{u}_i$ which have exactly $q + 1$ ones, and the rest zeros, then vectors with $q + 2$ ones and so on. Then build the polytope $\mathcal{C} = CH(\{\mathbf{u}_i\})$ and define $Mon_{\mathcal{C}}$. We obtain the set of α-monotone functions. Again we emphasize that the amounts by which the inputs increase satisfy certain proportions, derived from using the convex hull of vectors we just defined.

7.11.2 Majority and Preferential Inputs

Let us now consider slightly more complex situations by specifying different polytopes $\mathcal{C}$ using interesting subsets of vertices. We start with the following subset.

As in the case of simple majority, we use $m = \binom{n}{q}$ (multiples of) vectors $\mathbf{u}_i$ with exactly q ones and $n-q$ zeros. Let $\mathbf{u}_{m+1} = (1, 0, \ldots, 0)$. In the case of three variables, this set is illustrated in Fig. 7.7. Here as long as either the first input increases, or the majority of inputs increases, the value of f must not decrease. That is, the first

Fig. 7.7 The set C in Example 7.65

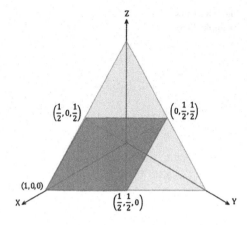

input has preferential treatment, perhaps due to its importance. But this input is not a dictator, because F is not directionally monotone with respect to every vector $\mathbf{u}$ with a positive first component, see Example 7.65. However, the first input acts almost as a veto player, as with the exception of just a few vectors, the first component of $\mathbf{u}$ must be non-negative, also illustrated in Example 7.65.

Example 7.65 Let $n = 3$. Take $\mathbf{u}_1 = \frac{1}{2}(1, 1, 0)$, $\mathbf{u}_2 = \frac{1}{2}(1, 0, 1)$, $\mathbf{u}_3 = \frac{1}{2}(0, 1, 1)$ and $\mathbf{u}_4 = (1, 0, 0)$, and construct the convex hull C of these vectors. The illustration is provided in Fig. 7.7. Note that vector $\mathbf{u} = (\frac{1}{4}, 0, \frac{3}{4}) \notin C$, hence the first input is not a dictator. On the other hand, with the exception of $\mathbf{u} = \mathbf{u}_3$, all $\mathbf{u} \in C$ have a nonzero first component, hence input one is almost a veto player, whose veto is overwritten only by the unanimity of the other two inputs.

7.11.3 Coalitions

Now we are ready to select any coalition $\mathcal{A}$ of inputs and specify monotonicity with respect to that coalition. Consider a 0-1 fuzzy measure $M : 2^{\mathcal{N}} \rightarrow \{0, 1\}$, $\mathcal{N} = \{1, 2, \ldots, n\}$. Those values of M that are ones specify coalitions of inputs with respect to which we need to ensure monotonicity. Let χ denote the characteristic function of a set and $|\mathcal{A}|$ denote the set cardinality.

Definition 7.66 (*Monotonicity with respect to coalition*) Let M be a 0-1 (Dirac) fuzzy measure. A function f is called monotone with respect to the coalitions in M, if $f(\mathbf{x} + a\mathbf{u}) \geq f(\mathbf{x})$ for any $\mathbf{u}$ whose components are defined by $u_i = \frac{1}{|\mathcal{A}|}\chi(\mathcal{A})M(\mathcal{A})$.

Remark 7.67 Monotonicity of fuzzy measures with respect subset inclusion ensures that coalition-monotone functions are weakly monotone. Also recall that the cone C is constructed using the convex hull of vectors $\mathbf{u}_i$. Since fuzzy measures are monotone, we have that f must be directionally monotone with respect to pointwise maxima of

Fig. 7.8 The set $\mathcal{C}$ in
Example 7.68

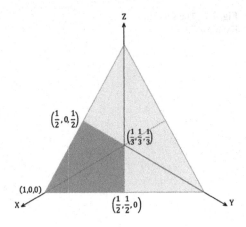

vectors defined by the non-zero values of M. That is, if $\mathbf{u}, \mathbf{v}$ correspond to subsets
$\mathcal{A}, \mathcal{B}$ for which the value of M is one, then we also have vector $\mathbf{w} = \mathbf{u} \vee \mathbf{v}$, which is
not necessarily in the $CH(\{\mathbf{u}, \mathbf{v}\})$.

Example 7.68 Let $n = 3$. Take $M(\{1\}) = 1$, $M(\{1, 2\}) = M(\{1, 3\}) = M$
$(\{1, 2, 3\}) = 1$ and the rest of the values are zeros. The illustration of the set $\mathcal{C}$ is
provided in Fig. 7.8. Note the difference to Fig. 7.7. Adding the value $M(\{2, 3\}) = 1$
to our fuzzy measure will change $\mathcal{C}$ to the one plotted in Fig. 7.7.

Example 7.69 Monotonicity with respect to majority can be obtained by specifying
a symmetric fuzzy measure such that $M(\mathcal{A}) = 1$ for any subset with cardinality $|\mathcal{A}| \geq$
$\lfloor \frac{n}{2} \rfloor + 1$. Similarly, α-monotonicity is obtained using symmetric fuzzy measures with
$|\mathcal{A}| \geq q = \lfloor \alpha n \rfloor + 1$.

One could think of many other examples of monotone functions with respect
to coalitions. For example, an input may not be important by itself but only in the
presence of some other inputs, in which case monotonicity with respect the corresponding singleton is not needed, but monotonicity with respect to certain coalitions
involving that input matters.

An important point made in this section is a formal technique to define (and verify)
monotonicity with respect to subsets of inputs, based on cone monotonicity. First
we identify a set of *"basis"* vectors $\mathbf{u}_i$ using the formula in Definition 7.66. Then
we need to construct the convex hull $\mathcal{C}$ of the basis vectors, an important step that
could be mistakenly omitted. After that the construction or verification process can
be completed using $\mathcal{C}$.

Monotonicity with respect to coalitions, and as its special cases, monotonicity
with respect to a majority of the inputs, has an attractive interpretation in the context
of decision making, voting and other problems. This type of monotonicity can be
converted to cone monotonicity, which is better suited for formal analysis. One key
result implies that only convex cones need to be considered. All coalition-monotone
functions are necessarily weakly monotone, which again justifies that notion as the
weakest type of monotonicity.

7.12 Directional Monotonicity

The concept of weak monotonicity can be further extended if we consider monotonicity along general rays. This idea leads to the notion of directional monotonicity presented in [BFKM15].

Definition 7.70 (*r-increasing function*) Let $\mathbf{r} = (r_1, \ldots, r_n)$ be a real n-dimensional vector, $\mathbf{r} \neq \mathbf{0}$. A function $f : \mathbb{I}^n \to \mathbb{I}$ is $\mathbf{r}$-increasing if for all points $(x_1, \ldots, x_n) \in \mathbb{I}^n$ and for all $c > 0$ such that $(x_1 + cr_1, \ldots, x_n + cr_n) \in [0, 1]^n$ it holds

$$f(x_1 + cr_1, \ldots, x_n + cr_n) \geq f(x_1, \ldots, x_n) .$$

That is, an $\mathbf{r}$-increasing function is a function which is increasing along the ray (direction) determined by the vector $\mathbf{r}$. For this reason, we say that f is directionally monotone, or, more specifically, directionally increasing. Note that in Definition 7.42 only the directions within the first octant were allowed. Every increasing function (in the usual sense) is, in particular, $\mathbf{r}$-increasing for every non-negative real vector $\mathbf{r}$. However, the class of directionally increasing functions is much wider than that of aggregation functions. For instance:

- Fuzzy implication functions (see [BBS03]) are $(-1, 1)$-increasing functions. This implies that many other functions, which are widely used in applications and which can be obtained from implication functions, are also directionally increasing. This is the case, for instance, of some subsethood measures (see [BPB07]);
- Many functions used for comparison of data are also directionally increasing. In particular, functions based on component-wise comparison by means of the Euclidean distance $|x - y|$, such as restricted equivalence functions [BBP06];
- Weakly monotone increasing functions [WB15] are a particular case of directionally increasing functions, with $\mathbf{r} = (1, \ldots, 1)$.

Note 7.71 (i) For any $\alpha > 0$ the increasing/decreasing monotonicity of a function f in the direction $\alpha \mathbf{r}$ is equivalent to the same property in the direction $\mathbf{r}$.

(ii) If a function f has continuous first-order partial derivatives with respect to all variables (which ensures the differentiability of f) then it has the directional derivative in the direction of any $\mathbf{r} \neq \mathbf{0}$ at each point $\mathbf{x} \in \mathbb{I}^n$ admitting a positive constant $c > 0$ such that $\mathbf{x} + c\mathbf{r} \in \mathbb{I}$. Such a point will be called an admissible point). Consequently, f is $\mathbf{r}$-increasing ($\mathbf{r}$-decreasing) if and only if its directional derivative in the direction of $\mathbf{r}$ is non-negative (non-positive) at each admissible point $x \in \mathbb{I}^n$.

Example 7.72 The Łukasiewicz implication $I_L(x, y) = \min\{1, 1 - x + y\}$, is $(1, 1)$-increasing. Indeed, for all points $(x, y) \in [0, 1]^2$ and all $c > 0$ such that $(x, y) + c(1, 1) = (x+c, y+c) \in [0, 1]^2$, we have $I_L(x+c, y+c) = \min\{1, 1-x-c+y+c\} = \min\{1, 1 - x + y\} = I_L(x, y)$.

Note that I_L is also a $(1, 1)$-decreasing function. However, the Reichenbach implication $I_R(x, y) = 1 - x + xy$, is neither $(1, 1)$-increasing nor $(1, 1)$-decreasing.

Example 7.73 Consider the function $f : [0, 1]^2 \rightarrow [0, 1]$ given by $f(x, y) = x - \max\{0, x - y\}^2$. f is a continuous function, but not an aggregation function (its monotonicity in the first coordinate fails). So f is $(1, 1)$-increasing, $(0, 1)$-increasing, but it is neither $(1, 0)$-increasing nor $(1, 0)$-decreasing.

Example 7.74 The weighted Lehmer mean $L_\lambda^1 : \mathbb{I}^2 \rightarrow \mathbb{I}$, given by

$$L_\lambda^1(x, y) = \frac{\lambda x^2 + (1 - \lambda)y^2}{\lambda x + (1 - \lambda)y}$$

with the convention $0/0 = 0$, where $\lambda \in]0, 1[$, is $(1 - \lambda, \lambda)$-increasing.

7.12.1 Properties of r-Monotone functions

Proposition 7.75 *A function f is $\mathbf{r}$-decreasing if and only if f is $(-\mathbf{r})$-increasing.*

Now we consider whether a function f is $\mathbf{u}$-increasing whenever it is $\mathbf{r}$- and $\mathbf{s}$-increasing and $\mathbf{u}$ is a linear combination of $\mathbf{r}$ and $\mathbf{s}$.

Theorem 7.76 *Let $\mathbf{r}$ and $\mathbf{s}$ be n-dimensional vectors such that for a given $a, b \geq 0$ with $a + b > 0$, it holds that for each point $x \in \mathbb{I}^n$ and $c > 0$ such that $\mathbf{x} + c\mathbf{u} \in \mathbb{I}^n$, where $\mathbf{u} = a\mathbf{r} + b\mathbf{s}$, the points $\mathbf{x} + ca\mathbf{r}$ or $\mathbf{x} + cb\mathbf{s}$ are also in $\mathbb{I}^n$. Then, if a function f is both $\mathbf{r}$- and $\mathbf{s}$-increasing then it is also $\mathbf{u}$-increasing.*

Corollary 7.77 *Let $\mathbf{r} = (r_1, \ldots, r_n)$ and $\mathbf{s} = (s_1, \ldots, s_n)$ be n-dimensional vectors that satisfy the condition*

$$d_{\mathbf{r},\mathbf{s}} := card\{i \in \{1, \ldots, n\} \mid r_i s_i \geq 0\} \geq n - 1,$$

and let a function f be $\mathbf{r}$-increasing and $\mathbf{s}$-increasing. Then f is $\mathbf{u}$-increasing for each $\mathbf{u} = a\mathbf{r} + b\mathbf{s}$ with $a, b \geq 0$ and $a + b > 0$.

The conditions in Theorem 7.76 may be difficult to check. The next result provides an easier way to study directional monotonicity.

Theorem 7.78 *Let a function f have continuous first-order partial derivatives with respect to each variable. If f is $\mathbf{r}$-increasing and $\mathbf{s}$-increasing for some n-dimensional directions $\mathbf{r}$ and $\mathbf{s}$, then f is also $\mathbf{u}$ increasing for each $\mathbf{u} = a\mathbf{r} + b\mathbf{s}$, where $a = 0$, $b = 0$, $a + b > 0$.*

Regarding the preservation of directional monotonicity, we have the following results.

Proposition 7.79 *Let f be an $\mathbf{r}$-increasing function. If $\varphi : \mathbb{I} \rightarrow \mathbb{I}$ is an increasing (decreasing) function then $g = \varphi \circ f$ is an $\mathbf{r}$-increasing (decreasing) function.*

Proposition 7.80 *Let $f_1, \ldots, f_k$ be $\mathbf{r}$-increasing functions and let $f : \mathbb{I}^k \to \mathbb{I}$ be an increasing function. Then the function $g = f(f_1, \ldots, f_k)$ is an $\mathbf{r}$-increasing function.*

The next result concerns the directional monotonicity of transformed increasing functions.

Proposition 7.81 *Let f be an increasing function and $f_i : \mathbb{I} \to \mathbb{I}$, $i = 1, \ldots, n$, be monotone functions. If a function $g : \mathbb{I}^n \to \mathbb{I}$ is defined by*

$$g(x_1, \ldots, x_n) = f(f_1(x_1), \ldots, f_n(x_n)),$$

then g is $\mathbf{r}$-increasing for each vector $\mathbf{r} = (r_1, \ldots, r_n)$ such that:

$$\forall i \in \{1, \ldots, n\}, r_i = 0 \text{ if } f_i \text{ is increasing and } r_i = 0 \text{ if } f_i \text{ is decreasing.}$$

Regarding duality, we have the following property.

Proposition 7.82 *If a function $f : \mathbb{I}^n \to \mathbb{I}$ is $\mathbf{r}$-increasing then its dual is also $\mathbf{r}$-increasing. Moreover, both functions are $\mathbf{r}$-increasing with respect to the same set of vectors.*

A function f which is $\mathbf{r}$-increasing and also $(-\mathbf{r})$-increasing is called an $\mathbf{r}$-constant function. These functions can be characterized as follows.

Proposition 7.83 *A function $f : \mathbb{I}^n \to \mathbb{I}$ is $\mathbf{r}$-constant if and only if $f(\mathbf{x}) = f(\mathbf{y})$ for all $\mathbf{x}, \mathbf{r} \in \mathbb{I}^n$ such that $\mathbf{x} = \mathbf{y} + k\mathbf{r}$ for some $k \in \mathbb{R}$.*

For $n = 2$, we have the following complete characterization of $\mathbf{r}$-constant functions.

Proposition 7.84 *Let $\mathbf{r} = (r_1, r_2)$ be a non-null two-dimensional vector. Then $g : \mathbb{I}^2 \to \mathbb{I}$ is an $\mathbf{r}$-constant function if and only if there is a function $f : \mathbb{I} \to \mathbb{I}$ such that*

$$g(x, y) = f(h_{(r_2, -r_1)}(x, y))$$
$$= f\left(\frac{r_2 x - r_1 y + \max\{0, r_1\} - \min\{0, r_2\}}{|r_1| + |r_2|} \right).$$

7.12.2 The Set of Directions of Increasingness

Given a function $f : \mathbb{I}^n \to \mathbb{I}$, let $\mathcal{C}(f)$ denote the set of all real n-dimensional vectors $\mathbf{r} = (r_1, \ldots, r_n)$ for which f is $\mathbf{r}$-increasing (compare to the cone of monotonicity in Definition 7.43). There are functions f with an empty set of directions of increasingness, $\mathcal{C}(f) = \emptyset$.

Proposition 7.85 *Let* $f : \mathbb{I}^n \to \mathbb{I}$ *be a function with a strict local extreme at a point* $\mathbf{x}_0$ *in the interior of* $\mathbb{I}$. *Then* $\mathcal{C}(f) = \emptyset$.

If we consider the Lehmer mean L^1, we have that $\mathcal{C}(L) = \{(a, a) \mid a > 0\}$. Moreover, for the weighted Lehmer mean L_λ it holds that $\mathcal{C}(L_\lambda^1) = \{(a(1 - \lambda), a\lambda) \mid a > 0\}$.

For continuous piecewise linear functions we have the following characterization of directional increasing monotonicity.

Lemma 7.86 *If* $f : \mathbb{I}^n \to \mathbb{I}$ *is a continuous piecewise linear function determined by linear functions* $B_1, \ldots, B_k$, *then* $\mathcal{C}(f) = \cap_{j=1}^k \mathcal{C}(B_j)$.

Proposition 7.87 *Let a function* $f : \mathbb{I}^n \to \mathbb{I}$ *be affine, i.e.,* $f(\mathbf{x}) = b + \sum_{i=1}^n a_i x_i$. *Then*

$$\mathcal{C}(f) = \{(r_1, \ldots, r_n) \in \mathbb{R}^n \backslash \{\mathbf{0}\} \mid \sum_{i=1}^n a_i r_i = 0\},$$

and f *is* $\mathbf{r}$-*constant if and only if* $\sum_{i=1}^n a_i r_i = 0$.

Corollary 7.88 *Let* $\mathbf{w} = (w_1, \ldots, w_n)$ *be a weighting vector. Then, for the weighted arithmetic mean* $M_{\mathbf{w}}$ *it holds that*

$$\mathcal{C}(M_{\mathbf{w}}) = \{(r_1, \ldots, r_n) \in \mathbb{R}^n \backslash \{\mathbf{0}\} \mid \sum_{i=1}^n w_i r_i = 0\}.$$

Note that in particular OWA functions and Choquet integrals are instances of piecewise linear aggregation functions.

7.13 Pre-aggregation Functions

7.13.1 Definitions and Properties

The notions of weak monotonicity and directional monotonicity are crucial in order to weaken the monotonicity conditions required to aggregation-like functions. In particular, the latter has led to the notion of pre-aggregation function [Luc+16].

Definition 7.89 (*Pre-aggregation function*) A function $f : [0, 1]^n \to [0, 1]$ is said to be an n-ary pre-aggregation function if the following conditions hold:

(PA1) There exists a real vector $\mathbf{r} \in [0, 1]^n$ $(\mathbf{r} \neq \mathbf{0})$ such that f is $\mathbf{r}$-increasing.
(PA2) f satisfies the boundary conditions: $f(0, \ldots, 0) = 0$ and $f(1, \ldots, 1) = 1$.

Example 7.90 Some examples of pre-aggregation functions are the following.

(i) Consider the mode, $Mod(x_1, \ldots, x_n)$, see p. 256. The mode is $(1, \ldots, 1)$-increasing, and it is a particular case of pre-aggregation function which is not an aggregation function.

(ii) $f(x, y) = x - (\max\{0, x - y\})^2$ is, for instance, $(0, 1)$-increasing, and it is an example of a pre-aggregation function which is not an aggregation function.

(iii) Weakly increasing functions satisfying the boundary conditions for aggregation functions (see Sect. 7.3) are also pre-aggregation functions which need not be aggregation functions.

(iv) Take $\lambda \in]0, 1[$. The weighted Lehmer mean is a pre-aggregation function.

(v) Define $A, B : [0, 1]^2 \to [0, 1]$ by

$$A(x, y) = \begin{cases} x(1 - x), & \text{if } y \leq 3/4 , \\ 1 & \text{otherwise,} \end{cases}$$

and

$$B(x, y) = \begin{cases} y(1 - y), & \text{if } x \leq 3/4 , \\ 1 & \text{otherwise.} \end{cases}$$

Then both A and B are pre-aggregation functions which are not aggregation functions. In fact, A is $(0, a)$-increasing for any $a > 0$ but for no other direction $\mathbf{r} = (a, b)$, $b > 0$, while B is $(b, 0)$-increasing for any $b > 0$ but for no other direction $\mathbf{r} = (a, b)$, $a > 0$. However, $C = (A + B)/2$ is not a pre-aggregation function, just illustrating the fact that the class of all pre-aggregation functions with a fixed dimension n is not a convex class.

If f is a pre-aggregation function with respect to a vector $\mathbf{r}$ we just say that f is an $\mathbf{r}$-pre-aggregation function.

Note 7.91 Note that if $A : [0, 1]^n \to [0, 1]$ is an aggregation function, then A is also a pre-aggregation function. Pre-aggregation functions which are not aggregation functions will be called proper pre-aggregation functions.

Proposition 7.92 *Let $A : [0, 1]^m \to [0, 1]$ be an aggregation function. Let $f_i : [0, 1]^n \to [0, 1]$ ($i \in \{1, \ldots, m\}$) be a family of m $\mathbf{r}$-pre-aggregation functions for the same vector $\mathbf{r} \in [0, 1]^n$. Then, the function $A(f_1, \ldots, f_m)$ is also an $\mathbf{r}$-pre-aggregation function.*

7.13.2 Construction of Pre-aggregation Functions by Composition

Fix $n \in \mathbb{N}$. Let I be a proper subset of $\mathcal{N} = \{1, \ldots, n\}$ and consider that $\mathcal{I} = \{i_1, \ldots, i_k\}$ with $i_1 < \cdots < i_k$. For an n-tuple $\mathbf{x} = (x_1, \ldots, x_n) \in [0, 1]^n$, its

I-projection is a k-tuple $\mathbf{x}_I = (x_{i_1}, \ldots, x_{i_k})$, where k is the cardinality of $\mathcal{I}$. We will use I-projections $\mathbf{x}_I$ of points $\mathbf{x} \in [0, 1]^n$ and I-projections $\mathbf{r}_I$ of (geometrical) vectors $\mathbf{r} \in [0, 1]^n$ as well. Finally, for a function $f: [0, 1]^n \to [0, 1]$, let $\mathcal{C}^\uparrow(f) = \{\mathbf{r} \in [0, 1]^n \mid f \text{ is } \mathbf{r}\text{-increasing}\}$. Note that the zero vector is not excluded now.

Proposition 7.93 *Let* $\{\mathcal{I}_1, \ldots, \mathcal{I}_k\}$ *be a partition of* $\mathcal{N}$, $k > 1$. *For* $j \in \{1, \ldots, k\}$, *let* $n_j = |\mathcal{I}_j|$ *and consider functions* $f_j: [0, 1]^{n_j} \to [0, 1]$ *such that* $f_j(1, \ldots, 1) = 1$. *Then, for any aggregation function* $g: [0, 1]^k \to [0, 1]$, *the composite function* $h: [0, 1]^n \to [0, 1]$ *defined by*

$$h(\mathbf{x}) = g\left(f_1\left(\mathbf{x}_{I_1}\right), \ldots, f_k\left(\mathbf{x}_{I_k}\right)\right)$$

is $\mathbf{r}$-increasing *for any vector* $\mathbf{r} \in [0, 1]^n$ *such that* $\mathbf{r}_{I_j} \in \mathcal{C}^\uparrow(f_j)$, $j = 1, \ldots, k$, *and* $h(\mathbf{1}) = 1$. *Moreover, if there is a* $j_0 \in \{1, \ldots, k\}$ *such that* f_{j_0} *is a pre-aggregation function, and* 0 *is an annihilator of* g, *then the function* h *is a pre-aggregation function.*

Proof Clearly, $h(\mathbf{1}) = g\left(f_1\left(\mathbf{1}_{I_1}\right), \ldots, f_k\left(\mathbf{1}_{I_k}\right)\right) = g(1, \ldots, 1) = 1$. Moreover, if $f_{j_0}(0, \ldots, 0) = 0$ for some $j_0 \in \{1, \ldots, k\}$ and 0 is an annihilator of g, then

$$h(\mathbf{0}) = g\left(f_1\left(\mathbf{0}_{I_1}\right), \ldots, f_{j_0}\left(\mathbf{0}_{I_{j_0}}\right), \ldots, f_k\left(\mathbf{0}_{I_k}\right)\right)$$
$$= g\left(f_1\left(\mathbf{0}_{I_1}\right), \ldots, 0, \ldots, f_k\left(\mathbf{0}_{I_k}\right)\right) = 0.$$

Next, consider a vector $\mathbf{r} \in [0, 1]^n$ such that $\mathbf{r}_{I_j} \in \mathcal{C}^\uparrow(f_j)$ for each $j = 1, \ldots, k$. Then, for any $c > 0$ and $\mathbf{x} \in [0, 1]^n$ such that also $\mathbf{x} + c\mathbf{r} \in [0, 1]^n$, it holds that

$$h(\mathbf{x} + c\mathbf{r}) = g\left(f_1\left(\mathbf{x}_{I_1} + c\mathbf{r}_{I_1}\right), \ldots, f_k\left(\mathbf{x}_{I_k} + c\mathbf{r}_{I_k}\right)\right)$$
$$\geq g\left(f_1\left(\mathbf{x}_{I_1}\right), \ldots, f_k\left(\mathbf{x}_{I_k}\right)\right) = h(\mathbf{x}),$$

where the inequality follows from the increasing monotonicity of the aggregation function g, and the fact that $f_j\left(\mathbf{x}_{I_j} + c\mathbf{r}_{I_j}\right) \geq f_j\left(\mathbf{x}_{I_j}\right), j = 1, \ldots, k$.

Now, suppose that f_{j_0} is a pre-aggregation function, i.e., $f_{j_0}(0, \ldots, 0) = 0$ and f_{j_0} is $\mathbf{v}$-increasing for some non-zero vector $\mathbf{v} \in [0, 1]^{n_{j_0}}$. Due to the above mentioned facts h satisfies the boundary conditions and is directionally increasing in the direction of a non-zero vector $\mathbf{r} \in [0, 1]^n$ such that $\mathbf{r}_{I_{j_0}} = \mathbf{v}$ and $\mathbf{r}_{\mathcal{N} \setminus I_{j_0}} = (0, \ldots, 0)$, which proves that h is a pre-aggregation function. $\square$

Example 7.94 Let $n = 2$ and $\mathbf{v} = (v_1, v_2) \in]0, 1]^2$. For obtaining a proper pre-aggregation function which is $\mathbf{v}$-increasing, it is enough to consider the weighted Lehmer mean $L_\lambda^1: [0, 1]^2 \to [0, 1]$ with $\lambda = \frac{v_2}{v_1 + v_2}$, see Example 7.90(iv), given by

$$L_\lambda^1(x, y) = \frac{v_2 x^2 + v_1 y^2}{v_2 x + v_1 y}.$$

This fact and Proposition 7.93 allow us to construct a pre-aggregation function h which is directionally increasing in the direction of any a-priori given vector $\mathbf{0} \neq \mathbf{r} \in [0, 1]^n$.

Consider, for example, $n = 4$ and $\mathbf{r} = (0.5, 0.4, 0.3, 0.7)$. Let $g = \min$, $\mathcal{I}_1 = \{1, 3\}$, $\mathcal{I}_2 = \{2, 4\}$, $f_1 = L_{3/8}^1$, $f_2 = L_{7/11}^1$. Then $h \colon [0, 1]^4 \to [0, 1]$ given by

$$h(x_1, x_2, x_3, x_4) = \min \left\{ \frac{3x_1^2 + 5x_3^2}{3x_1 + 5x_3}, \frac{7x_2^2 + 4x_4^2}{7x_2 + 4x_4} \right\}$$

is an $\mathbf{r}$-increasing proper pre-aggregation function.

7.13.3 Choquet-Like Construction Method of Pre-aggregation Functions

By replacing the product operation in the definition of the Choquet integrals by other aggregation functions, we get a method to build pre-aggregation functions.

Let $v : 2^N \to [0, 1]$ be a fuzzy measure and $M : [0, 1]^2 \to [0, 1]$ be a function such that $M(0, x) = 0$ for every $x \in [0, 1]$. Taking as the basis the Choquet integral, we define the function $C_v^M : [0, 1]^n \to [0, n]$ by

$$C_v^M(\mathbf{x}) = \sum_{i=1}^{n} M\left(x_{(i)} - x_{(i-1)}, v\left(A_{(i)} \right) \right), \tag{7.36}$$

where $\mathcal{N} = \{1, \ldots, n\}$, $(x_{(1)}, \ldots, x_{(n)})$ is an increasing permutation on the input $\mathbf{x}$ with the convention that $x_{(0)} = 0$, and $A_{(i)} = \{(i), \ldots, (n)\}$ is the subset of indices of $n - i + 1$ largest components of $\mathbf{x}$.

We have the following result.

Theorem 7.95 Let $M : [0, 1]^2 \to [0, 1]$ be a function such that for all $x, y \in [0, 1]$ it satisfies $M(x, y) \leq x$, $M(x, 1) = x$, $M(0, y) = 0$ and M is (1,0)-increasing. Then, for any fuzzy measure v, C_v^M is a pre-aggregation function which is idempotent and averaging, i.e.,

$$\min(x_1, \ldots, x_n) \leq C_v^M(x_1, \ldots, x_n) \leq \max(x_1, \ldots, x_n).$$

Note 7.96 Under the constraints of Theorem 7.95, we cannot ensure the monotonicity of C_v^M, i.e., C_v^M is in general a proper pre-aggregation function. To see it, observe the following:

Table 7.3 Some pre-aggregation functions obtained using the t-norms

T-norm	Resulting pre-aggregation function
Minimum	$C_v^{\min}(\mathbf{x}) = \sum_{i=1}^{n} \min \left\{ x_{(i)} - x_{(i-1)}, v\left(A_{(i)}\right) \right\}$
Łukasiewicz	$C_v^{TL}(\mathbf{x}) = \sum_{i=1}^{n} \max \left\{ 0, x_{(i)} - x_{(i-1)} + v\left(A_{(i)}\right) - 1 \right\}$
Drastic product	$C_v^{D}(\mathbf{x}) = \sum_{i=1}^{n} \begin{cases} x_{(1)}, & \text{if } i = 1 \\ v\left(A_{(i)}\right), & \text{if } x_{(i)} - x_{(i-1)} = 1 \\ 0 & \text{otherwise} \end{cases}$
Nilpotent minimum	$C_v^{NM}(\mathbf{x}) = \sum_{i=1}^{n} \begin{cases} \min \left\{ x_{(i)} - x_{(i-1)}, v\left(A_{(i)}\right) \right\}, \\ \qquad \text{if } x_{(i)} - x_{(i-1)} + v\left(A_{(i)}\right) > 1 \\ 0 \qquad \text{otherwise} \end{cases}$
Hamacher	$C_v^{TH}(\mathbf{x}) = \sum_{i=1}^{n} \begin{cases} 0, \text{ if } x_{(i)} = x_{(i-1)} \text{ and } v\left(A_{(i)}\right) = 0 \\ \dfrac{(x_{(i)} - x_{(i-1)}) \cdot v(A_{(i)})}{x_{(i)} - x_{(i-1)} + v(A_{(i)}) - (x_{(i)} - x_{(i-1)}) \cdot v(A_{(i)})} \\ \qquad\qquad\qquad\qquad \text{otherwise} \end{cases}$

(i) Take $M(x, y) = \min(x, y)$. Consider $\mathcal{N} = \{1, 2, 3, 4\}$ and the cardinality measure v. Then, we have that

$$C_v^{\min}(0.05, 0.1, 0.7, 0.9) = 0.8, \quad \text{whereas} \quad C_v^{\min}(0.05, 0.1, 0.8, 0.9) = 0.7,$$

so $C_v^{\min}$ is not an increasing function.

(ii) Consider the Łukasiewicz t-norm $T_L(x, y) = \max\{0, x + y - 1\}$. Again, for $\mathcal{N} = \{1, 2, 3, 4\}$ and the cardinality measure v we have that

$$C_v^{TL}(0.05, 0.1, 0.7, 0.9) = 0.15, \quad \text{whereas} \quad C_v^{TL}(0.05, 0.2, 0.7, 0.9) = 0.05,$$

so C_v^{TL} is not an increasing function. Analogous counterexamples can be found for the cases of the drastic product, the Hamacher product or the nilpotent minimum t-norms taken as M.

Consider $\mathcal{N} = \{1, \ldots, n\}$ and a fuzzy measure $v : 2^N \to [0, 1]$. In Table 7.3, we present the value of C_v^T, which are pre-aggregation functions but not aggregation functions for the different t-norms.

7.13.4 Sugeno-Like Construction Method of Pre-aggregation Functions

Recall that the formula for the discrete Sugeno integral $S_v : [0, 1]^n \to [0, 1]$ can be written as

$$S_v(\mathbf{x}) = \bigvee_{i=1}^{n} \min\left\{x_{(i)}, v\left(A_{(i)}\right)\right\}.$$

Inspired by this formula, for any function $M: [0, 1]^2 \to [0, 1]$, we define the function $S_v^M: [0, 1]^n \to [0, 1]$ by the formula

$$S_v^M(\mathbf{x}) = \bigvee_{i=1}^{n} M\left(x_{(i)}, v\left(A_{(i)}\right)\right). \tag{7.37}$$

We prove a sufficient condition for M ensuring that S_v^M is a pre-aggregation function for any fuzzy measure v.

Proposition 7.97 *Let $M: [0, 1]^2 \to [0, 1]$ be a function increasing in the first variable and let for each $y \in [0, 1]$, $M(0, y) = 0$ and $M(1, 1) = 1$. Then S_v^M defined in (7.37) is a pre-aggregation function for any fuzzy measure v.*

Proof It is easy to check that, for any v,

$$S_v^M(\mathbf{0}) = \bigvee_{i=1}^{n} M\left(0, v\left(A_{(i)}\right)\right) = 0$$

and

$$S_v^M(\mathbf{1}) = \bigvee_{i=1}^{n} M\left(1, v\left(A_{(i)}\right)\right) = M(1, v(A_{(1)})) = M(1, 1) = 1.$$

Moreover, for vector $\mathbf{1} = (1, \dots, 1)$ we get

$$S_v^M(\mathbf{x} + c\mathbf{1}) = \bigvee_{i=1}^{n} M\left(x_{(i)} + c, v\left(A_{(i)}\right)\right) \geq \bigvee_{i=1}^{n} M\left(x_{(i)}, v\left(A_{(i)}\right)\right) = S_v^M(\mathbf{x}),$$

i.e., S_v^M is $\mathbf{1}$-increasing, which completes the proof that S_v^M is a pre-aggregation function. $\qquad\square$

Note that any function M satisfying the constraints of Proposition 7.97 is, in fact, a binary $(1, 0)$-increasing pre-aggregation function which satisfies $M(0, y) = 0$ for each $y \in [0, 1]$.

Example 7.98 (i) Let $M: [0, 1]^2 \to [0, 1]$ be any aggregation function. Then $S_v^M: [0, 1]^n \to [0, 1]$ is also an aggregation function, independently of v.
(ii) Consider the function $f(x, y) = x|2y - 1|$. Note that f is a proper pre-aggregation function which satisfies the constraints of Proposition 7.97, and thus, for any v, the function $S_v^f: [0, 1]^n \to [0, 1]$, $S_v^f(\mathbf{x}) = \bigvee_{i=1}^{n} f\left(x_{(i)}, v\left(A_{(i)}\right)\right)$ is a pre-aggregation function (even an aggregation function thought f is not).

For example, for $n = 2$, $v(\{1\}) = 1/3$, $v(\{2\}) = 3/4$, we get

$$S_v^f(x, y) = \begin{cases} x \vee \frac{y}{2}, & \text{if } x \leq y, \\ y \vee \frac{x}{3}, & \text{if } x > y. \end{cases}$$

References

[And+72]   D. Andrews et al., *Robust Estimates of Location: Surveys and Advances* (Princeton University Press, Princeton, 1972)
[AY13]   P. Angelov, R.R. Yager, Density-based averaging—a new operator for data fusion. Inf. Sci. **222**, 163–174 (2013)
[Baj58]   M. Bajraktarević, Sur une équation fonctionelle aux valeurs moyennes. Glasnik Mat.-Fiz. i Astronom. Drustvo Mat. Fiz. Hrvatske. Ser. II **13**, 243–248 (1958)
[Baj63]   M. Bajraktarević, Sur une genéralisation des moyennes quasilineaire. Publ. Inst. Math. Beograd. **3**(17), 69–76 (1963)
[BC93]   T. Bednarski, B.R. Clarke, Trimmed likelihood estimation of location and scale of the normal distribution. Aust. J. Stat. **35**, 141–153 (1993)
[Bel05]   G. Beliakov, Monotonicity preserving approximation of multivariate scattered data. BIT **45**, 653–677 (2005)
[Bel06]   G. Beliakov, Interpolation of Lipschitz functions. J. Comp. Appl. Math. **196**, 20–44 (2006)
[Bel07]   G. Beliakov, Construction of aggregation operators for automated decision making via optimal interpolation and global optimization. J. Ind. Manage. Optim. **3**, 193–208 (2007)
[BBP12]   G. Beliakov, H. Bustince, D. Paternain, Image reduction using means on discrete product lattices. IEEE Trans. Image Process. **21**, 1070–1083 (2012)
[BC07]   G. Beliakov, T. Calvo, Construction of aggregation operators with noble reinforcement. IEEE Trans. Fuzzy Syst. **15**, 1209–1218 (2007)
[BCL07]   G. Beliakov, T. Calvo, J. Lázaro, Pointwise construction of Lipschitz aggregation operators with specific properties. Int. J. Uncertain. Fuzziness Knowl.-Based Syst. **15**, 193–223 (2007)
[BCW14]   G. Beliakov, T. Calvo, T. Wilkin, Three types of monotonicity of averaging aggregation. Knowl.-Based Syst. **72**, 114–122 (2014)
[BCW15]   G. Beliakov, T. Calvo, T. Wilkin, On the weak monotonicity of Gini means and other mixture functions. Inf. Sci. **300**, 70–84 (2015)
[BLVW14]   G. Beliakov, G. Li, H.Q. Vu, T. Wilkin, Fuzzy measures of pixel cluster compactness, in *Proceedings of the 11th IEEE International Conference on Fuzzy Systems (FUZZ-IEEE)*. Beijing, China (2014)
[BLVW15]   G. Beliakov, G. Li, H.Q. Vu, T. Wilkin, Characterizing compactness of geometrical clusters using fuzzy measures. IEEE Trans. Fuzzy Syst. **23**, 1030–1043 (2015)
[BS15]   G. Beliakov, J. Spirková, Weak monotonicity of mixture functions with special weighting functions, under review (2015)
[BW14]   G. Beliakov, T. Wilkin, On some properties of weighted averaging with variable weights. Inf. Sci. **281**, 1–7 (2014)
[Bul03]   P.S. Bullen, *Handbook of Means and Their Inequalities* (Kluwer, Dordrecht, 2003)
[BBP06]   H. Bustince, E. Barrenechea, M. Pagola, Restricted equivalence functions. Fuzzy Sets Syst. **157**, 2333–2346 (2006)
[BBS03]   H. Bustince, P. Burillo, F. Soria, Automorphisms, negations and implication operators. Fuzzy Sets Syst. **134**, 209–229 (2003)
[BFKM15]   H. Bustince, J. Fernandez, A. Kolesárová, R. Mesiar, Directional monotonicity of fusion functions. Eur. J. Oper. Res. **244**, 300–308 (2015)

[BPB07]   H. Bustince, M. Pagola, E. Barrenechea, Construction of fuzzy indices from fuzzy
             DI-subsethood measures: application to the global comparison of images. Inf. Sci.
             177, 906–929 (2007)
[CBW14]   T. Calvo, G. Beliakov, T. Wilkin, On weak monotonicity of some mixture functions,
             in *Proceedings of ESTYLF'2014 Conference*. Zaragoza, Spain (2014)
[DAPT97]   P.L. De Angelis, P.M. Pardalos, G. Toraldo, Quadratic programming with box con-
             straints, in *Developments in Global Optimization*, ed. by I.M. Bomze, et al. (Kluwer
             Academic Publishers, Dordrecht, 1997)
[Ela02]   M. Elad, On the origin of the bilateral filter and ways to improve it. IEEE Trans.
             Image Process. **11**, 1141–1151 (2002)
[Gin58]   C. Gini, *Le Medie*. Milan (Russian translation, Srednie Velichiny, Statistica,
             Moscow, 1970): Unione Tipografico-Editorial Torinese (1958)
[HLn97]   A.S. Hadi, A. Luceño, Maximum trimmed likelihood estimators: a unified approach,
             examples, and algorithms. Comput. Stat. Data Anal. **25**, 251–272 (1997)
[HO99]   D. Hawkins, D. Olive, Applications and algorithms for least trimmed sum of absolute
             deviations regression. Comput. Stat. Data Anal. **32**, 119–134 (1999)
[KMM07]   A. Kolesárová, G. Mayor, R. Mesiar, Weighted ordinal means. Inf. Sci. **177**,
             3822–3830 (2007)
[Luc+16]   G. Lucca et al., Pre-aggregation functions: construction and anapplication. IEEE
             Trans. Fuzzy Syst. (in press). doi:10.1109/TFUZZ.2015.2453020 (2016)
[MPR03]   R.A. Marques Pereira, R.A. Ribeiro, Aggregation with generalized mixture opera-
             tors using weighting functions. Fuzzy Sets Syst. **137**, 43–58 (2003)
[MS06]   R. Mesiar, J. Spirková, Weighted means and weighting functions. Kybernetika **42**,
             151–160 (2006)
[MSV08]   R. Mesiar, J. Spirková, L. Vavríková, Weighted aggregation operators based on
             minimization. Inf. Sci. **178**, 1133–1140 (2008)
[PFBMB15]   D. Paternain, J. Fernandez, H. Bustince, R. Mesiar, G. Beliakov, Construction of
             image reduction operators using averaging aggregation functions. Fuzzy Sets Syst.
             261, 81–111 (2015)
[Ped09]   W. Pedrycz, Statistically grounded logic operators in fuzzy sets. Eur. J. Oper. Res.
             193, 520–529 (2009)
[PM90]   P. Perona, J. Malik, Scale space and edge detection using anisotropic diffusion.
             IEEE Trans. Pattern Anal. Mach. Intell. **12**(7), 629–639 (1990)
[Rez83]   B. Reznick, Some inequalities for products of power sums. Pac. J. Math. **104**,
             443–450 (1983)
[RMP03]   R. Ribeiro, R. Marques Pereira, Generalized mixture operators using weighting
             functions. Eur. J. Oper. Res. **145**, 329–342 (2003)
[Rou84]   P.J. Rousseeuw, Least median of squares regression. J. Am. Stat. Assoc. **79**, 871–880
             (1984)
[RL03]   P.J. Rousseeuw, A.M. Leroy, *Robust Regression and Outlier Detection* (Wiley, New
             York, 2003)
[š08]   J. Špirková, Weighted aggregation operators and their applications, in *Dissertation
             Thesis* (2008)
[TS92]   G.R. Terrell, D.W. Scott, Variable kernel density estimation. Ann. Stat. **20**,
             1236–1265 (1992)
[TM98]   C. Tomasi, R. Manduchi, Bilateral filtering for gray and color images, in *Proceedings
             of the 6th International Conference on Computer Vision*. Bombay, India (1998), pp.
             839–846
[TW80]   J.F. Traub, H. Wozniakowski, *A General Theory of Optimal Algorithms* (Academic
             Press, New York, 1980)
[WB01]   J. van de Weijer, R. van den Boomgaard, Local mode filtering, in *Computer Vision
             and Pattern Recognition*, vol. 2 (IEEE Computer Society, 2001), pp. 428–433
[Wil13]   T. Wilkin, Image reduction operators based on non-monotonic averaging functions,
             in *Proceedings of the 10th IEEE International Conference on Fuzzy Systems*. doi:10.
             1109/FUZZ-IEEE.2013.6622458. Hyderabad, India (2013)

[WB14]   T. Wilkin, G. Beliakov, Weakly monotone averaging functions. arXiv:1408.0328, (2014)

[WB15]   T. Wilkin, G. Beliakov, Weakly monotone averaging functions. Int. J. Intell. Syst. **30**, 144–169 (2015)

[WBC14]   T. Wilkin, G. Beliakov, T. Calvo, Weakly monotone averaging functions, in *Proceedings of IPMU 2014 Conference*. Montpellier, France (2014)

[Wol76]   P.H. Wolfe, Finding the nearest point in a polytope. Math. Progr. **11**, 128–149 (1976)

[YB10]   R.R. Yager, G. Beliakov, OWA operators in regression problems. IEEE Trans. Fuzzy Syst. **18**, 106–113 (2010)

Chapter 8
Averages on Lattices

Abstract This chapter covers averages defined on product lattices, in particular in the intuitionistic fuzzy sets and interval-valued fuzzy sets setting. Some general construction methods and special cases are outlined. The medians on lattices are discussed in detail and several alternative forms of the median function are presented.

8.1 Aggregation of Intervals and Intuitionistic Fuzzy Values

Many decision making applications involve aggregation of several fuzzy sets, particularly where fuzzy sets are used to express preferences [KY97, Yag88]. It has also been argued that extensions of fuzzy sets capture various uncertainties better [CM08, Men07, WM11]. Such extensions include the introduction of type-2 fuzzy sets and interval-valued fuzzy sets by Zadeh in 1973 [Zad73], vague sets [GB93], L-fuzzy sets [Gog67] see also [Bus10, BMPBG08, Li10a, MJL06, Zad08, Zad09]. Later in [Ata86], Atanassov introduced the idea of intuitionistic fuzzy sets (AIFS), see also [Ata99, Ata12]. In [Ata89, BB96, DBR00] the authors advanced the theory of operators and relations for Atanassov's intuitionistic fuzzy sets and interval-valued intuitionistic fuzzy sets.

As noticed in [AG89] and later developed in [BB96, DK03, DK07], AIFS are mathematically equivalent to interval-valued fuzzy sets (IVFS). Hence one can concentrate on the mathematical theories in one or the other representation without loss of generality.

Aggregation methods on AIFS or IVFS are an important and dynamic area, with many new applications to decision making problems and reaching consensus, see e.g., [BBCJB14, Li10b, NSMK11] and image processing [BBP12, GFBB11]. Recent results in this area relate to extensions of triangular norms and conorms [BBP08a, BMPBG08, Des07, DK05, DK08] and averaging functions [BBGMP11, TC10, Wei10, Xu07, Xu10, XY06, Yag09, ZXNL10]. The recent books [Ata12, Li14, Xu12, Xu14] are particularly good sources.

Besides AIFS and IVFS and some of their extensions, averaging on lattices arises in color image processing where the colors are coded by using red, green and blue (RGB) components (or alternative representations) [BBP12, Gal+13].

© Springer International Publishing Switzerland 2016
G. Beliakov et al., *A Practical Guide to Averaging Functions*,
Studies in Fuzziness and Soft Computing 329,
DOI 10.1007/978-3-319-24753-3_8

Similarly, averaging of multichannel signals in signal processing requires aggregation on lattices techniques.

8.1.1 Preliminary Definitions

We review several relevant concepts and highlight the correspondence between notions in AIFS and IVFS [Ata86].

Definition 8.1 (*Atanassov's intuitionistic fuzzy set*) An AIFS $\mathcal{A}$ on X is defined as $\mathcal{A} = \{\langle x, \mu_{\mathcal{A}}(x), \nu_{\mathcal{A}}(x)\rangle | x \in X\}$, where $\mu_{\mathcal{A}}(x)$ and $\nu_{\mathcal{A}}(x)$ are the degrees of membership and nonmembership of x in $\mathcal{A}$, which satisfy $\mu_{\mathcal{A}}(x), \nu_{\mathcal{A}}(x) \in [0, 1]$ and $0 \le \mu_{\mathcal{A}}(x) + \nu_{\mathcal{A}}(x) \le 1$.

Definition 8.2 (*Interval-valued fuzzy set*) An IVFS $\mathcal{B}$ on X is defined as $\mathcal{B} = \{\langle x, [l_{\mathcal{B}}(x), r_{\mathcal{B}}(x)]\rangle | x \in X\}$, where $l_{\mathcal{B}}(x)$ and $r_{\mathcal{B}}(x)$ are the lower and upper ends of the membership interval, and satisfy $0 \le l_{\mathcal{B}}(x) \le r_{\mathcal{B}}(x) \le 1$.

Obviously an ordinary fuzzy set with a membership function $\mu(x)$ can be written as $\{\langle x, \mu(x), 1 - \mu(x)\rangle | x \in X\}$, or as $\{\langle x, [\mu(x), \mu(x)]\rangle | x \in X\}$. AIFS can be represented by means of IVFS and vice versa. To ease the notation we suppress the dependence of the membership/non-membership values on x, i.e., we will just consider *Atanassov's intuitionistic fuzzy values* (AIFV), which are the membership and non-membership pairs for a fixed x denoted $A = \langle \mu_A, \nu_A\rangle$. For the interval-valued fuzzy sets, we will shorten the notation to consider only the *interval valued fuzzy values* (IVFV), which are the left and right terminals of the interval of membership, $B = [l_B, r_B]$.

Several indices are used to characterize AIFV.

Definition 8.3 (*Score, accuracy, indeterminacy*) The Score, Accuracy and degree of indeterminacy $\pi(A)$ of an AIFV A are defined by

$$Score(A) = \mu_A - \nu_A,$$
$$Accuracy(A) = \mu_A + \nu_A,$$
$$\pi(A) = 1 - (\mu_A + \nu_A).$$

We list the indices in Table 8.1 together with their counterparts in the IVFV representation. We use $\frac{l+r}{2}$ to denote the center of the IVFV interval $[l, r]$.

8.1.2 Aggregation on Product Lattices

The set of AIFV or IVFV is a special case of a product lattice. For this reason it makes sense to present the following results in the framework of aggregation on lattices in

Table 8.1 Various indices and operations on AIFV and IVFV

Index	AIFV representation	IVFV representation
Membership	$\langle \mu, \nu \rangle$	$[l, r] = [\mu, 1 - \nu]$
Degree of indeterminancy	$\pi = 1 - (\mu + \nu)$	$r - l = length([l, r])$
Score	$\mu - \nu$	$r + l - 1 = 2\left(\frac{l+r}{2}\right) - 1$
Accuracy	$\mu + \nu = 1 - \pi$	$l - r + 1 = 1 - length([l, r])$

general, and specify the instances of product lattices when necessary. We state some general definitions following [Bir67].

Definition 8.4 (*Poset*) A partial order is a binary relation $\leq$ over a set P which is reflexive, antisymmetric and transitive. A poset $(P, \leq)$ is a set P with a partial order $\leq$.

Definition 8.5 (*Chain*) A chain in a poset P is a subset $C \subseteq P$ in which each pair of elements is comparable. The *length* of a chain is given by the cardinality of the chain minus one.

Observe that any finite chain of k elements is order isomorphic to the poset $\{0, \ldots, k - 1\}$ with the usual order between integers.

Definition 8.6 (*Lattice*) A lattice $\mathcal{L} = (P, \leq, \wedge, \vee)$ is a poset in which every two elements have both meet and join, denoted by $\wedge, \vee$ respectively. The bottom and top elements of the lattice will be denoted by $0_{\mathcal{L}}$ and $1_{\mathcal{L}}$.

Note 8.7 In fact, if $\mathcal{L} = (P, \leq, \wedge, \vee)$ is a lattice, the following two items hold for any two elements $a, b \in \mathcal{L}$.

 (i) $a \leq b$ if and only if $a \wedge b = a$;
(ii) $a \leq b$ if and only if $a \vee b = b$.

Definition 8.8 (*Product lattice*) If $\mathcal{L}_1 = (P_1, \leq_1, \wedge_1, \vee_1)$ and $\mathcal{L}_2 = (P_2, \leq_2, \wedge_2, \vee_2)$ are two lattices, their Cartesian product is the lattice $\mathcal{L}_1 \times \mathcal{L}_2 = (P_1 \times P_2, \leq, \wedge, \vee)$ with $\leq$ defined by

$$(x_1, y_1) \leq (x_2, y_2) \Leftrightarrow x_1 \leq_1 x_2 \text{ and } y_1 \leq_2 y_2, \text{ and} \tag{8.1}$$

$$(x_1, y_1) \wedge (x_2, y_2) = (x_1 \wedge_1 x_2, y_1 \wedge_2 y_2), \tag{8.2}$$

$$(x_1, y_1) \vee (x_2, y_2) = (x_1 \vee_1 x_2, y_1 \vee_2 y_2). \tag{8.3}$$

Definition 8.9 (*Distributive lattice*) A lattice $\mathcal{L}$ is called distributive if $(x \vee y) \wedge z = (x \wedge z) \vee (y \wedge z)$ for all $x, y, z \in \mathcal{L}$.

Note 8.10 A Cartesian product of distributive lattices is distributive [Bir67].

Let us denote by $\mathcal{L}$ the lattice of non-empty intervals $\mathcal{L} = \{[a, b] | (a, b) \in [0, 1]^2, a \leq b\}$ with the partial order $\leq_{\mathcal{L}}$ defined as $[a, b] \leq_{\mathcal{L}} [c, d] \Leftrightarrow a \leq c$ and $b \leq d$. Observe that, from Note 8.7, this order defines the operations $\wedge$ and $\vee$ in $\mathcal{L}$. The top and bottom elements are respectively $1_{\mathcal{L}} = [1, 1], 0_{\mathcal{L}} = [0, 0]$.

The corresponding partial order in the AIFV representation $\leq_{\mathcal{L}_{AIFV}}$ is defined as $\langle \mu_1, \nu_1 \rangle \leq_{\mathcal{L}_{AIFV}} \langle \mu_2, \nu_2 \rangle \Leftrightarrow \mu_1 \leq \mu_2$ and $\nu_2 \leq \nu_1$. The top and bottom elements are $1_{\mathcal{L}_{AIFV}} = \langle 1, 0 \rangle$ and $0_{\mathcal{L}_{AIFV}} = \langle 0, 1 \rangle$ respectively. Note that $\mathcal{L}$ is a sublattice of the Cartesian product $[0, 1]^2$, and is therefore distributive.

An extension of Definition 1.5 to lattices can be stated as follows [DK05].

Definition 8.11 (*Aggregation function on a lattice*) An aggregation function on a lattice $f_L : \mathcal{L}^n \to \mathcal{L}$ is a function non-decreasing with respect to $\leq_{\mathcal{L}}$ and satisfying $f_L(0_{\mathcal{L}}, \ldots, 0_{\mathcal{L}}) = 0_{\mathcal{L}}$ and $f_L(1_{\mathcal{L}}, \ldots, 1_{\mathcal{L}}) = 1_{\mathcal{L}}$.

This directly leads to the following definition of aggregation functions for Atannassov intuitionistic fuzzy values.

Definition 8.12 (*Aggregation function for AIFV*) An aggregation function on AIFV $f_{AIFV} : \mathcal{L}_{AIFV}^n \to \mathcal{L}_{AIFV}$ is a function non-decreasing with respect to $\leq_{\mathcal{L}_{AIFV}}$ and satisfying $f_{AIFV}(0_{\mathcal{L}_{AIFV}}, \ldots, 0_{\mathcal{L}_{AIFV}}) = 0_{\mathcal{L}_{AIFV}}$ and $f_{AIFV}(1_{\mathcal{L}_{AIFV}}, \ldots, 1_{\mathcal{L}_{AIFV}}) = 1_{\mathcal{L}_{AIFV}}$.

There are many ways to construct such aggregation functions for AIFVs (and hence IVFVs). One way is to apply the usual aggregation functions to the ends of the intervals. Such functions are called representable in [DK08].

Definition 8.13 (*Representable aggregation function*) f_L is a *representable* aggregation function if there are aggregation functions f_1, f_2 such that $f_1 \leq f_2$, and f_L can be expressed as

$$f_L(B_1, \ldots, B_n) = [f_1(l_{B_1}, \ldots, l_{B_n}), f_2(r_{B_1}, \ldots, r_{B_n})] \tag{8.4}$$

where $B_i = [l_{B_i}, r_{B_i}], i = 1, \ldots, n$.

Definition 8.14 (*Natural extension*) A natural extension of an aggregation function f to $\mathcal{L}$ is a representable aggregation function f_L with $f_1 = f_2 = f$.

Various constructions of IV aggregation functions, including t-norms and t-conorms, uninorms and OWA are discussed in [DK08]. In the special case of OWA, Yager [Yag09] uses $f_1 = f_2 = OWA$.

Definition 8.15 (*Idempotent function for AIFV*) An aggregation function on AIFV $f_{AIFV} : \mathcal{L}_{AIFV}^n \to \mathcal{L}_{AIFV}$ is idempotent if $f_{AIFV}(A, \ldots, A) = A$ for every $x \in \mathcal{L}$.

Definition 8.16 (*Averaging function for AIFV*) An aggregation function on AIFV $f_{AIFV} : \mathcal{L}_{AIFV}^n \to \mathcal{L}_{AIFV}$ is averaging if

$$\inf\{A_1, \ldots, A_n\} \leq_{\mathcal{L}_{AIFV}} f_{AIFV}(A_1, \ldots, A_n) \leq_{\mathcal{L}_{AIFV}} \sup\{A_1, \ldots, A_n\}$$

for every $A_1, \ldots, A_n \in \mathcal{L}$.

The following proposition will be useful in the sequel.

Proposition 8.17 *For intuitionistic aggregation functions defined on $\mathcal{L}_{AIFV}$, idempotency (Definition 8.15) is equivalent to averaging behavior (Definition 8.16).*

Proof 1. Idempotency implies averaging behaviour : Let $B = \inf\{A_1, \ldots, A_n\} \Rightarrow B \leq_{\mathcal{L}_{AIFV}} A_i, \forall i$. From the monotonicity and idempotency of f_{AIFV}, we have

$$B = f_{AIFV}(B, B, \ldots, B) \leq_{\mathcal{L}_{AIFV}} f_{AIFV}(A_1, \ldots, A_n).$$

Hence f_{AIFV} is bounded from below by $\inf\{A_1, \ldots, A_n\}$.
On the other hand, if $B = \sup\{A_1, \ldots, A_n\}$, we have $A_i \leq_{\mathcal{L}_{AIFV}} B, \forall i$. Then

$$f_{AIFV}(A_1, \ldots, A_n) \leq_{\mathcal{L}_{AIFV}} f_{AIFV}(B, B, \ldots, B) = B.$$

Hence f_{AIFV} is bounded from above by $\sup\{A_1, \ldots, A_n\}$.
2. Averaging behaviour implies idempotency: Suppose $A_1 = A_2 = \cdots = A_n = B$. It follows that $\inf\{A_1, \ldots, A_n\} = \sup\{A_1, \ldots, A_n\} = B \Rightarrow f_{AIFV}(A_1, \ldots, A_n) = B$. $\qquad\qquad\square$

8.1.3 Arithmetic Means and OWA for AIFV

The Atanassov's intuitionistic weighted means and OWA functions were defined in [Xu07, XY06] based on the following arithmetic operations on AIFV [Ata94].

$$A \oplus B = \langle \mu_A + \mu_B - \mu_A\mu_B, \nu_A\nu_B \rangle, \tag{8.5}$$

$$A \otimes B = \langle \mu_A\mu_B, \nu_A + \nu_B - \nu_A\nu_B \rangle. \tag{8.6}$$

From these formulas one obtains the following equations [DBR00]

$$n\,A = \underbrace{A \oplus \ldots \oplus A}_{n\ times} = \langle 1 - (1 - \mu_A)^n, \nu_A^n \rangle, \tag{8.7}$$

$$A^n = \underbrace{A \otimes \ldots \otimes A}_{n\ times} = \langle \mu_A^n, 1 - (1 - \nu_A)^n \rangle, \tag{8.8}$$

for any $n = 1, 2, \ldots$, which can then be extended for positive real n.
Since OWA requires the sort operation, one needs to define a *total order* on AIFVs. The following total order on AIFVs was proposed in [Xu07, XY06].

$$A < B$$

if and only if   (8.9)

(i) $Score(A) < Score(B)$ or

(ii) $Score(A) = Score(B)$ and

$$Accuracy(A) < Accuracy(B).$$

Clearly, in the IVFV representation, this corresponds to ordering according to the center of the membership interval, and then (if the centers coincide) according to the length of the interval $(length([l_A, r_A]) > length([l_B, r_B]) \Rightarrow A < B)$.

An alternative *lexicographic* total order is: $A < B$ if and only if $l_A < l_B$ or $l_A = l_B$ and $r_A < r_B$. Obviously, many other total orders are possible. In particular, score and accuracy allow one to define a specific instance of the so-called admissible orders, that is, linear orders which extend the usual partial order $[a, b] \leq [c, d]$ if and only if $a \leq c$ and $b \leq d$. In particular, in [BFKM13] the following result appears.

Proposition 8.18 *Let* $f, g : [0, 1]^2 \to [0, 1]$ *be two continuous aggregation functions, such that for all* $(x, y), (u, v)$ *with* $x \leq y$ *and* $u \leq v$ *the equalities* $f(x, y) = f(u, v)$ *and* $g(x, y) = g(u, v)$ *can only hold if* $(x, y) = (u, v)$. *Define the relation* $[x, y] \leq_{f,g} [u, v]$ *if and only if* $f(x, y) < f(u, v)$ *or* $f(x, y) = f(u, v)$ *and* $g(x, y) \leq g(u, v)$. *Then* $\leq_{f,g}$ *is an admissible order between intervals.*

This result can be applied directly in the Atanassov intuitionistic fuzzy setting. Some applications of these admissible orders in decision making can be found in [BGBKM13].

Let us mention one undesirable property of the ordering (8.9): it is not preserved under multiplication by a scalar: $A < B$ does not necessarily imply $\lambda A < \lambda B$ where λ is a scalar, as can be seen from the Example 8.19 below. This has a profound implication on the lack of monotonicity of aggregation functions for AIFV defined below with respect to the chosen total ordering, as shown in Proposition 8.23.

Example 8.19 Take $A = \langle 0.5, 0.4 \rangle$, $B = \langle 0.3, 0.2 \rangle$ and $\lambda = 0.6$. Since $Score(A) = Score(B) = 0.1$ and $Accuracy(A) = 0.9$, $Accuracy(B) = 0.5$, then $B < A$. But $\lambda A = \langle 1 - (1 - 0.5)^{0.6}, 0.4^{0.6} \rangle \cong \langle 0.3402, 0.5771 \rangle$, $\lambda B = \langle 1 - (1 - 0.3)^{0.6}, 0.2^{0.6} \rangle \cong \langle 0.1927, 0.3807 \rangle$, and $Score(\lambda A) \cong -0.2369$, $Score(\lambda B) \cong -0.188$, so $\lambda A < \lambda B$. Thus $B < A$ does not imply $\lambda B < \lambda A$.

The following aggregation functions were defined on AIFV in [Xu07, XY06] using operations of addition and multiplication by a scalar.

Definition 8.20 (*Intuitionistic weighted arithmetic mean*) The intuitionistic weighted arithmetic mean with respect to a weighting vector **w**, $IWAM_{\mathbf{w}}$, is defined as

$$IWAM_{\mathbf{w}}(A_1, \ldots, A_n) = w_1 A_1 \oplus w_2 A_2 \oplus \cdots \oplus w_n A_n$$

$$= \langle 1 - \prod_{i=1}^{n}(1 - \mu_{A_i})^{w_i}, \prod_{i=1}^{n} \nu_{A_i}^{w_i} \rangle, \qquad (8.10)$$

with the convention $0^0 = 1$.

Definition 8.21 (*Intuitionistic OWA*) The intuitionistic ordered weighted averaging operator with respect to a weighting vector $\mathbf{w}$, $IFOWA_{\mathbf{w}}$, is defined as

$$IFOWA_{\mathbf{w}}(A_1, \ldots, A_n) = w_1 A_{\sigma(1)} \oplus w_2 A_{\sigma(2)} \oplus \cdots \oplus w_n A_{\sigma(n)}$$

$$= \langle 1 - \prod_{i=1}^{n}(1 - \mu_{A_{\sigma(i)}})^{w_i}, \prod_{i=1}^{n} \nu_{A_{\sigma(i)}}^{w_i} \rangle, \qquad (8.11)$$

where $A_{\sigma(i)}$ is the ith largest value according to the total order (8.9): $A_{\sigma(1)} \geq \cdots \geq A_{\sigma(n)}$.

One problem with these definitions is that they are not consistent with aggregation operations on the ordinary fuzzy sets (when $\mu = 1 - \nu$), nor with the natural extension of aggregation to IVFV (i.e. using representable aggregation functions, see Definition 8.13) [BBGMP11]. Consider Example 8.22.

Example 8.22 Three pieces of software, x_1, x_2, x_3 are examined by three malicious software detection systems $\mathcal{A}_1, \mathcal{A}_2, \mathcal{A}_3$ for security threats. The evaluations $\mu_{\mathcal{A}_i}(x_j)$ indicate to what degree the software contains code which is *potentially* malicious, while the values $\nu_{\mathcal{A}_i}(x_j)$ indicate to what extent the software is definitively not malicious. An AIFV of $\langle 0, 1 \rangle$ hence indicates with certainty that the software is safe, $\langle 1, 0 \rangle$ means that the software contains only code which is potentially malicious, while $\langle 0, 0 \rangle$ denotes complete ignorance. The evaluations given by each of the three systems are shown in Table 8.2.

It is desired that the evaluations of each system be aggregated to give an overall evaluation. A weighting vector $\mathbf{w} = (0.5, 0.3, 0.2)$ is used, in the case of the IWAM denoting the relative importance of the detection systems, and in the case of the IFOWA giving more importance to higher evaluations. The aggregated AIFV evaluations using IWAM and IFOWA are shown in Table 8.3, as well as the respective aggregations $\mu_{\mathcal{A}_i}(x_j)$ values only using the standard aggregation functions (i.e. for ordinary fuzzy sets).

Table 8.2 AIFV evaluations for software x_1, x_2, x_3 given by three malicious software detection systems (Example 8.22)

Software	$\mathcal{A}_1$	$\mathcal{A}_2$	$\mathcal{A}_3$
x_1	$\langle 0.3, 0.7 \rangle$	$\langle 0.5, 0.5 \rangle$	$\langle 0.9, 0.1 \rangle$
x_2	$\langle 0, 0 \rangle$	$\langle 0, 1 \rangle$	$\langle 0.8, 0.1 \rangle$
x_3	$\langle 1, 0 \rangle$	$\langle 0.3, 0.3 \rangle$	$\langle 0.5, 0.3 \rangle$

Table 8.3 Aggregation of AIFV evaluations using IWAM, IFOWA and standard evaluations using WAM and OWA (Example 8.22)

Software	$IWAM$	$IFOWA$	$WAM(\mu)$	$OWA(\mu)$
x_1	$\langle 0.571, 0.429 \rangle$	$\langle 0.761, 0.239 \rangle$	0.48	0.66
x_2	$\langle 0.275, 0 \rangle$	$\langle 0.553, 0 \rangle$	0.16	0.4
x_3	$\langle 1, 0 \rangle$	$\langle 1, 0 \rangle$	0.69	0.71

The overall evaluations for x_1 help illustrate the differences between these aggregation functions in their standard form and their definitions on AIFV. Note that for x_1 individual evaluations satisfy $\mu = 1 - \nu$. The aggregated AIFVs using $IWAM$ and $IFOWA$ also satisfy $\mu = 1 - \nu$, however the aggregated μ values are different to those obtained using the standard WAM and OWA contrary to our expectations.

For x_2, the evaluation of $\nu = 1$ given by $\mathcal{A}_2$ suggests that the software is definitively not malicious. The detection system $\mathcal{A}_1$ provides an evaluation which suggests complete ignorance. The evaluation of $\mu = 0$ indicates that no code has been found that is potentially malicious, however $\nu = 0$ also suggests that no code is confirmed to be safe. We see that the certainty of $\mathcal{A}_2$ is absorbed by $\mathcal{A}_1$ when it comes to the aggregated ν values, which are both zero.

In the case of x_3, we see that a single evaluation given by $\mathcal{A}_1$ dominates the others when aggregation is performed using $IWAM$ and $IFOWA$.

Another undesirable feature of (8.10) and (8.11) is that whenever one of the arguments $A_i = \langle 1, 0 \rangle$ and the corresponding weight is not 0, we have $IWAM_w(A_1, \ldots, A_n) = IOWA_w(A_1, \ldots, A_n) = \langle 1, 0 \rangle$, which is rather counterintuitive for an averaging operation. On the other hand, if one of the arguments is $\langle 0, 1 \rangle$ and its weight is not zero, it is not accounted for at all, which is again counterintuitive. The latter feature helps to prove that IWAM and IOWA in (8.10) and (8.11) are not monotone with respect to the order based on the score and accuracy.

Proposition 8.23 *Aggregation operators for AIFV defined by (8.10) and (8.11) are not monotone with respect to the ordering in (8.9).*

Proof Take $A_1 = \langle 0, 1 \rangle$, $A_2 = \langle 0.5, 0.4 \rangle$, $A_3 = \langle 0.3, 0.2 \rangle$ and $w = (0.4, 0.6)$. We know from Example 8.19 that $A_3 < A_2$. Monotonicity requires that $IWAM(A_1, A_3) < IWAM(A_1, A_2)$, but $IWAM(A_1, A) = 0.6A$ for any intuitionistic value A. From Example 8.19, $0.6A_2 < 0.6A_3$, so monotonicity does not hold. For IOWA the same argument applies. $\square$

Of course, both (8.10) and (8.11), and the functions defined in the next section are monotone with respect to the *partial order* $\leq_{\mathcal{L}_{AIFV}}$. However it raises an interesting question of whether there are alternative definitions of aggregation functions which are monotone with respect to the total order (8.9). The affirmative answer was provided in [BBGMP11], and we will present it in the next subsection.

8.1.4 Alternative Definitions of Aggregation Functions on AIFV

This section follows the work [BBGMP11]. Let us write the operation $A \oplus B$ on AIFV as follows

$$A \oplus B = \langle S(\mu_A, \mu_B), T(\nu_A, \nu_B) \rangle, \tag{8.12}$$

where $T = T_P$ is the product t-norm and $S = S_P$ is its dual t-conorm (called probabilistic sum), defined by $S_P(x, y) = 1 - T_P(1-x, 1-y)$. Here we can use any pair of dual t-norm and t-conorm. We first concentrate on continuous Archimedean t-norms and t-conorms, and in particular on the product t-norm, because it was used in many existing definitions. An Archimedean t-norm is strict if it is continuous and strictly increasing.

It is well known (see [KMP00]) that a strict Archimedean t-norm is expressed via its additive generator g as follows

$$T(x, y) = g^{-1}(g(x) + g(y)),$$

and the same applies to its dual t-conorm,

$$S(x, y) = h^{-1}(h(x) + h(y)),$$

with $h(t) = g(1 - t)$. We remind that an additive generator of a continuous Archimedean t-norm is a strictly decreasing function $g : [0, 1] \to [0, \infty]$ such that $g(1) = 0$. For nilpotent operations the inverse changes to the pseudoinverse. The additive generator is not unique, it is defined up to an arbitrary positive multiplier [GMMP09, KMP00]. For T_P, an additive generator is $g(t) = -\log(t)$.

Now, let $C = A \oplus B$. Then $g(\nu_C) = g(\nu_A) + g(\nu_B)$ and $h(\mu_C) = h(\mu_A) + h(\mu_B)$. Further, if $D = \lambda A$, then $g(\nu_D) = \lambda g(\nu_A)$, $h(\mu_D) = \lambda h(\mu_A)$.

Let us now denote by $\langle \hat{\mu}, \hat{\nu} \rangle = \langle h(\mu), g(\nu) \rangle$, the transformed membership/non-membership pair. In this notation, for AIFV $C = A \oplus B$ and $D = \lambda A$,

$$\langle \hat{\mu}_C, \hat{\nu}_C \rangle = \langle \hat{\mu}_A + \hat{\mu}_B, \hat{\nu}_A + \hat{\nu}_B \rangle,$$

$$\langle \hat{\mu}_D, \hat{\nu}_D \rangle = \langle \lambda \hat{\mu}_A, \lambda \hat{\nu}_A \rangle.$$

We deduce that if $C = IWAM_w(A_1, \ldots, A_n)$ then

$$\langle \hat{\mu}_C, \hat{\nu}_C \rangle = \langle \sum_{i=1}^{n} w_i \hat{\mu}_{A_i}, \sum_{i=1}^{n} w_i \hat{\nu}_{A_i} \rangle,$$

and explicitly,

$$IWAM_w(A_1, \ldots, A_n) = \langle h^{-1}\left(\sum_{i=1}^{n} w_i h(\mu_{A_i})\right), g^{-1}\left(\sum_{i=1}^{n} w_i g(\nu_{A_i})\right)\rangle. \quad (8.13)$$

If $C = IOWA_w(A_1, \ldots, A_n)$ then

$$IOWA_w(A_1, \ldots, A_n) = \langle h^{-1}\left(\sum_{i=1}^{n} w_i h(\mu_{A_{\sigma(i)}})\right), g^{-1}\left(\sum_{i=1}^{n} w_i g(\nu_{A_{\sigma(i)}})\right)\rangle. \quad (8.14)$$

By taking $g = -\log$ we recover Eqs. (8.10), (8.11) from (8.13) and (8.14) respectively.

In fact Eqs. (8.13), (8.14) show that the degrees of membership and non-membership of the combined AIFVs are simply weighted quasi-arithmetic means of the respective degrees of the components (additionally, in (8.14) the arguments are reordered). The properties of these functions discussed in several publications (e.g., [Xu07, Xu10, XY06]) are now simple consequences of this fact.

Proposition 8.24 *The IWAM and IOWA functions in (8.13) and (8.14) are idempotent, monotone with respect to the partial order $\leq_{L_{AIFV}}$, and are bounded by*

$$A_- \leq_{\mathcal{L}_{AIFV}} IWAM_w(A_1, \ldots, A_n) \leq_{\mathcal{L}_{AIFV}} A_+,$$

$$A_- \leq_{\mathcal{L}_{AIFV}} IOWA_w(A_1, \ldots, A_n) \leq_{\mathcal{L}_{AIFV}} A_+,$$

with $A_- = \langle\min \mu_{A_i}, \max \nu_{A_i}\rangle$, $A_+ = \langle\max \mu_{A_i}, \min \nu_{A_i}\rangle$. The IOWA is symmetric.

Proof The proof follows from the fact that weighted quasi-arithmetic means in both components of (8.13) and (8.14) are idempotent monotone functions. The symmetry in (8.14) is obvious. □

Next we consider the weighted quasi-arithmetic mean (WQAM)

$$M_{\mathbf{w}}(x_1, \ldots, x_n) = \varphi^{-1}(\sum_{i=1}^{n} w_i \varphi(x_i))$$

with a generator φ. We remind that $\varphi : [0, 1] \rightarrow [-\infty, \infty]$ is a continuous strictly monotone function.

The intuitionistic WQAM was defined in [ZXNL10] as follows (the authors considered only power functions $\varphi(t) = t^p$, $p > 0$, and called it the generalized intuitionistic fuzzy weighted averaging (GIFWA))

$$IWQAM_{w,p}(A_1, \ldots, A_n)$$

$$= \langle\left(1 - \prod(1 - \mu_{A_i}^p)^{w_i}\right)^{1/p}, 1 - \left(1 - \prod(1 - (1 - \nu_{A_i})^p)^{w_i}\right)^{1/p}\rangle. \quad (8.15)$$

It can be written in our notation in this way

$$IWQAM_w(A_1, \ldots, A_n)$$

$$= \langle \varphi^{-1}(h^{-1}\left(\sum w_i h(\varphi(\mu_{A_i}))\right)), 1 - \varphi^{-1}(h^{-1}\left(\sum w_i h(\varphi(1 - \nu_{A_i}))\right))\rangle \quad (8.16)$$

$$= \langle u^{-1}\left(\sum w_i u(\mu_{A_i})\right), v^{-1}\left(\sum w_i v(\nu_{A_i})\right)\rangle,$$

with $u = h \circ \varphi$ and $v = h \circ \varphi \circ N$, where N is the standard negation. Note that this formula is similar to (8.13), just with a different pair of additive generators; note $u(t) = v(1 - t)$. In fact, it gives the same expression as (8.13), as long as we use a different pair of t-norm and t-conorm in (8.12) instead of the product, namely a strict t-norm with the generator v and its dual. If φ is the power function as in [ZXNL10], it is the pair of dual t-norm and t-conorm defined by using the generator $g(t) = -\log(1 - (1 - t)^{\lambda})$.

In general, the function (8.15) inherits the same problems of inconsistency with the operations on ordinary fuzzy sets mentioned at the end of Sect. 8.1.3 (to see this, it is sufficient to take $\varphi = Id$ to get IWAM as the special case we already considered). Later we will show that by using a different pair of t-norm and t-conorm in (8.12), we can achieve consistency.

Let us now extend these constructions to other aggregation functions. Take any aggregation function f, and following the same approach, define an aggregation function for AIFV f_{AIFV}. There are several ways of doing this. First, directly by using the following definition:

Definition 8.25 (*Intuitionistic aggregation function*) Let f be an aggregation function, and g is an additive generator of the t-norm in (8.12). f_{AIFV} is an aggregation function on AIFV corresponding to f if $f_{AIFV}(A_1, \ldots, A_n) = C$ with

$$\langle \hat{\mu}_C, \hat{\nu}_C \rangle = \langle f_{\sigma}(\hat{\mu}_{A_1}, \ldots, \hat{\mu}_{A_n}), f_{\sigma}(\hat{\nu}_{A_1}, \ldots, \hat{\nu}_{A_n}) \rangle, \quad (8.17)$$

where the index σ indicates that f may depend on a permutation σ of the participating $A_1, \ldots, A_n$.

The dependence on σ is needed specifically for IOWA and Choquet integrals by (8.11), where the ordering is done with respect to AIFV and not with respect to μ and ν separately.

Proposition 8.26 *The expression (8.17) is well defined, i.e., f_{AIFV} is monotone with respect to $\leq_{\mathcal{L}_{AIFV}}$ and its result is always an AIFV.*

Proof Monotonicity is straightforward by noticing that both components of (8.17) are monotone with respect to all μ_{A_i} and ν_{A_i}, $i = 1, \ldots, n$. To show that $\mu_C + \nu_C \leq 1$ it is convenient to write (8.17) in the IVFV representation

$$[\hat{l}_C, \hat{r}_C] = [f_{\sigma}(\hat{l}_{A_1}, \ldots, \hat{l}_{A_n}), f_{\sigma}(\hat{r}_{A_1}, \ldots, \hat{r}_{A_n})],$$

where $\hat{l} = h(l)$ and $\hat{r} = h(r)$. Since $l_{A_i} \leq r_{A_i}$, $h(l_{A_i}) \leq h(r_{A_i})$ for all i, and because f_σ is non-decreasing in all components, we get $\hat{l}_C \leq \hat{r}_C$ and therefore $l_C \leq r_C$. □

By taking the aggregation functions from [TC10, Wei10, Xu07, Xu10, ZXNL10] (WAM, OWA, Choquet integrals), and the generator $g = -\log$ we obtain the respective formulas as special cases.

However, WQAM and generalized OWA as defined in [ZXNL10] by (8.15) do not fit this definition. Equation (8.15) is consistent with (8.16) whereas (8.17) generally leads to a different formula

$$IWQAM_w(A_1, \ldots, A_n)$$
$$= \langle h^{-1}(\varphi^{-1}\left(\sum_{i=1}^n w_i \varphi(h(\mu_{A_i}))\right)), g^{-1}(\varphi^{-1}\left(\sum_{i=1}^n w_i \varphi(g(\nu_{A_i}))\right))\rangle.$$

An alternative, which accommodates and extends definitions in [Xu10, ZXNL10] to a special class of generated aggregation functions, is the following:

Definition 8.27 (*Intuitionistic generated aggregation function*) Let f be a generated aggregation function with a generator ϕ, so that f is defined by $\varphi(f(x_1, \ldots, x_n)) = \sum_{i=1}^n w_i \varphi(x_i)$. f_{AIFV_w} is an aggregation function on AIFV if $f_{AIFV_w}(A_1, \ldots, A_n) = C$ with

$$\langle \hat{\mu}_C, \hat{\nu}_C \rangle = \langle \sum_{i=1}^n w_{\sigma(i)}\hat{\mu}_i, \sum_{i=1}^n w_{\sigma(i)}\hat{\nu}_i \rangle, \tag{8.18}$$

and $\hat{\mu} = u(\mu)$, $\hat{\nu} = v(\nu)$, $u = h \circ \varphi$, $v = h \circ \varphi \circ N$, $h = g \circ N$ and g is an additive generator of a strict Archimedean t-norm. $w_{\sigma(i)}$; $i = 1, \ldots, n$, denote weights which depend on the ordering of pairs $\langle \mu_i, \nu_i \rangle$; $i = 1, \ldots, n$.

By specifying how $w_{\sigma(i)}$ depends on the ordering of the arguments, we get weighted quasi-arithmetic means, generalized OWA and generalized Choquet integrals as special cases. Note that if $\varphi = Id$, Eq. (8.18) is equivalent to Eq. (8.17).

8.1.5 Consistency with Operations on Ordinary Fuzzy Sets

So far we have presented constructions that generalize those in [TC10, Wei10, Xu07, Xu10, ZXNL10], however the problem of consistency with the operations on ordinary fuzzy sets remains. To ensure consistency, let us now look at the following alternative. Take in (8.12) S_L, T_L, as the Łukasiewicz t-conorm and t-norm respectively (see Example 1.82), $S_L(x, y) = \min(1, x + y)$, $T_L(x, y) = \max(0, x + y - 1)$ with $g(t) = 1 - t$. Then we have for $\lambda \in [0, 1]$

$$\lambda A = \langle \lambda \mu_A, 1 - \lambda(1 - \nu_A) \rangle.$$

We immediately obtain that the ordering of AIFV described in Sect. 8.1.3 is invariant under multiplication by a scalar $\lambda \in [0, 1]$, because:

$Score(A) = Score(B) \Rightarrow Score(\lambda A) = Score(\lambda B),$
$Score(A) < Score(B) \Rightarrow Score(\lambda A) < Score(\lambda B),$
$Accuracy(A) = Accuracy(B) \Rightarrow Accuracy(\lambda A) = Accuracy(\lambda B),$
$Accuracy(A) < Accuracy(B) \Rightarrow Accuracy(\lambda A) < Accuracy(\lambda B).$

It is sufficient to observe that such an operation can be written in the IVFS representation as $\lambda[l, r] = [\lambda l, \lambda r]$.

Taking into account that $h^{-1} = h = Id$ and $g^{-1} = g = 1 - Id$ on $[0, 1]$, and that $\sum_{i=1}^{n} w_i = 1$ ensures the argument of g^{-1} is in $[0, 1]$, we consequently obtain from (8.13), (8.14)

$$IWAM_w(A_1, \ldots, A_n) = \langle \sum_{i=1}^{n} w_i \mu_{A_i}, 1 - \sum_{i=1}^{n} w_i(1 - \nu_{A_i}) \rangle \qquad (8.19)$$

$$= \langle \sum_{i=1}^{n} w_i \mu_{A_i}, \sum_{i=1}^{n} w_i \nu_{A_i} \rangle,$$

and

$$IOWA_w(A_1, \ldots, A_n) = \langle \sum_{i=1}^{n} w_i \mu_{A_{\sigma(i)}}, \sum_{i=1}^{n} w_i \nu_{A_{\sigma(i)}} \rangle. \qquad (8.20)$$

The expression (8.19) is a natural extension of the WAM according to the Definition 8.13, and hence is a representable f_L. However (8.20) cannot be written as a representable f_L, because the order in which $\mu_{A_{\sigma(i)}}$ and $\nu_{A_{\sigma(i)}}$ are arranged depends on both μ and ν according to (8.9). In contrast, representable f_L imply that aggregation of μ values does not depend on ν values and vice versa.

Consequently $IOWA_w$ is not the same as the function OWA_L in [Yag09], which is a natural extension of OWA according to Definition 8.13 with $f_1 = f_2 = OWA$.

What is remarkable is that the use of Łukasiewicz t-norm and t-conorm in (8.12) is the only way consistency with the ordinary fuzzy sets can be achieved [BBGMP11].

Proposition 8.28 *The operations defined by (8.13) and (8.14) are consistent with the operations on ordinary fuzzy sets if and only if the t-norm and t-conorm in (8.12) are Łukasiewicz ones.*

Proof The necessity is straightforward, by letting $\nu_i = 1 - \mu_i$ in (8.19), (8.20). Sufficiency: for consistency of (8.13) with operations on ordinary fuzzy sets we need to ensure

$$h^{-1}\left(\sum w_i h(\mu_{A_i})\right) = \sum w_i \mu_{A_i} \text{ for all } \mu_{A_i} \in [0, 1], \qquad (8.21)$$

(since $\nu = 1 - \mu$, the respective condition for non-membership is automatically verified). The unique solution h to the functional equation (8.21) is an affine function [Acz66]. Because h is a generator of a t-conorm $h(0) = 0$, and hence it is a linear function. The multiplier is irrelevant since additive generators are defined up to an arbitrary positive multiplier. Hence $h = Id$ on $[0,1]$ or its multiple.

The same result holds for (8.14), since the order (8.9) coincides with the usual order on real numbers in that case. □

Corollary 8.29 *The operations defined by (8.13) and (8.16) are consistent with the natural extensions of WAM and WQAM to f_L if and only if the t-norm and t-conorm in (8.12) are Łukasiewicz ones.*

Proof By using Proposition 8.28 and noticing that $u = h \circ \varphi$ in (8.16). □

However, this result does not hold for natural extensions of OWA and other operations that require sorting based on order (8.9). Since the order (8.9) uses both μ and ν, the operations (8.14) are not representable.

Another useful feature of this construction that uses Łukasiewicz t-norm and t-conorm, is that the resulting aggregation functions are monotone not only with respect to the partial order $\leq_L$ ($\leq_{L_{AIFV}}$), but also with respect to the total order in (8.9).

Proposition 8.30 *Operations (8.16), (8.19) and (8.20) are monotone with respect to the order (8.9).*

Proof Immediate for (8.16) and (8.19), since these operations are natural extensions of WAM and WQAM to f_L. For (8.20) note that in both sums the ordering is the same, and the function can be written as

$$IOWA_w(A_1, \ldots, A_n) = \sum w_i A_{\sigma(i)} = \sum w_i \langle \mu_{A_{\sigma(i)}}, \nu_{A_{\sigma(i)}} \rangle,$$

which is monotone with respect to (8.9), because now multiplication by a scalar and addition preserve the ordering (8.9). □

This results does not hold for a natural extension of OWA, as can be seen from the following example.

Example 8.31 Take $A = [0, 1]$, $B = [0.1, 0.9]$ in the IVFS representation, so that $A < B$ according to (8.9). Let $w = (1, 0)$, in which case $OWA_L(A, B) = [0.1, 1]$. Now if $C = [0.1, 0.95]$ then $A < B < C$ but $IOWA_L(B, C) = [0.1, 0.95] < OWA_L(A, B)$.

Finally, this approach can be extended to other averaging aggregation functions, such as quasi-arithmetic means, Choquet integral and induced OWA. We define the aggregation functions by using Definition 8.25, which in the case of generated functions is equivalent to Definition 8.27.

Let us illustrate this construction on two examples.

Example 8.32 Let f be a weighted quasi-arithmetic mean with a generator φ. We have

$$IWQAM_w(A_1, \ldots, A_n) = \langle \varphi^{-1}(\sum w_i \varphi(\mu_{A_i})), 1 - \varphi^{-1}(\sum w_i \varphi(1 - \nu_{A_i})) \rangle$$
$$= \langle f(\mu_{A_1}, \ldots, \mu_{A_n}), f^d(\nu_{A_1}, \ldots, \nu_{A_n}) \rangle,$$

where f^d denotes the dual of f with respect to the standard negation. When $\varphi = Id$ we get (8.19) as a special case (we remind that a WAM is self-dual). In the IVFS representation we have

$$IWQAM_w(A_1, \ldots, A_n) = [f(l_{A_1}, \ldots, l_{A_n}), f(r_{A_1}, \ldots, r_{A_n})].$$

This means that $IWQAM_w$ is representable and a natural extension of WQAM. The choices $\varphi(t) = t^p, p \neq 0$, $\varphi(t) = \log(t)$ $\varphi(t) = p^t, p \neq 1$ and $\varphi(t) = \exp(\exp(t))$ lead to natural extensions of weighted power means, geometric mean, exponential mean and double exponential means. Other types of quasi-arithmetic means in Chap. 2 can also be used.

Example 8.33 Consider the logarithmic mean $f(x, y) = \frac{y - x}{\log y - \log x}$, which is not a quasi-arithmetic mean. Using Definition 8.25, we obtain the logarithmic mean for AIFS

$$ILM(A_1, A_2) = \langle \frac{\mu_2 - \mu_1}{\log \mu_2 - \log \mu_1}, 1 - \frac{\nu_1 - \nu_2}{\log(1 - \nu_2) - \log(1 - \nu_1)} \rangle.$$

Formulas for generalized logarithmic, Bonferroni, Heronian and other means can be found in Chap. 2. We note that the constructions for "generalized" aggregation functions for AIFS in [TC10, Wei10, Xu07, Xu10, XY06, ZXNL10] do not work in these cases.

8.1.6 Medians for AIFV

The median is suitable for modeling consensus in group decision making [GLMP09] as it discards the extremal values and serves as a robust estimate of a "typical" input. It can be expressed as a special case of the OWA functions, see Chap. 3. A straightforward way to define the median for IVFS is by applying it to the ends of the membership interval independently. Such a construction is called a representable aggregation function, see Definition 8.13.

Another way of defining the median is by using IFOWA functions as presented in the previous section, following [TC10, Xu07, Xu10, ZXNL10]. Let us instantiate the median function from the above mentioned constructions. Using (8.11), with all weights except those pertaining to the middle observations equal to zero, we obtain expression (8.22), where G is the geometric mean $G(x, y) = \sqrt{xy}$ and G^d is its dual $G^d(x, y) = 1 - \sqrt{(1 - x)(1 - y)}$, and $A_{\sigma(i)} = \langle \mu_{A_{\sigma(i)}}, \nu_{A_{\sigma(i)}} \rangle$ is the ith largest input A_i, according to the ordering (8.9).

Using the alternative formulation (8.20) we obtain expression (8.23). Note that Med_1 and Med_2 only differ where n is even, with the former averaging the middle two arguments using the geometric mean and its dual while the latter uses the arithmetic mean. Finally the natural extension of the median is given by (8.24).

Let us now make several observations about equations (8.22)–(8.24). First, notice that neither (8.22) nor (8.23) coincide with the natural extension Med_L (and in fact are not representable) because the ordering is different. The medians in (8.22) and (8.23) are internal for odd n, and Med_L is generally not internal. Furthermore, (8.22) is not consistent with the usual median (when $1 - \mu = \nu$), whereas both (8.23) and (8.24) are (and they coincide in this case since we would have $Score(\langle \mu, 1 - \mu \rangle) = 2\mu - 1$).

It should be noted too that the median Med_1 is not monotone with respect to the ordering (8.9), which is used in its definition. To illustrate this, consider Example 8.34.

Example 8.34 The AIFVs, $A_1 = \langle 0, 1 \rangle$, $A_2 = \langle 0.3, 0.2 \rangle$, $A_3 = \langle 0.5, 0.4 \rangle$ are ordered $A_1 < A_2 < A_3$ according to (8.9). Therefore we expect $Med(A_1, A_2) < Med(A_1, A_3)$. However, we have

$$Med_1(A_1, A_2) \approx \langle 0.163, 0.447 \rangle > \langle 0.293, 0.632 \rangle \approx Med_1(A_1, A_3),$$

because

$$Score(Med_1(A_1, A_2)) \approx -0.284 > -0.339 \approx Score(Med_1(A_1, A_3)).$$

In contrast, monotonicity of Med_2 with respect to (8.9) holds.

Finally, let us write (8.22)–(8.24) in the IVFV representation for completeness, which can be determined using $\mu_{A_i} = l_{B_i}$, $1 - \nu_{A_i} = r_{B_i}$. Equations (8.25)–(8.27) provide these expressions.

As with the definitions on AIFV, $B_{\sigma(i)}$ denotes the ith largest input according to (8.9), while Med_L considers the l_{B_i} and r_{B_i} independently.

8.1.7 Bonferroni Means

This section is based on the work by Beliakov and James [BJ12].

Yager and Xu construction

First we present Xu and Yager's extension [XY11] of the Bonferroni mean, and highlight some characteristics inconsistent with the original definition of the Bonferroni mean. We then present some alternatives. In [XY11], the Bonferroni mean was generalized to AIFS by replacing the sum and the product operations with those defined on fuzzy sets, i.e. we have

$$Med_1(A_1, \ldots, A_n) = \begin{cases} A_{\sigma\left(\frac{n+1}{2}\right)}, & n \text{ is odd}, \\ \langle G^d(\mu_{A_{\sigma\left(\frac{n}{2}\right)}}, \mu_{A_{\sigma\left(\frac{n}{2}+1\right)}}), G(\nu_{A_{\sigma\left(\frac{n}{2}\right)}}, \nu_{A_{\sigma\left(\frac{n}{2}+1\right)}}) \rangle, & n \text{ is even}. \end{cases}$$

$$(8.22)$$

$$Med_2(A_1, \ldots, A_n) = \begin{cases} A_{\sigma\left(\frac{n+1}{2}\right)}, & n \text{ is odd,} \\ \langle \frac{1}{2}(\mu_{A_{\sigma\left(\frac{n}{2}\right)}} + \mu_{A_{\sigma\left(\frac{n}{2}+1\right)}}), \frac{1}{2}(\nu_{A_{\sigma\left(\frac{n}{2}\right)}} + \nu_{A_{\sigma\left(\frac{n}{2}+1\right)}}) \rangle, & n \text{ is even.} \end{cases}$$
(8.23)

$$Med_L(A_1, \ldots, A_n) = \langle Med(\mu_{A_1}, \ldots, \mu_{A_n}), Med(\nu_{A_1}, \ldots, \nu_{A_n}) \rangle.$$
(8.24)

$$Med_1(B_1, \ldots, B_n) = \begin{cases} B_{\sigma\left(\frac{n+1}{2}\right)}, & n \text{ is odd,} \\ \left[G^d(l_{B_{\sigma\left(\frac{n}{2}\right)}}, l_{B_{\sigma\left(\frac{n}{2}+1\right)}}), G^d(r_{B_{\sigma\left(\frac{n}{2}\right)}}, r_{B_{\sigma\left(\frac{n}{2}+1\right)}}) \right], & n \text{ is even.} \end{cases}$$
(8.25)

$$Med_2(B_1, \ldots, B_n) = \begin{cases} B_{\sigma\left(\frac{n+1}{2}\right)}, & n \text{ is odd,} \\ \left[\frac{1}{2}(l_{B_{\sigma\left(\frac{n}{2}\right)}} + l_{B_{\sigma\left(\frac{n}{2}+1\right)}}), \frac{1}{2}(r_{B_{\sigma\left(\frac{n}{2}\right)}} + r_{B_{\sigma\left(\frac{n}{2}+1\right)}}) \right], & n \text{ is even.} \end{cases}$$
(8.26)

$$Med_L(B_1, \ldots, B_n) = [Med(l_{B_1}, \ldots, l_{B_n}), Med(r_{B_1}, \ldots, r_{B_n})].$$
(8.27)

$$IFB^{p,q}(A_1, \ldots, A_n) = \left(\frac{1}{n(n-1)} \bigoplus_{i,j=1, i\neq j}^{n} A_i^p \otimes A_j^q \right)^{\frac{1}{p+q}}.$$
(8.28)

Based on the sum and probabilistic sum operations to define the arithmetic operations (8.5), (8.6), the following definition is obtained [XY11].

Definition 8.35 (*Intuitionistic Bonferroni mean*) Let $p, q \geq 0$ and A_i an AIFV, $i = 1, \ldots, n$. The intuitionistic Bonferroni mean is the function

$$IFB^{p,q}(A_1, \ldots, A_n) = \left\langle \left(1 - \prod_{i,j=1, i\neq j}^{n} (1 - \mu_{A_i}^p \mu_{A_j}^q)^{\frac{1}{n(n-1)}} \right)^{\frac{1}{p+q}},$$
(8.29)

$$1 - \left(1 - \prod_{i,j=1, i\neq j}^{n} (1 - (1 - \nu_{A_i})^p (1 - \nu_{A_j})^q)^{\frac{1}{n(n-1)}} \right)^{\frac{1}{p+q}} \right\rangle.$$

If we compare the Bonferroni mean in its original form, concentrating for the moment on the aggregation of the μ_{A_i} values, we can better understand the behavior of this function. For $p = q = 1$ the inner function of the original Bonferroni mean can be interpreted as an arithmetic mean of all product pairs $x_i x_j$.

Table 8.4 A comparison of the outputs of the intuitionistic Bonferroni mean ($IFB^{1,1}$) and the standard Bonferroni mean ($B^{1,1}$) for standard fuzzy sets

A_1	A_2	A_3	A_4	$IFB^{1,1}(A_1,\ldots,A_4)$	$B^{1,1}(A_1,\ldots,A_4)$
$\langle 1,0 \rangle$	$\langle 0,1 \rangle$	$\langle 0,1 \rangle$	$\langle 0,1 \rangle$	$\langle 0,1 \rangle$	$\langle 0,1 \rangle$
$\langle 1,0 \rangle$	$\langle 0.5,0.5 \rangle$	$\langle 0,1 \rangle$	$\langle 0,1 \rangle$	$\langle 0.3303,0.6697 \rangle$	$\langle 0.2887,0.7113 \rangle$
$\langle 1,0 \rangle$	$\langle 0.9,0.1 \rangle$	$\langle 0,1 \rangle$	$\langle 0,1 \rangle$	$\langle 0.5646,0.4355 \rangle$	$\langle 0.3873,0.6127 \rangle$
$\langle 1,0 \rangle$	$\langle 1,0 \rangle$	$\langle 0,1 \rangle$	$\langle 0,1 \rangle$	$\langle 1,0 \rangle$	$\langle 0.4082,0.5918 \rangle$
$\langle 1,0 \rangle$	$\langle 1,0 \rangle$	$\langle 1,0 \rangle$	$\langle 0,1 \rangle$	$\langle 1,0 \rangle$	$\langle 0.7071,0.2929 \rangle$

Note that $IFB^{1,1}$ is $\langle 1,0 \rangle$ as soon as two arguments are $\langle 1,0 \rangle$, while in its original form even three evaluations of $\langle 1,0 \rangle$ does not have this effect

$$\left(\frac{1}{n(n-1)} \sum_{i,j=1,i\neq j}^{n} x_i x_j \right)^{\frac{1}{2}}. \tag{8.30}$$

On the other hand, Eq. (8.29) takes the dual of the geometric mean G^d of the product pairs $\mu_{A_i}\mu_{A_j}$.

As a consequence, this definition does not coincide with the original definition in the case of ordinary fuzzy sets, nor is it able to express any of the Bonferroni mean's special cases (for ordinary fuzzy sets, the aggregation of μ_{A_i} should correspond with that of the crisp values). Further to this, by duality, G^d has an absorbent element $e = 1$, i.e. $G^d(x_1,\ldots,x_n) = 1$ if $\exists i : x_i = 1$. This means that if there exists a single pair $\mu_{A_i} = \mu_{A_j} = 1$, the aggregated value of the membership values will be 1. This is somewhat antithetical to the requirement of the Bonferroni mean to have at least two non-zero outputs to give an output greater than zero (which can be used to model mandatory criteria). Consider the values in Table 8.4 which shows the input and output AIFVs with $n = 4$ for Eq. (8.29) and the standard Bonferroni mean (assuming $x_i = \mu_{A_i} = 1 - \nu_{A_i}$).

We note that, for these values, it behaves similarly to the original function except when the two values approach 1.

A weighted version of the function is also expressed in [XY11]. The following multiplication by a scalar rule is used:

$$nA = \langle 1 - (1 - \mu_A)^n, \nu_A^n \rangle.$$

Note 8.36 The above equation for nA is consistent with the operation used for $A \oplus B$ (i.e. it follows logically from $A \oplus A \oplus \cdots \oplus A$. The decision to use this as opposed to $nA = \langle n\mu_A, 1 - n(1 - \nu_A) \rangle$, however, is discussed in Xu and Yager's paper.

The result (for standard AIFVs) will be the function,

$$IFB_{\mathbf{w}}^{p,q}(A_1, \ldots, A_n)$$

$$= \left\langle \left(1 - \prod_{i,j=1, i\neq j}^{n} (1 - (1 - (1 - \mu_{A_i})^{w_i})^p (1 - (1 - \mu_{A_j})^{w_i})^q)^{\frac{1}{n(n-1)}}\right)^{\frac{1}{p+q}},\right.$$

$$\left. 1 - \left(1 - \prod_{i,j=1, i\neq j}^{n} (1 - (1 - \nu_{A_i}^{w_i})^p (1 - \nu_{A_j}^{w_i})^q)^{\frac{1}{n(n-1)}}\right)^{\frac{1}{p+q}}\right\rangle. \qquad (8.31)$$

Applying the weights to the Bonferroni mean in this way results in a function which can no longer be considered an averaging aggregation function. This is easily seen by looking at the aggregation of the μ_{A_i} values. The expression differs to the unweighted version by replacing each μ_{A_i} with $1 - (1 - \mu_{A_i})^{w_i}$ which is strictly less than μ_{A_i} for $w_i < 1$ on the open interval $]0, 1[$. It would be more appropriate to replace $\frac{1}{n(n-1)}$ (i.e. equal weights for each pair) with $\frac{w_i u_j}{1-u_i}$ where $\mathbf{w}$ is the weighting vector associated with the p index and $\mathbf{u}$ is the weighting vector associated with the q index.

We now present some alternative definitions which are more consistent with the original Bonferroni mean and its weighted extensions.

Application of generator transformations

One way to approach the definition of the Bonferroni mean for AIFVs is by direct application of Definition 8.25 to either its original or generalized form. It can also be expressed as a composed intuitionistic aggregation function. Beliakov and James [BJ12] show that both of these constructions result in an averaging intuitionistic aggregation function. The main advantage of these constructions is their conceptual simplicity, as the construction is broken down into manageable blocks, and automatic verification of the desired properties, which can be time consuming if explicit expressions are used as in [XC11, XY11].

By using Definition 8.25, we can define the AIFV aggregation function corresponding to the Bonferroni mean in a straightforward manner, which will be consistent with the original function.

Definition 8.37 (*Intuitionistic generalized Bonferroni mean*) Let $B_{\mathbb{M}}$ denote the generalized Bonferroni mean defined by (6.11) and g be the additive generator of an Archimedean t-norm with its dual t-conorm generated by $h(t) = g(1-t)$. $IFB_{\mathbb{M}}$ is the generalized Bonferroni mean on AIFV with,

$$IFB_{\mathbb{M}}(A_1, \ldots, A_n)$$

$$= \left\langle h^{-1} \circ B_{\mathbb{M}}\big(h(\mu_{A_1}), \ldots, h(\mu_{A_n})\big), g^{-1} \circ B_{\mathbb{M}}\big(g(\mu_{A_1}), \ldots, g(\mu_{A_n})\big)\right\rangle. \qquad (8.32)$$

Proposition 8.38 *The generalized Bonferroni mean on AIFV, $IFB_{\mathbb{M}}$ is an averaging intuitionistic aggregation function.*

Proof We need only show that $IFB_{\mathbb{M}}$ is monotone and idempotent.

1. monotonicity: Since $B_{\mathbb{M}}$, h are both monotone increasing, it follows that the composition $h^{-1} \circ B_{\mathbb{M}}(h(t))$ will also be monotone increasing. Similarly, since g is monotone decreasing, $g^{-1} \circ B_{\mathbb{M}}(g(t))$ is monotone decreasing.

Therefore $IFB_{\mathbb{M}}$ will be monotone with respect to $\leq_{LAIFV}$ (as indeed will any aggregation function defined in this way).

2. idempotency: The idempotency of $B_{\mathbb{M}}$ guarantees $h^{-1} \circ B_{\mathbb{M}}(h(\mu_A), \ldots,$ $h(\mu_A)) = h^{-1}(h(\mu_A)) = \mu_A$ and $g^{-1} \circ B_{\mathbb{M}}(g(\nu_A), \ldots, g(\nu_A)) = g^{-1}(g(\mu_A))\nu_A$.

Therefore $IFB_{\mathbb{M}}(A, \ldots, A) = A$. $\qquad\qquad\qquad\qquad\qquad\qquad\qquad\qquad\square$

The following example shows that the original Bonferroni mean is a special case.

Example 8.39 We let $g(t) = 1 - t$ (and hence $h(t) = t$), and use the arithmetic mean, the power mean and product with powers respectively for M_1, M_2, C i.e.,

$$\mathbb{M} = \left\langle \sum_{i=1}^{n} \frac{1}{n} x_i \, , \, \left(\sum_{j=1, j\neq i}^{n} \frac{1}{n-1} x_j^q \right)^{\frac{1}{q}} \, , \, x^p y^q \right\rangle.$$

Explicitly, we will then have

$$IFB_{\mathbb{M}}(A_1, \ldots, A_n) \quad = \left\langle \left(\sum_{i,j=1, i\neq j}^{n} \tfrac{1}{n(n-1)} \mu_{A_i}^p \mu_{A_j}^q \right)^{\frac{1}{p+q}} \, , \right. \qquad (8.33)$$

$$\left. 1 - \left(\sum_{i,j=1, i\neq j}^{n} \tfrac{1}{n(n-1)} (1 - \nu_{A_i})^p (1 - \nu_{A_j})^q \right)^{\frac{1}{p+q}} \right\rangle.$$

The expression in Example 8.39 corresponds with the natural extension of the original Bonferroni mean. It will therefore have the same outputs as those given in Table 8.4 for $B^{1,1}$. The requirement to have at least two non-zero (non $\langle 0, 1 \rangle$) inputs is upheld, and we no longer have the problem of only two $\langle 1, 0 \rangle$ inputs resulting in a $\langle 1, 0 \rangle$ output.

In the case of $g(t) = -\ln t$, the function will not correspond with Xu and Yager's extension, but will still have the undesirable behavior with any two $\langle 1, 0 \rangle$ input values since $g(0) = h(1) = \infty$. However, alternative choices of g and h can transform the function without this drawback. Example 8.40 shows one such construction.

Example 8.40 If we let $g(t) = (1 - t)^2$, we then have $h(t) = t^2$ and the inverse functions, $g^{-1}(t) = 1 - t^{\frac{1}{2}}, h^{-1}(t) = t^{\frac{1}{2}}$. The resulting function will be bounded between Eq. (8.33) and the maximum, i.e. give higher values in general but still hold the averaging property. It can be expressed explicitly as

$$IFB_{\mathbb{M},g}(A_1,\ldots,A_n) = \left\langle \left(\sum_{i,j=1,i\neq j}^{n} \frac{1}{n(n-1)} \mu_{A_i}^{2p}\mu_{A_j}^{2q} \right)^{\frac{1}{2(p+q)}}, \right. \tag{8.34}$$

$$\left. 1 - \left(\sum_{i,j=1,i\neq j}^{n} \frac{1}{n(n-1)}(1-v_{A_i}^2)^p(1-v_{A_j}^2)^q \right)^{\frac{1}{2(p+q)}} \right\rangle.$$

We note that we still require at least two non-zero membership values for the membership output to be greater than zero, but that we cannot obtain a full membership unless all four inputs are equal to 1.

The intuitionistic Bonferroni mean as a composed aggregation function

Based on Eq. (6.10), Xu and Chen consider a generalized Bonferroni mean in the context of interval valued AIFS [XC11]. Using $p = q = 1$, the function is given in terms of the A_i as

$$IVIFB(A_1,\ldots,A_n) = \left(\frac{1}{n}\bigoplus_{i=1}^{n} \left(\frac{1}{n-1} \left(\bigoplus_{j=1,j\neq i}^{n} A_j \right) \otimes A_i \right) \right)^{\frac{1}{2}}. \tag{8.35}$$

Note 8.41 While the expression replaces the standard operations in Eq. (6.10) with intuitionistic ones, for the usual choice of $\oplus$, $\otimes$ (Eqs. (8.5), (8.6)), distributivity does not hold, i.e. $A \otimes (B \oplus C) \neq (A \otimes B) \oplus (A \otimes C)$, and the function will not be a consistent with Eq. (8.28).

Alternative definitions of the arithmetic $\oplus$, $\otimes$ operations, however will allow this rearrangement, so we will use this expression as the basis of our construction.

As with the generalized Bonferroni mean, Beliakov and James [BJ12] define the intuitionistic Bonferroni mean as a composition of intuitionistic aggregation functions. We denote by $\mathbf{A}_{j\neq i}$ the input set of all AIFVs except the ith, i.e. $(A_1,\ldots,A_{i-1},A_{i+1},\ldots,A_n)$.

Definition 8.42 (*Intuitionistic generalized Bonferroni mean*) Let $i\mathbb{M} = \langle iM_1, iM_2, iC\rangle$, with $iM_1 : \mathcal{L}_{AIFV}^n \to \mathcal{L}_{AIFV}, iM_2 : \mathcal{L}_{AIFV}^{n-1} \to \mathcal{L}_{AIFV}$ and $iC : \mathcal{L}_{AIFV} \times \mathcal{L}_{AIFV} \to \mathcal{L}_{AIFV}$ denote intuitionistic aggregation functions, with $d_{iC}(A) = iC(A,A)$ invertible such that $d_{iC}^{-1}(d_{iC}(A)) = A$. The generalized intuitionistic Bonferroni mean is given by,

$$iB_{\mathbb{M}}(A_1,\ldots,A_n) = d_{iC}^{-1} \circ iM_1\Big(iC\Big(A_1, iM_2(\mathbf{A}_{j\neq 1})\Big),\ldots, iC\Big(A_n, iM_2(\mathbf{A}_{j\neq n})\Big)\Big). \tag{8.36}$$

Note 8.43 We use the existing (intuitionistic) aggregation functions as building blocks, and do not require their extensive expressions to verify the desired properties of the resulting function, as illustrated in the following proposition.

Proposition 8.44 *Where iC is an aggregation function on $\mathcal{L}_{AIFV}$ and iM_1, iM_2 are averaging aggregation functions, $iB_{\mathbb{M}}$ will be an averaging intuitionistic aggregation function, regardless of iC.*

Proof It is sufficient to show that $iB_{\mathbb{M}}$ is monotone and averaging.

1. monotonicity: Given the monotonicity of the components $d_{iC}^{-1}, iM_1, iM_2, iC$ with respect to the partial order $\leq_{\mathcal{L}_{AIFV}}$, it is straightforward that $iB_{\mathbb{M}}$ will also be monotone.

2. idempotency: Given the input vector $(A, A, \ldots, A)$ we have

$$iM_2(A, A, \ldots, A) = A, \quad \text{hence} \quad iC\big(A, iM_2(A, A, \ldots, A)\big) = iC(A, A),$$

which implies

$$iM_1\Big(iC\big(A, iM_2(A, A, \ldots, A)\big)\ldots iC\big(A, iM_2(A, A, \ldots, A)\big)\Big) = iC(A, A) = d_C(A).$$

Therefore $iB_{\mathbb{M}}(A, \ldots, A) = d_C^{-1}(d_C(A)) = A$. $\qquad\qquad\qquad\qquad\qquad\square$

The averaging behavior of the composed function, which follows from the choice of functions iM_1, iM_2, implies the closure, monotonicity and idempotency. We also note that we automatically obtain weighted intuitionistic Bonferroni means if we use weighted intuitionistic means as iM_1, iM_2. As with the generalized Bonferroni mean for real inputs, further properties will depend on the components. We mention some of these here; the proofs are in [BJ12].

Proposition 8.45 *The intuitionistic generalized Bonferroni mean will be symmetric if iM_1, iM_2 are symmetric.*

Proposition 8.46 *If E is an idempotent element of iM_1 and an absorbing element of both iM_2 and iC, $iB_{\mathbb{M}}$ will have absorbing element E, independent of iM_1.*

Whilst we can use any averaging intuitionistic aggregation functions for iM_1, iM_2 and any intuitionistic aggregation function for iC (provided the reverse diagonal exists), we may wish to ensure that our choices are consistent with the underlying intuitionistic arithmetic operations. Beliakov and James [BJ12] proposed the use of generated functions which can help us identify special cases of Eq. (8.36). We refer the reader to [BJ12] for details.

Concluding this section we should mention another relevant work related to averaging of AIFS. Recently Yager [Yag13, YA13] introduced Pythagorean fuzzy values, where instead of the condition $\mu_A + \nu_A \leq 1$ which relates the membership–non-membership pairs in the AIFS framework, we have the condition $\mu_A^2 + \nu_A^2 \leq 1$. A number of constructions of averaging functions for Pythagorean fuzzy values was presented in [BJ14]. The authors used polar representation of membership and non-membership pairs to obtain various averaging functions, in particular the WAM.

8.2 Medians on Lattices

8.2.1 Medians as Penalty Based Functions

The definitions of the median based on IFOWA are problematic because of the sorting operation. Since intervals form a lattice which is not a chain, there is no unique ordering, and sorting the inputs in IFOWA becomes problematic. In this section the median will be defined on distributive lattices not using the sort operation, as in [Bar80, Mon08, MH80]. Furthermore, the median can also be calculated as a penalty-based function [Bel09] (or equivalently with respect to a dissimilarity function, e.g. [MSV08]), in a manner consistent with its definition on lattices. Again, no sorting is required. The results in this section are based on [BBF11, BBJCF11].

We already know from Chap. 5 that the arithmetic means and the median are solutions to simple optimization problems, in which a measure of disagreement between the inputs is minimized. The arithmetic mean is the solution to

$$\arg \min_y \sum_{i=1}^{n} (x_i - y)^2, \tag{8.37}$$

whereas the median is a solution to

$$\arg \min_y \sum_{i=1}^{n} |x_i - y|. \tag{8.38}$$

Since we are interested in medians on the lattice $\mathcal{L}$, we need to generalize Eq. (8.38) to the case where x_i and $y \in \mathcal{L}$. Notice that no sorting of the inputs is required in (8.38), and this penalty-based construction will allow us to circumvent the problem of defining a total order on the lattice $\mathcal{L}$. The next section deals with this issue in detail.

Definition 8.47 (*Natural distance*) A graph $G(V, E)$ consists of a set of vertices V (finite or infinite) joined by edges E. A path in a graph is a sequence of vertices linked by edges. A natural distance d between two vertices in a graph is the shortest path distance, i.e., the minimum number of edges on a path linking two vertices.

Definition 8.48 (*Covering graph*) The covering graph $G(P)$ of a poset P is a graph whose vertices are P's elements and $u_{a,b}$ with $a \neq b$ is an edge if and only if $a \leq b$ and there is no $c \neq a, b$ such that $a \leq c \leq b$ or $b \leq a$.

Definition 8.49 (*Natural distance on a lattice*) The natural distance d between elements of a distributive lattice is the shortest path distance on its covering graph. Endowed with this distance, a distributive lattice becomes a metric space.

We now consider the Cartesian product of finite chains $\mathcal{C}$ of length K (and hence their covering graphs are finite). All the finite chains of the same length are isomorphic to each other. So we can always assume that we are working with chains $C_K = 0 \leq 1 \leq 2 \leq \cdots \leq K - 1$. We now instantiate the natural distance on a lattice for a Cartesian product of finite chains.

Definition 8.50 (*Natural distance on product*) Let $\mathcal{L} = (X, \leq, \wedge, \vee)$ be a product of finite chains. The natural distance between $x, y \in \mathcal{L}$ is defined as the length minus 1 of a chain C joining the least element $a = x \wedge y$ and the greatest element $b = x \vee y$,

$$d(x, y) = length(C) - 1. \tag{8.39}$$

We note that all the chains with the least element a and the greatest element b on a product lattice in Definition 8.50 have the same length. This definition of the distance is equivalent to the following

$$d(x, y) = \sum_{i=1}^{m} d_i(x_i, y_i) = \sum_{i=1}^{m} |x_i - y_i|, \tag{8.40}$$

where d_i is the distance on the ith chain in the product of m chains, which is equal to the absolute value of the difference of the corresponding integers (as each chain is isomorphic to $\{0, 1, \ldots, K - 1\}$), which we also denoted by x_i, y_i.

8.2.2 Median Graphs and Distributive Lattices

Let us now recall various results that appeared mainly in the context of distributive lattices and social choice theory [MMR98, Mon08, MH80].

Definition 8.51 (*Median on a graph*) A median of three vertices a, b, c of a graph G is a vertex $Med(a, b, c)$ that belongs to the shortest path between any two of a, b, c.

Not every graph allows the median of every three vertices.

Definition 8.52 (*Median graph*) A *median graph* G is a connected graph in which every three vertices have a unique median.

Medians of three elements on lattices were defined algebraically in [Bir67], namely
$$Med(x, y, z) = (x \wedge y) \vee (y \wedge z) \vee (z \wedge x). \tag{8.41}$$

This coincides with the usual definition of the median of real numbers.

Barbut [Bar80] has extended the notion of the median on a lattice to any set I of elements of odd cardinality $n = |I| = 2p + 1$ using

$$Med(x_1, \ldots, x_n) = \bigvee_{K \subset I, |K| = p+1} \left(\bigwedge_{i \in K} x_i \right) \tag{8.42}$$
$$= \bigwedge_{K \subset I, |K| = p+1} \left(\bigvee_{i \in K} x_i \right).$$

For an even number of elements, the median is not unique, but the medians form the median interval whose bounds are given by the algebraic formulas in [MH80]. An interval $I(a, b)$ between $a, b \in V$ in a median graph $G(V, E)$ is defined as $I(a, b) = \{v \in V | d(a, b) = d(a, v) + d(v, b)\}$.

The existence of a median is a characteristic property of distributive lattices:

Theorem 8.53 [Bar80] *A lattice is distributive if and only if every odd number of elements admits the median $Med(\mathbf{x})$.*

Furthermore, Barbut [Bar80] shows that a lattice is distributive if and only if one can construct a distance function $d : \mathcal{L} \times \mathcal{L} \to \mathbb{R}$, such that (a) d is positive and $d(x, y) = 0$ if and only if $x = y$, (b) $d(x, y) = d(x, t) + d(t, y)$ if $t \in [x \wedge y, x \vee y]$, and $d(x, y) = d(y, x)$ if $x \le y$.

With the help of such a distance function, the median is expressed as a minimizer of

$$Med_L(x_1, \ldots, x_n) = \arg\min_y \sum_{i=1}^{n} d(x_i, y). \tag{8.43}$$

If n is odd, the minimizer is unique, otherwise the minimizers form an interval in $\mathcal{L}$.

Besides distributive lattices, median graphs include trees and some other graphs. A structure which contains the trees and distributive lattices is called a median semi-lattice or a median graph [Mon08]. A Cartesian product of median graphs is a median graph, and medians of the product can be computed by finding the medians of the factors.

Equipped with the tools from the theory of median lattices, we can state the main result as follows: *The median on $\mathcal{L}$ is a Cartesian product of the medians, i.e., it is the natural extension of the Median.*

This result coincides with the definition of the median as a penalty based function (8.38). Indeed, by using the natural distance (8.59), we write the penalty on $\mathcal{L}^n \times \mathcal{L}$ as follows

$$P(B_1, \ldots, B_n, Y) = \sum_{i=1}^{n} d(B_i, Y) = \sum_{i=1}^{n} |l_{B_i} - l_Y| + |r_{B_i} - r_Y|, \tag{8.44}$$

where as earlier $B_i = [l_{B_i}, r_{B_i}]$ and $Y = [l_Y, r_Y]$ (we use IVFV notation here since the lattice of intervals is more easily interpreted). The median is the minimizer of

$$Med_L(B_1, \ldots, B_n) = \arg\min_{Y \in L} \sum_{i=1}^{n} |l_{B_i} - l_Y| + |r_{B_i} - r_Y|$$

$$= \arg\min_{Y \in L} \sum_{i=1}^{n} |l_{B_i} - l_Y| + \sum_{i=1}^{n} |r_{B_i} - r_B|. \tag{8.45}$$

Each term in the first sum in the latter expression depends on l_Y only, and each term in the second sum depends on r_Y only. Therefore minimization with respect to l_Y

and r_Y can be performed separately. Since the solutions to the optimization problems with respect to l_Y and r_Y are the usual medians, the result is the Cartesian product of the medians, i.e., a natural extension of the median.

8.2.3 Medians on Infinite Lattices and Fermat Points

The main result of the previous section, that the median on a Cartesian product of lattices is the Cartesian product of the medians, provides a unique solution for finite lattices. In the case of IVFV and AIFV, the lattice $\mathcal{L}$ is infinite. Following Monjardet [Mon08], we now consider the Cartesian product $\mathbb{R}^2$. Here, there are other ways to define the distance, besides the natural distance (8.59). The usual Euclidean distance is one such choice. Then the median is defined as a solution to

$$Med_F(B_1, \ldots, B_n) = \arg\min_{Y \in L} \sum_{i=1}^{n} \sqrt{|l_{B_i} - l_Y|^2 + |r_{B_i} - r_Y|^2}. \qquad (8.46)$$

The definition of the median by (8.46) was advocated in the 1950s by Gini in [Gin58], who writes that the usual definition of the median based on sorting has no sense for pairs, triples and n-tuples, and it is expression (8.46) that should be used.

This time the solution, called the geometric median, *is not the Cartesian product of the medians*. In fact this is the generalized Fermat problem, and its solution is known as the Fermat-Weber point [Bri95, Kuh73]. It is unique if the points are not collinear (in the collinear case at least one of the points $B_1, \ldots, B_n$ is optimal and it can be found in linear time by a selection algorithm). Fermat points were considered recently in the context of multidistances in [MM11].

It has been shown that no general solution or explicit formula (using standard operations) exists [Baj88]. However, iterative procedures which converge toward the solution more accurately at each step have been developed. This follows from the convexity of the distance to each point B_i and the convexity of the function giving the sum of distances. Since the sum of convex functions is convex, such procedures aimed at decreasing the total distance at each step cannot become trapped in a local optimum [BMM03, Kuh73].

Weiszfeld's algorithm [CT89, Kuh73, Wei37], is one such approach and is a form of iteratively re-weighted least squares [Bis09]. Each new estimate Y_{k+1} is calculated from a weighted average of the inputs, where the weights are defined according to the inverse distances between each input B_i and the current estimate Y_k. The sequence of the estimates $Y_1, Y_2, \ldots$ is given by

$$Y_{k+1} = \left(\sum_{i=1}^{n} \frac{B_i}{d(B_i, Y_k)} \right) \Big/ \left(\sum_{i=1}^{n} \frac{1}{d(B_i, Y_k)} \right). \qquad (8.47)$$

The initial estimate Y_1 is chosen arbitrarily. Other procedures have also been described, for instance, Bose et al. [BMM03] and the algorithms presented in [Qi00, Xue97].

We should mention that the restriction $Y \in \mathcal{L}$ (as opposed to $Y \in \mathbb{R}^2$ in the Fermat problem) does not affect the solution or the algorithms, as an optimal Y is necessarily in the convex hull of B_i, $i = 1, \ldots, n$. Since all $B_i \in \mathcal{L}$ then necessarily the optimum $Y \in \mathcal{L}$ [Kuh73].

We see that in the continuous case, the median of IVFV is not defined uniquely, but depends on the chosen distance. Only for the natural distance in the Cartesian product $\mathbb{R}^2$ (the Manhattan distance) does it give the Cartesian product of the medians, which is the natural extension of the usual median. Otherwise the median is the Fermat point of the inputs. In that case it does not coincide with the definitions arising in AIFS theory, presented in Sect. 8.1.6. Furthermore, in general such a median is not internal.

When the distance is not Euclidean, but is based on a p-norm, $p > 1$, the solution is also numerical [Xue00]. However, for the Chebyshev norm $p = \infty$ there is an exact solution to the Fermat problem, and hence a geometric median, similar to that of the natural distance. This will become apparent in the next section, where we examine the Hausdorff distance between intervals.

If one desires to define an internal median, then a simple solution is to minimize

$$Med_F(B_1, \ldots, B_n) = \arg \min_{Y \in \{B_1, \ldots, B_n\}} \sum_{i=1}^{n} \sqrt{|l_{B_i} - l_Y|^2 + |r_{B_i} - r_Y|^2}, \quad (8.48)$$

which is trivially done by exhaustive search.

8.2.4 Medians Based on Distances Between Intervals

Hausdorff distance

In this section we look at the definition of the median as a penalty based function (Definition 5.2). Following Chavent et al. [CS08], we extend this definition to $\mathcal{L}$, by using

$$Med_H(B_1, \ldots, B_n) = \arg \min_{Y \in L} \sum_{i=1}^{n} d(B_i, Y), \quad (8.49)$$

and using the Hausdorff distance between intervals. We remind that the Hausdorff distance between sets X and Y is defined by

$$d(X, Y) = \max\{\sup_{x \in X} \inf_{y \in Y} d(x, y), \sup_{y \in Y} \inf_{x \in X} d(x, y)\}. \quad (8.50)$$

In the case of the intervals $A = [l_A, r_A]$, $B = [l_B, r_B]$, the Hausdorff distance is expressed as

$$d(A, B) = \max\{|l_A - l_B|, |r_A - r_B|\}, \tag{8.51}$$

i.e., it is the Chebyshev distance between vectors (l_A, r_A) and (l_B, r_B). Using the identity $\max\{|x - y|, |x + y|\} = |x| + |y|$, Chavent et al. [CS08] show that

$$d(A, B) = |m_A - m_B| + |hl_A - hl_B|, \tag{8.52}$$

where m_A, m_B are the midpoints and hl_A, hl_B are the radii (or half-lengths) of the intervals A, B, $m_A = \frac{r_A - l_A}{2}$, $hl_A = \frac{r_A + l_A}{2}$. Thus the Hausdorff distance between the intervals is at the same time the Chebyshev distance between the vectors (l_A, r_A), (l_B, r_B) and the Manhattan distance between the vectors (m_A, hl_A), (m_B, hl_B). In this case we express the median as

$$Med_H(B_1, \ldots, B_n) = \arg\min_{Y \in L} \sum_{i=1}^{n} |m_{B_i} - m_Y| + |hl_{B_i} - hl_Y|. \tag{8.53}$$

We immediately see (compare to (8.45)) that the result is the Cartesian product of the medians of the interval centers and radii, i.e.,

Theorem 8.54 *If d is the Hausdorff distance, then the median on $\mathcal{L}$ is expressed by $Y = Med_H(B_1, \ldots, B_n)$ with $Y = [l_Y, r_Y]$, $l_Y = m_Y - hl_Y$, $r_Y = m_Y + hl_Y$, and $m_Y = Med(m_{B_1}, \ldots, m_{B_n})$, $hl_Y = Med(hl_{B_1}, \ldots, hl_{B_n})$.*

The result in Theorem 8.54 can be stated in AIFS representation. Since the median is invariant with respect to monotone transformations, and since the accuracy of AIFV $Accuracy(A) = 1 - 2hl_A$ and its score $Score(A) = 2m_A - 1$, we have that the median of AIFVs $A_1, \ldots, A_n$ is the AIFV Y, such that its accuracy and score are the medians of the accuracies and scores of $A_1, \ldots, A_n$. This formula is different from the natural extension of the median.

Wasserstein-based distance

The Hausdorff distance is essentially based on the ends of the compared intervals. Irpino and Verde [IV08] reviewed other interval distances that take into account a measure of overlap between the intervals. One such distance appeared in the context of fuzzy numbers in Bertoluzza et al. [BCS95]. Bertoluzza's distance is in fact a family of functions

$$d_{Ber}^2(A, B) = \tag{8.54}$$

$$\int_0^1 (t|l_A - l_B| + (1 - t)|r_A - r_B|)^2 \gamma(t) dt + \sum_{s=1}^{S} k_s (a_s - b_s)^2,$$

with $a_s = t_s l_A + (1 - t_s) r_A$, $b_s = t_s l_B + (1 - t_s) r_B$ parameterized by a function γ and S, and in the simplest non-trivial case $\gamma = 0$, $S = 3$ and $t_1 = 0$, $t_2 = 0.5$, $t_3 = 1$, it yields

$$d_{Ber}^2(A, B) = (m_A - m_B)^2 + \frac{2}{3}(hl_A - hl_B)^2. \tag{8.55}$$

Irpino and Verde [IV08] introduced Wasserstein-based distance using the following reasoning. Think of two independent random variables uniformly distributed on the intervals A and B respectively. In general if F and G are the distribution functions of two random variables, the Wasserstein L_2 metric is

$$d_{Wass}(F, G) = \left(\int_0^1 (F^{-1}(t) - G^{-1}(t))^2 dt \right)^{1/2}, \tag{8.56}$$

where F^{-1} and G^{-1} are quantile functions of the two distributions. When both F and G are uniform, this equation simplifies to

$$d^2_{Wass}(A, B) = (m_A - m_B)^2 + \frac{1}{3}(hl_A - hl_B)^2. \tag{8.57}$$

This distance is also known as Mallow's distance.

Following, Irpino and Verde expressed various distances between the intervals, such as Bertoluzza, Wasserstein, Coppi and D'Urso and the L_2 distance using a generic formula [IV08]

$$d^2(A, B) = \alpha(m_A - m_B)^2 + \beta(hl_A - hl_B)^2, \tag{8.58}$$

in which parameters $\alpha, \beta > 0$ determine the relative importance of the position component and the size of the interval respectively. There is no explicit solution to the median problem (8.49) in all these cases (the argument is the same as for the Euclidean distance [Baj88]). A numerical solution involves minimization of a convex function, and methods discussed in Sect. 8.2.3 are applicable.

Table 8.5 provides a summary of the different medians we studied. We will now illustrate the use of these medians in the following section.

8.2.5 Numerical Comparison

The following example gives some indication as to the behavior of these medians when aggregating the membership values of interval valued fuzzy sets and Atanassov intuitionistic fuzzy sets.

Example 8.55 Four experts A, B, C, D provide the membership and non-membership pair evaluations for four objects $x_1, \ldots, x_4$. The resulting Atanassov intuitionistic fuzzy values are given in Table 8.6. We are interested in aggregating these evaluations for each object x_i separately by using the median.

For all objects x_i, it holds that the evaluations of the experts satisfy $C_i < B_i < A_i < D_i$ according to the order defined in (8.9) based on the score and accuracy.

For object x_1, the relative ordering of the expert evaluations seems straightforward as there is little overlap between the intervals. For x_2 and x_3 however, the ordering

Table 8.5 Medians in the IVFV representation

Median based on		$Med(B_1, \ldots, B_n)$
Operations on AIFV with the product t-norm	Med_1	$\begin{cases} B_{(\frac{n+1}{2})}, & n \text{ is odd} \\ [G^d(l_{\sigma(\frac{n}{2})}, l_{\sigma(\frac{n}{2}+1)}), G^d(r_{\sigma(\frac{n}{2})}, r_{\sigma(\frac{n}{2}+1)})], & n \text{ is even} \end{cases}$
Operations on AIFV with the Łukasiewicz t-norm	Med_2	$\begin{cases} B_{(\frac{n+1}{2})}, & n \text{ is odd} \\ [A(l_{\sigma(\frac{n}{2})}, l_{\sigma(\frac{n}{2}+1)}), A(r_{\sigma(\frac{n}{2})}, r_{\sigma(\frac{n}{2}+1)})], & n \text{ is even} \end{cases}$
Natural extension	Med_L	$[Med(l_1, \ldots, l_n), Med(r_1, \ldots, r_n)]$
Natural distance in a lattice	Med_L	$[Med(l_1, \ldots, l_n), Med(r_1, \ldots, r_n)]$
Fermat points: • in Manhattan distance	Med_L	$[Med(l_1, \ldots, l_n), Med(r_1, \ldots, r_n)]$
• in Euclidean distance (geometric median)	Med_F	$\arg\min_{B \in L} \sum_{i=1}^{n} \sqrt{\mid l_i - y_l \mid^2 + \mid r_i - y_r \mid^2}$
• in Chebyshev distance	Med_H	same as for Hausdorff distance
Distance between intervals: • Hausdorff distance	Med_H	$[m_B - hl_B, m_B + hl_B]$ where $m_B = Med(m_{B_1}, \ldots, m_{B_n})$, $\quad hl_B = Med(hl_{B_1}, \ldots, hl_{B_n})$
• Wasserstein distance	Med_{Wass}	$\arg\min_{B \in L} \sum_{i=1}^{n} \sqrt{\mid m_i - m_B \mid^2 + \frac{1}{3} \mid hl_i - hl_B \mid^2}$
• Bertoluzza distance	Med_{Ber}	$\arg\min_{B \in L} \sum_{i=1}^{n} \sqrt{\mid m_i - m_B \mid^2 + \frac{2}{3} \mid hl_i - hl_B \mid^2}$

Table 8.6 Expert AIFV evaluations from Example 8.55

Expert	x_1	x_2	x_3	x_4
A	$\langle 0.3, 0.6 \rangle$	$\langle 0.6, 0.3 \rangle$	$\langle 0, 0 \rangle$	$\langle 0.8, 0 \rangle$
B	$\langle 0.2, 0.7 \rangle$	$\langle 0.5, 0.2 \rangle$	$\langle 0.3, 0.4 \rangle$	$\langle 0.3, 0.6 \rangle$
C	$\langle 0.1, 0.7 \rangle$	$\langle 0.55, 0.3 \rangle$	$\langle 0.2, 0.3 \rangle$	$\langle 0, 0.4 \rangle$
D	$\langle 0.4, 0.5 \rangle$	$\langle 0.6, 0 \rangle$	$\langle 0.6, 0.4 \rangle$	$\langle 0.9, 0 \rangle$

is not so intuitive, as intervals overlap. For x_3, expert A's evaluation is the entire interval, which would usually be interpreted as complete ignorance pertaining to the item, however this evaluation is ranked higher than those given by experts C and B.

We now compute the different medians and resulting AIFVs. To illustrate the differences for odd and even inputs, we evaluate the medians of each triplet A, B, C and show the results in Table 8.7, while Table 8.8 gives the median AIFV evaluations of all four experts A, B, C, D (accuracy is to 3 d.p.).

Firstly, we consider the median of three evaluations. We can see that for x_1, the medians are identical, corresponding to the evaluation given by expert B, which is very intuitive, as $C < B < A$. For x_3, Med_1 and Med_2 also give Expert B's evaluation as per the defined ordering of the intervals, however the remaining medians return the evaluation of expert C, which is not in the middle. For x_4, B's evaluation is

Table 8.7 The medians of three evaluations A, B, C in Example 8.55

Median	x_1	x_2	x_3	x_4
$Med_1(A, B, C)$	$\langle 0.2, 0.7 \rangle$	$\langle 0.5, 0.2 \rangle$	$\langle 0.3, 0.4 \rangle$	$\langle 0.3, 0.6 \rangle$
$Med_2(A, B, C)$	$\langle 0.2, 0.7 \rangle$	$\langle 0.5, 0.2 \rangle$	$\langle 0.3, 0.4 \rangle$	$\langle 0.3, 0.6 \rangle$
$Med_L(A, B, C)$	$\langle 0.2, 0.7 \rangle$	$\langle 0.55, 0.3 \rangle$	$\langle 0.2, 0.3 \rangle$	$\langle 0.3, 0.4 \rangle$
$Med_F(A, B, C)$	$\langle 0.2, 0.7 \rangle$	$\langle 0.551, 0.298 \rangle$	$\langle 0.2, 0.3 \rangle$	$\langle 0.274, 0.485 \rangle$
(Internal F)	$\langle 0.2, 0.7 \rangle$	$\langle 0.55, 0.3 \rangle$	$\langle 0.2, 0.3 \rangle$	$\langle 0.3, 0.6 \rangle$
$Med_H(A, B, C)$	$\langle 0.2, 0.7 \rangle$	$\langle 0.575, 0.275 \rangle$	$\langle 0.2, 0.3 \rangle$	$\langle 0.25, 0.55 \rangle$
$Med_{Wass}(A, B, C)$	$\langle 0.2, 0.7 \rangle$	$\langle 0.557, 0.286 \rangle$	$\langle 0.2, 0.3 \rangle$	$\langle 0.267, 0.533 \rangle$
(Internal Wass)	$\langle 0.2, 0.7 \rangle$	$\langle 0.55, 0.3 \rangle$	$\langle 0.2, 0.3 \rangle$	$\langle 0.3, 0.6 \rangle$
$Med_{Ber}(A, B, C)$	$\langle 0.2, 0.7 \rangle$	$\langle 0.554, 0.293 \rangle$	$\langle 0.2, 0.3 \rangle$	$\langle 0.267, 0.502 \rangle$
(Internal Ber)	$\langle 0.2, 0.7 \rangle$	$\langle 0.55, 0.3 \rangle$	$\langle 0.2, 0.3 \rangle$	$\langle 0.3, 0.6 \rangle$

Table 8.8 The medians of four evaluations in Example 8.55

Median	x_1	x_2	x_3	x_4
$Med_1(A, B, C, D)$	$\langle 0.252, 0.648 \rangle$	$\langle 0.553, 0.245 \rangle$	$\langle 0.163, 0 \rangle$	$\langle 0.626, 0 \rangle$
$Med_2(A, B, C, D)$	$\langle 0.25, 0.65 \rangle$	$\langle 0.55, 0.25 \rangle$	$\langle 0.15, 0.2 \rangle$	$\langle 0.55, 0.3 \rangle$
$Med_L(A, B, C, D)$	$\langle 0.25, 0.65 \rangle$	$\langle 0.575, 0.25 \rangle$	$\langle 0.25, 0.35 \rangle$	$\langle 0.55, 0.2 \rangle$
$Med_F(A, B, C, D)$	$\langle 0.3, 0.6 \rangle$	$\langle 0.557, 0.257 \rangle$	$\langle 0.216, 0.288 \rangle$	$\langle 0.741, 0.071 \rangle$
(Internal F)	$\langle 0.3, 0.6 \rangle$	$\langle 0.55, 0.3 \rangle$	$\langle 0.2, 0.3 \rangle$	$\langle 0.8, 0 \rangle$
$Med_H(A, B, C, D)$	$\langle 0.25, 0.65 \rangle$	$\langle 0.538, 0.238 \rangle$	$\langle 0.275, 0.325 \rangle$	$\langle 0.55, 0.3 \rangle$
$Med_{Wass}(A, B, C, D)$	$\langle 0.3, 0.6 \rangle$	$\langle 0.557, 0.257 \rangle$	$\langle 0.216, 0.288 \rangle$	$\langle 0.741, 0.071 \rangle$
(Internal Wass)	$\langle 0.3, 0.6 \rangle$	$\langle 0.6, 0.3 \rangle$	$\langle 0.2, 0.3 \rangle$	$\langle 0.8, 0 \rangle$
$Med_{Ber}(A, B, C, D)$	$\langle 0.3, 0.6 \rangle$	$\langle 0.557, 0.257 \rangle$	$\langle 0.216, 0.288 \rangle$	$\langle 0.741, 0.071 \rangle$
(Internal Ber)	$\langle 0.3, 0.6 \rangle$	$\langle 0.55, 0.3 \rangle$	$\langle 0.2, 0.3 \rangle$	$\langle 0.8, 0 \rangle$

the output of the internal Med_F, Med_{Wass} and Med_{Ber} while for x_2 the output is C. The median Med_L is not internal and in the case of x_4 its output is neither A, B or C.

For the median of 4 evaluations, clearly the results are much more varied as we no longer have equivalence between Med_1 and Med_2. The Med_{Wass} and Med_{Ber} are the same to this level of accuracy given that the optimization problems are quite similar, however they do actually differ. Whilst the outputs are quite similar for x_1, there is much more disagreement for x_3 and x_4, where the difference for the ν values differ by up to 0.3 and the μ values by up to 0.25. For example compare the outputs of Med_1 and Med_L, which do not even overlap. Given such significant differences, it suggests that the best median to use will strongly depend on the context and interpretation of the AIFV pairs.

It turns out that there is no unique definition of the median on lattices, such as the sets of AIFV and IVFV, and that different mathematical theories lead to quite distinct medians. First, medians based on intuitionistic OWA functions require sorting the inputs, and hence need a complete order. Algebraic operations on AIFS based on the product t–norm lead to the median Med_1 being inconsistent with the chosen complete order. This can be remedied by using the Łukasiewicz t–norm, Med_2, although the need to sort the inputs does not go away.

The second result is that the natural extension of the median Med_L coincides with the median on distributive lattices when the natural distance on the lattice is used. It also coincides with the Fermat point in the Manhattan distance. Calculation of the geometric median (in the Euclidean distance) is rather complicated, and can be done only by using numerical algorithms.

The third view of the median, which is different, yet related to the first two, is based on minimizing a penalty for disagreement between the inputs. Here various distances between the intervals can be used, because the lattice is infinite. The Hausdorff distance leads to the median Med_H, which is computed exactly by finding the medians of the midpoints and half-lengths of the intervals. Wasserstein and Bertoluzza distances, on the other hand, lead to minimization problems which need to be solved numerically.

We should mention that medians on distributive lattices Med_L are also applicable to a more general case of Goguen's L-fuzzy sets [Gog67], as long as the lattice $\mathcal{L}$ is distributive. Here one needs to define a distance on $\mathcal{L}$, which will be the shortest path distance on the corresponding median graph.

8.3 Penalty Functions on Cartesian Products of Lattices

In Chap. 5 we have seen that the averaging functions of real variables can be constructed by means of penalty functions. An analogous construction can be defined for general lattices. In particular, in [BBCJB14, BJPMB13] this construction is performed by using restricted dissimilarity functions [BBP06, BBP08b] (see Definition 5.20) as the first step.

Restricted dissimilarity functions can be used to define distances between fuzzy sets.

Definition 8.56 (*Distance between fuzzy sets*) [Liu92] A mapping $\mathcal{D} : \mathcal{FS}(U)^2 \to [0, 1]$ is a distance over $\mathcal{FS}(U)$ if

1. $\mathcal{D}(A, B) = \mathcal{D}(B, A)$ for every $A, B \in \mathcal{FS}(U)$;
2. $\mathcal{D}(A, B) = 0$ if and only if $A = B$;
3. $\mathcal{D}(A, B) = 1$ if and only if A and B are complementary crisp sets;
4. if $A \leq A' \leq B' \leq B$, then $\mathcal{D}(A, B) \geq \mathcal{D}(A', B')$.

In this section the distances between fuzzy sets are understood in the sense of Liu.

Theorem 8.57 [BBP08b, BMBP07] *Let M be an aggregation function such that it satisfies*

(A1) $M(x_1, \ldots, x_n) = 1$ *if and only if* $x_1 = \cdots = x_n = 1;$
(A2) $M(x_1, \ldots, x_n) = 0$ *if and only if* $x_1 = \cdots = x_n = 0,$

and let $d_R : [0, 1]^2 \to [0, 1]$ *be a restricted dissimilarity function. Then*

$$\mathcal{D}(A, B) = \overset{n}{\underset{i=1}{M}} (d_R(A(u_i), B(u_i)))$$

for all $A, B \in \mathcal{FS}(U)$ *defines a distance in the sense of Liu.*

From the point of view of applications we consider lattices which are the Cartesian product of finite chains $\mathcal{C}$. Moreover, and since all the finite chains of the same length are isomorphic to each other, we can always assume that we are working with chains of the type $\mathcal{C} = 0 \leq 1 \leq 2 \leq \cdots \leq n - 1$.

Note 8.58 The Cartesian product needs not be a totally ordered set, i.e., a chain.

Theorem 8.59 [Bir67] *Let* $\mathcal{L}_k = \{\mathcal{C}_1 \times \cdots \times \mathcal{C}_k, \leq, \wedge, \vee\}$. *Let* a, b *be two elements in* $\mathcal{L}_k$ *such that* $a \leq b$. *Then all the maximal chains joining* a *and* b *have the same length.*

Corollary 8.60 *Take* $a, b \in \mathcal{L}_k = \{\mathcal{C}_1 \times \cdots \times \mathcal{C}_k, \leq, \wedge, \vee\}$. *Then all the maximal chains joining* $\wedge(a, b)$ *and* $\vee(a, b)$ *are of the same length.*

So the distance between $x, y \in \mathcal{L}$ can be defined as the length of the chain $\mathcal{C}$ with minimal element $a = \wedge(x, y)$ and maximal element $b = \vee(x, y)$, minus one, see Definition 8.50. That is,

$$d(x, y) = length(\mathcal{C}) - 1.$$

This definition is equivalent to the following.

$$d(x, y) = \sum_{i=1}^{m} d_i(x_i, y_i) = \sum_{i=1}^{m} |x_i - y_i|, \tag{8.59}$$

where d_i is the distance in the ith chain. It is easy to see that Eq. (8.59) defines a distance, called the natural distance.

In order to build penalty functions over a Cartesian product of lattices, we first extend the definition of the distance to L-fuzzy sets using the notion of restricted dissimilarity function.

Note 8.61 In this work we will only consider chains $\mathcal{C}_i$ with supremum and infimum, which are lattices.

Consider the lattice $\mathcal{L}_m = \{\mathcal{C}_1 \times \cdots \times \mathcal{C}_m, \leq, \wedge, \vee\}$. For each chain $\mathcal{C}_i$ we denote by $\vee(\mathcal{C}_i)$ and $\wedge(\mathcal{C}_i)$ its top and bottom elements, respectively. We also denote

$$1_{\mathcal{L}_m} = (\vee(\mathcal{C}_1), \ldots, \vee(\mathcal{C}_m)),$$
$$0_{\mathcal{L}_m} = (\wedge(\mathcal{C}_1), \ldots, \wedge(\mathcal{C}_m)).$$

Definition 8.62 (*Lattice restricted dissimilarity function*) Take $\mathcal{L}_m = \{\mathcal{C}_1 \times \cdots \times \mathcal{C}_m, \leq, \wedge, \vee\}$. The mapping

$$\delta_R : \mathcal{L}_m \times \mathcal{L}_m \to \mathcal{L}_m$$

is a lattice restricted dissimilarity function if

1. $\delta_R(x, y) = \delta_R(y, x)$ for any $x, y \in \mathcal{L}_m$;
2. $\delta_R(x, y) = 1_{\mathcal{L}_m}$ if and only if for any $i = 1, \ldots, m$,

$$x_i = \vee(\mathcal{C}_i) \text{ and } y_i = \wedge(\mathcal{C}_i),$$

or

$$x_i = \wedge(\mathcal{C}_i) \text{ and } y_i = \vee(\mathcal{C}_i);$$

3. $\delta_R(x, y) = 0_{\mathcal{L}_m}$ if and only if $x = y$;
4. If $x \leq y \leq z$ then $\delta_R(x, y) \leq \delta_R(x, z)$ and $\delta_R(y, z) \leq \delta_R(x, z)$.

Proposition 8.63 *Let each* $\delta_{R_i} : \mathcal{C}_i \times \mathcal{C}_i \to \mathcal{C}_i$ *be a lattice restricted dissimilarity function. Then the mapping defined as*

$$\delta_R(x, y) = (\delta_{R_1}(x_1, y_1), \ldots, \delta_{R_m}(x_m, y_m)) \qquad (8.60)$$

for every $x, y \in \mathcal{L}_m$ *is a lattice restricted dissimilarity function.*

Let $\mathcal{FS}(U)^m$ denote the class $\mathbf{A} = (A_1, \ldots, A_m)$ with $A_i : U \to \mathcal{C}_i$ such that $\mathbf{A}(u_i) = (A_1(u_i), \ldots, A_m(u_i))$ for every $u_i \in U$. Notice that each A_i is an L-fuzzy set in the sense of Goguen [Gog67]; i.e., each A_i is a fuzzy set defined over the lattice $\{\mathcal{C}_i, \leq_i, \wedge_i, \vee_i\}$.

The construction methods for restricted dissimilarity functions described in [BBP08b] can be easily adapted to lattice restricted dissimilarity functions, so we do not develop them here.

Definition 8.64 (*Lattice distance between fuzzy sets*) Take $\mathcal{L}_m = \{\mathcal{C}_1 \times \cdots \times \mathcal{C}_m, \leq, \wedge, \vee\}$. The mapping

$$\Omega : \mathcal{FS}(U)^m \times \mathcal{FS}(U)^m \to \mathcal{L}_m$$

is a lattice distance in $\mathcal{FS}(U)^m$ if

1. $\Omega(\mathbf{A}, \mathbf{B}) = \Omega(\mathbf{B}, \mathbf{A})$ for every $\mathbf{A}, \mathbf{B} \in \mathcal{FS}(U)^m$;
2. $\Omega(\mathbf{A}, \mathbf{B}) = 0_{\mathcal{L}_m}$ if and only if $A_i = B_i$ for every $i = 1, \ldots, m$;
3. $\Omega(\mathbf{A}, \mathbf{B}) = 1_{\mathcal{L}_m}$ if and only if for every $i = 1, \ldots, m$, A_i and B_i are sets such that for every u_j

$$A_i(u_j) = \vee(C_i) \text{ and } B_i(u_j) = \wedge(C_i)$$

or

$$A_i(u_j) = \wedge(C_i) \text{ and } B_i(u_j) = \vee(C_i);$$

4. If $\mathbf{A} \leq \mathbf{A}' \leq \mathbf{B}' \leq \mathbf{B}$, then $\Omega(\mathbf{A}, \mathbf{B}) \geq \Omega(\mathbf{A}', \mathbf{B}')$ where $\mathbf{A} = (A_1, \ldots, A_m) \leq (A_1', \ldots, A_m') = \mathbf{A}'$ if $A_i \leq A_i'$ for every i.

We have the following result.

Proposition 8.65 *Let $M_1, \ldots, M_m$ be aggregation functions*

$$M_i : C_i \times C_i \rightarrow C_i.$$

Then the mapping

$$F : \mathcal{L}_m \times \mathcal{L}_m \rightarrow \mathcal{L}_m \text{ given by}$$
$$F(\mathbf{x}, \mathbf{y}) = (M_1(x_1, y_1), \ldots, M_m(x_m, y_m))$$

is an aggregation function over $\mathcal{L}_m$.

Now we can introduce a method of building lattice distances.

Proposition 8.66 *Let $\delta_{R_1}, \ldots, \delta_{R_m}$ be a lattice restricted dissimilarity function $\delta_{R_i} : C_i \times C_i \rightarrow C_i$. Let $M_1, \ldots, M_m$ be aggregation functions $M_i : C_i \times \cdots \times C_i \rightarrow C_i$ such that*

(L1) $M_i(x_1, \ldots, x_n) = 1_{\mathcal{L}}$ if and only if $x_i = \vee(C_i)$ for every $i = 1, \ldots, n$,
(L2) $M_i(x_1, \ldots, x_n) = 0_{\mathcal{L}}$ if and only if $x_i = \wedge(C_i)$ for every $i = 1, \ldots, n$.

Then

$$\Omega(\mathbf{A}, \mathbf{B}) = \left(\overset{n}{\underset{i=1}{M_1}} (R_1(A_1(u_i), B_1(u_i))), \ldots, \overset{n}{\underset{i=1}{M_m}} (R_m(A_m(u_i), B_m(u_i))) \right) \quad (8.61)$$

defines a lattice distance in $\mathcal{FS}(U)^m$.

We know that the arithmetic mean of convex functions is also a convex function. Next we consider other aggregation functions such that when applied to convex functions result in another convex function, similarly to the arithmetic mean.

Theorem 8.67 *Let $Y = (y_1, \ldots, y_m) \in \mathcal{L}_m$. For each y_i ($i = 1, \ldots, m$) we consider the set*

$$B_{y_i}(u_j) = y_i \text{ for all } u_j \in U, \quad (8.62)$$

and let $\mathbf{B}_Y = (B_{y_1}, \ldots, B_{y_m}) \in \mathcal{FS}(U)^m$. Let $M_1, \ldots, M_m$ be aggregation functions $M_i : C_i \times \cdots \times C_i \rightarrow C_i$ such that each of them when composed with convex *functions is also convex. Take the lattice restricted dissimilarity function* $\delta_R(x, y) = (\delta_{R_1}(x_1, y_1), \ldots, \delta_{R_m}(x_m, y_m))$ *such that each* δ_{R_i} *with* $i = 1, \ldots, m$ *is convex in one variable. Then*

$P_\Omega : \mathcal{FS}(U)^{m+1} \to \mathcal{L}_m$ *given by*

$$P_\Omega(\mathbf{A}, Y) = \Omega(\mathbf{A}, \mathbf{B}_Y) = \left(\overset{n}{\underset{i=1}{M_1}}(\delta_{R_1}(A_1(u_i), y_1)), \ldots, \overset{n}{\underset{i=1}{M_m}}(\delta_{R_m}(A_m(u_i), y_m)) \right)$$

(8.63)

satisfies:

1. $P_\Omega(\mathbf{A}, Y) \geq 0_{\mathcal{L}_m}$;
2. $P_\Omega(\mathbf{A}, Y) = 0_{\mathcal{L}_m}$ *if* $A_k(u_j) = y_k$ *for every* k *and for every* j;
3. *Each of its components is convex with respect to the corresponding* y_k *(*$k = 1, \ldots, m$*).*

Analogously to the case of the real variables (see [CMY04, Mes07] and Sect. 5.2), we use the term **lattice faithful restricted dissimilarity functions** to denote the following lattice restricted dissimilarity functions:

$$\delta_R(x, y) = K(d(x, y)) = K\left(\sum_{i=1}^{m} |x_i - y_i| \right) \qquad (8.64)$$

with $K : \mathbb{R} \to \mathbb{R}$ a convex function with the unique minimum at $K(0) = 0$.

Theorem 8.68 *In the setting of Theorem 8.67, if* $\delta_{R_1}, \ldots, \delta_{R_m}$ *are lattice faithful restricted dissimilarity functions, then the mapping*

$F_{\mathcal{L}_m} : \mathcal{FS}(U)^m \to \mathcal{L}_m$ *given by*

$$F_{\mathcal{L}_m}(\mathbf{A}) = \arg\min_Y P_\Omega(\mathbf{A}, Y) = \arg\min_Y \Omega(\mathbf{A}, \mathbf{B}_Y)$$

$$= \left(\arg\min_{y_1} (\overset{n}{\underset{i=1}{M_1}}(K_1(d(A_1(u_i), y_1)))), \ldots, \arg\min_{y_m} (\overset{n}{\underset{i=1}{M_m}}(K_m(d(A_m(u_i), y_m)))) \right)$$

$$= \left(\arg\min_{y_1} (\overset{n}{\underset{i=1}{M}}(K_1(|A_1(u_i) - y_1|))), \ldots, \arg\min_{y_m} (\overset{n}{\underset{i=1}{M}}(K_m(|A_m(u_i) - y_m|))) \right)$$

is such that each of its components is an averaging aggregation function over $\mathcal{FS}(U)$ *and* $F_{\mathcal{L}_m}(\mathbf{A})$ *is an averaging aggregation function over the Cartesian product* $\mathcal{FS}(U)^m$.

From now on we will denote by B_{y_q} the fuzzy set over U such that all its membership values are equal to $y_q \in [0, 1]$; that is, $B_{y_q}(u_i) = y_q \in [0, 1]$ for all $u_i \in U$.

Let $Y = (y_1, \ldots, y_m)$ and $\mathbf{B}_Y = (B_{y_1}, \ldots, B_{y_m}) \in \mathcal{FS}(U)^m$. We will denote by C^* a chain whose elements belong to $[0, 1]$ and by $\mathcal{L}_m^*$ the product such that $\mathcal{L}_m^* = C^* \times \cdots \times C^*$.

We can build penalty functions over a Cartesian product of lattices as follows.

Theorem 8.69 *Let $K_i : \mathbb{R} \to \mathbb{R}^+$ be convex functions with the unique minimum at $K_i(0) = 0$ $(i = 1, \ldots, m)$, and take the distance between fuzzy sets defined as*

$$\mathcal{D}(A, B) = \sum_{i=1}^{n} |A(u_i) - B(u_i)|, \tag{8.65}$$

where $A, B \in \mathcal{FS}(U)$ and $Cardinal(U) = n$. Then the mapping

$P_\nabla : \mathcal{FS}(U)^m \times \mathcal{L}_m^* \to \mathbb{R}^+$ *given by*

$$P_\nabla(\mathbf{A}, Y) = \sum_{q=1}^{m} K_q(\mathcal{D}(A_q, B_{y_q})) = \sum_{q=1}^{m} K_q \left(\sum_{p=1}^{n} |A_q(u_p) - y_q| \right) \tag{8.66}$$

satisfies

1. *$P_\nabla(\mathbf{A}, Y) \geq 0$;*
2. *$P_\nabla(\mathbf{A}, Y) = 0$ if and only if $A_q = y_q$ for every $q = 1, \ldots, m$;*
3. *is convex in y_q for every $q = 1, \ldots, m$.*

Observe that P_∇ is a penalty function defined over the Cartesian product of lattices $\mathcal{L}_m^{*n+1}$.

Example 8.70 • In the setting of Theorem 8.69, if we take $K_q(x) = x^2$ for all $q \in \{1, \ldots, m\}$, then

$$P_\nabla(\mathbf{A}, Y) = \sum_{q=1}^{m} \left(\sum_{p=1}^{n} |A_q(u_p) - y_q| \right)^2. \tag{8.67}$$

• If $K_q(x) = x$ for all $q \in \{1, \ldots, m\}$, then

$$P_\nabla(\mathbf{A}, Y) = \sum_{q=1}^{m} \sum_{p=1}^{n} |A_q(u_p) - y_q|. \tag{8.68}$$

Theorem 8.71 *In the setting of Theorem 8.69, the mapping*

$$F(\mathbf{A}) = \mu = \arg\min_{Y} P_\nabla(\mathbf{A}, Y), \tag{8.69}$$

where μ is the rounding to the smallest closest element, is an averaging aggregation function.

Penalty functions on Cartesian products of lattices have shown themselves very useful in decision making and consensus, see [BBCJB14, BJPMB13].

References

[Acz66]   J. Aczél, *Lectures on Functional Equations and Their Applications* (Academic Press, New York, 1966)
[Ata86]   K. Atanassov, Intuitionistic fuzzy sets. Fuzzy Sets Syst. **80**, 87–96 (1986)
[Ata89]   K. Atanassov, More on intuitionistic fuzzy sets. Fuzzy Sets Syst. **33**, 37–46 (1989)
[Ata94]   K. Atanassov, New operations defined over the intuitionistic fuzzy sets. Fuzzy Sets Syst. **61**, 137–142 (1994)
[Ata99]   K. Atanassov, *Intuitionistic Fuzzy Sets* (Physica-Verlag, Heidelberg, 1999)
[Ata12]   K. Atanassov, *On Intuitionistic Fuzzy Sets Theory* (Springer, Heidelberg, 2012)
[AG89]   K. Atanassov, G. Gargov, Interval valued intuitionistic fuzzy sets. Fuzzy Sets Syst. **31**, 343–349 (1989)
[Baj88]   C. Bajaj, The algebraic degree of geometric optimization problems. Discret. Comput. Geom. **3**, 177–191 (1988)
[Bar80]   M. Barbut, Médiane, distribuitivité, éloighements. Mathématiques et sciences humaines **70**, 5–31 (1980)
[Bel09]   G. Beliakov, Construction of aggregation functions from data using linear programming. Fuzzy Sets Syst. **160**, 65–75 (2009)
[BBF11]   G. Beliakov, H. Bustince, J. Fernandez, The median and its extensions. Fuzzy Sets Syst. **175**, 36–47 (2011)
[BBGMP11] G. Beliakov, H. Bustince, D.P. Goswami, U.K. Mukherjee, N.R. Pal, On averaging operators for Atanassov's intuitionistic fuzzy sets. Inf. Sci. **181**, 1116–1124 (2011)
[BBJCF11] G. Beliakov, H. Bustince, S. James, T. Calvo, J. Fernandez, Aggregation for Atanassov's intuitionistic and interval valued fuzzy sets: the median operator. IEEE Trans. Fuzzy Syst. **20**, 487–498 (2011)
[BBP12]   G. Beliakov, H. Bustince, D. Paternain, Image reduction using means on discrete product lattices. IEEE Trans. Image Process. **21**, 1070–1083 (2012)
[BJ12]   G. Beliakov, S. James, On extending generalized Bonferroni means to Atanassov orthopairs in decision making contexts. Fuzzy Sets Syst. **211**, 84–98 (2012)
[BJ14]   G. Beliakov, S. James, Averaging aggregation functions for preferences expressed as pythagorean membership grades and fuzzy orthopairs, in *Proceedings of the 11th IEEE InternationalConference on Fuzzy Systems (FUZZ-IEEE)*. Beijing, China (2014), pp. 298–305
[BCS95]   C. Bertoluzza, N. Corral, A. Salas, On a new class of distances between fuzzy numbers. Mathware Soft Comput. **2**, 71–84 (1995)
[Bir67]   G. Birkhoff, *Lattice Theory*, 3rd edn. (American Mathematical Society Colloquium Publications, New York, 1967)
[Bis09]   N. Bissantz, Convergence analysis of generalized iteratively reweighted least squares algorithms on convex function spaces. SIAM J. Optim. **19** (2009)
[BMM03]   P. Bose, A. Maheshwari, P. Morin, Fast approximations for sums of distances, clustering and the Fermat-Weber problem. Comput. Geom.: Theory Appl. **24**, 135–146 (2003)
[Bri95]   J. Brimberg, The Fermat-Weber location problem revisited. Math. Program. **71**, 71–76 (1995)
[Bus10]   H. Bustince, Interval-valued fuzzy sets in soft computing. Int. J. Comput. Intell. Syst. **3**, 215–222 (2010)
[BBCJB14] H. Bustince, E. Barrenechea, T. Calvo, S. James, G. Beliakov, Consensus in multi-expert decision making problems using penalty functions defined over a Cartesian product of lattices. Inf. Fusion **17**, 56–64 (2014)
[BBP06]   H. Bustince, E. Barrenechea, M. Pagola, Restricted equivalence functions. Fuzzy Sets Syst. **157**, 2333–2346 (2006)
[BBP08a]   H. Bustince, E. Barrenechea, M. Pagola, Generation of interval-valued fuzzy and Atanassovs intuitionistic fuzzy connectives from fuzzy connectives and from K_α operators: laws for conjunctions and disjunctions, amplitude. Int. J. Intell. Syst. **23**, 680–714 (2008)

[BBP08b]   H. Bustince, E. Barrenechea, M. Pagola, Relationship between restricted dissimi-
            larity functions, restricted equivalence functions and normal EN-functions: image
            thresholding invariant. Pattern Recognit. Lett. **29**, 525–536 (2008)
[BB96]   H. Bustince, P. Burillo, Structures on intuitionistic fuzzy relations. Fuzzy Sets Syst.
            78, 293–303 (1996)
[BFKM13]   H. Bustince, J. Fernández, A. Kolesásová, R. Mesiar, Generation of linear orders for
            intervals by means of aggregation functions. Fuzzy Sets Syst. **220**, 69–77 (2013)
[BGBKM13]   H. Bustince, M. Galar, B. Bedregal, A. Kolesásová, R. Mesiar, A new approach
            to interval-valued Choquet integrals and the problem of ordering in interval-valued
            fuzzy setapplications. IEEE Trans. Fuzzy Syst. **21**, 1150–1162 (2013)
[BJPMB13]   H. Bustince, A. Jurio, A. Pradera, R. Mesiar, G. Beliakov, Generalization of the
            weighted voting method using penalty functions constructed via faithful restricted
            dissimilarity functions. Eur. J. Oper. Res. **225**, 472–478 (2013)
[BMBP07]   H. Bustince, J. Montero, E. Barrenechea, M. Pagola, Semiautoduality in a restricted
            family of aggregation operators. Fuzzy Sets Syst. **158**, 1360–1377 (2007)
[BMPBG08]   H. Bustince, J. Montero, M. Pagola, E. Barrenechea, D. Gómez, A survey of interval-
            valued fuzzy sets, in *Handbook of Granular Computing* (Wiley, 2008), pp. 489–515
[CMY04]   T. Calvo, R. Mesiar, R.R. Yager, Quantitative weights and aggregation. IEEE Trans.
            Fuzzy Syst. **12**, 62–69 (2004)
[CM08]   O. Castillo, P. Melin, *Type-2 Fuzzy Logic Theory and Applications* (Springer, Berlin,
            2008)
[CT89]   R. Chandrasekaran, A. Tamir, Open questions concerning Weiszfeld's algorithm for
            the Fermat-Weber location problem. Math. Program., Ser. A **44**, 293–295 (1989)
[CS08]   M. Chavent, J. Saracco, On central tendency and dispersion measures for intervals
            and hypercubes. Commun. Stat.—Theory Methods **37**, 1471–1482 (2008)
[Des07]   G. Deschrijver, Arithmetic operators in interval-valued fuzzy set theory. Inf. Sci.
            177, 2906–2924 (2007)
[DK03]   G. Deschrijver, E.E. Kerre, On the relationship between some extensions of fuzzy
            set theory. Fuzzy Sets Syst. **133**, 227–235 (2003)
[DK05]   G. Deschrijver, E.E. Kerre, Implicators based on binary aggregation operators in
            interval-valued fuzzy set theory. Fuzzy Sets Syst. **153**, 229–248 (2005)
[DK07]   G. Deschrijver, E.E. Kerre, On the position of intuitionistic fuzzy set theory in the
            framework of theories modelling imprecision. Inf. Sci. **177**, 1860–1866 (2007)
[DK08]   G. Deschrijver, E.E. Kerre, Aggregation operators in interval-valued fuzzy and
            Atanassov's intuitionistic fuzzy set theory, in *Fuzzy Sets and Their Extensions: Rep-
            resentation, Aggregation and Models*, ed. by H. Bustince, F. Herrera, J. Montero
            (Springer, Heidelberg, 2008), pp. 183–203
[DBR00]   S.K. Dey, R. Biswas, A.R. Roy, Some operations on intuitionistic fuzzy sets. Fuzzy
            Sets Syst. **114**, 477–484 (2000)
[GFBB11]   M. Galar, J. Fernandez, G. Beliakov, H. Bustince, Interval-valued fuzzy sets applied
            to stereo matching of color images. IEEE Trans. Image Process. **20**, 1949–1961
            (2011)
[Gal+13]   M. Galar et al., Aggregation functions to combine RGB color channels in stereo
            matching. Opt. Express **21**, 1247–1257 (2013)
[GLMP09]   J.-L. Garcia-Lapresta, M. Martinez-Panero, Linguisticbased voting through centered
            OWA operators. Fuzzy Optim. Decis. Mak. **8**, 381–393 (2009)
[GB93]   W.L. Gau, D.J. Buehrer, Vague sets. IEEE Trans. Syst., Man Cybern. **23**, 610–614
            (1993)
[Gin58]   C. Gini, *Le Medie*. Milan (Russian translation, Srednie Velichiny, Statistica, Moscow,
            1970): Unione Tipografico- Editorial Torinese (1958)
[Gog67]   J. Goguen, L-fuzzy sets. J. Math. Anal.: Appl. **18**, 143–174 (1967)
[GMMP09]   M. Grabisch, J.-L. Marichal, R. Mesiar, E. Pap, *Aggregation Functions*, Encyclopedia
            of Mathematics and Its Foundations (Cambridge University Press, Cambridge, 2009)

[IV08]   A. Irpino, R. Verde, Dynamic clustering of interval data using a Wasserstein-based distance. Pattern Recognit. Lett. **29**, 1648–1658 (2008)

[KMP00]   E.P. Klement, R. Mesiar, E. Pap, *Triangular Norms* (Kluwer, Dordrecht, 2000)

[KY97]   G.J. Klir, B. Yuan, *Fuzzy Sets and Fuzzy Logic Theory and Application* (Prentice-Hall of India Pvt. Ltd., New Delhi, 1997)

[Kuh73]   H.W. Kuhn, A note on Fermat's problem. Math. Program. **4**, 98–107 (1973)

[Li10a]   D.-F. Li, Mathematical-programming approach to matrix games with payoffs represented by Atanassov's interval-valued intuitionistic fuzzy sets. IEEE Trans. Fuzzy Syst. **18**, 1112–1128 (2010)

[Li10b]   D.-F. Li, TOPSIS-based nonlinear-programming methodology for multiattribute decision making with interval-valued intuitionistic fuzzy sets. IEEE Trans. Fuzzy Syst. **18**, 299–311 (2010)

[Li14]   D.-F. Li, *Decision and Game Theory in Management With Intuitionistic Fuzzy Sets* (Springer, Berlin, 2014)

[Liu92]   X. Liu, Entropy, distance measure and similarity measure of fuzzy sets and their relations. Fuzzy Sets Syst. **52**, 305–318 (1992)

[MM11]   J. Martin, G. Mayor, Multi-argument distances. Fuzzy Sets Syst. **167**, 92–100 (2011)

[MMR98]   F.R. McMorris, H.M. Mulder, F.S. Roberts, The median procedure on median graphs. Discret. Appl. Math. **84**, 165–181 (1998)

[Men07]   J.M. Mendel, Type-2 fuzzy sets and systems: an overview. IEEE Comput. Intell. Mag. **2**, 20–29 (2007)

[MJL06]   J.M. Mendel, R.I. John, F. Liu, Interval type-2 fuzzy logic systems made simple. IEEE Trans. Fuzzy Syst. **14**, 808–821 (2006)

[Mes07]   R. Mesiar, Fuzzy set approach to the utility, preference relations, and aggregation operators. Eur. J. Oper. Res. **176**, 414–422 (2007)

[MSV08]   R. Mesiar, J. Spirková, L. Vavríková, Weighted aggregation operators based on minimization. Inf. Sci. **178**, 1133–1140 (2008)

[Mon08]   B. Monjardet, Mathématique Sociale and Mathematics. A case study: condorcets effect and medians. Electron. J. Hist. Probab. Stat. **4**, 1–26 (2008)

[MH80]   B. Monjardet, P.L. Hammer, Theorie de la mediane dans les treillis distributes finis et applications. Ann. Discret. Math. **9**, 87–91 (1980)

[NSMK11]   M. Nachtegael, P. Sussner, T. Melange, E.E. Kerre, On the role of complete lattices in mathematical morphology: from tool to uncertainty model. Inf. Sci. **181**, 1971–1988 (2011)

[Qi00]   L. Qi, A smoothing Newton method for minimizing a sum of Euclidean norms. SIAM J. Optim. **11**, 389–410 (2000)

[TC10]   C. Tan, X. Chen, Intuitionistic fuzzy Choquet integral operator for multi-criteria decision making. Expert Syst. Appl. **37**, 149–157 (2010)

[Wei10]   G. Wei, Some induced geometric aggregation operators with intuitionistic fuzzy information and their application to group decision making. Appl. Soft Comput. **10**, 423–431 (2010)

[Wei37]   E. Weiszfeld, Sur le point pour lequel la somme des distances de n points donnes est minimum. Tohoku Math. J. **43**, 355–386 (1937)

[WM11]   D. Wu, J.M. Mendel, Linguistic summarization using IF—THEN rules and interval type-2 fuzzy sets. IEEE Trans. Fuzzy Syst. **19**, 136–151 (2011)

[Xu07]   Z. Xu, Intuitionistic fuzzy aggregation operations. IEEE Trans. Fuzzy Syst. **15**, 1179–1187 (2007)

[Xu10]   Z. Xu, Choquet integrals of weighted intuitionistic fuzzy information. Inf. Sci. **180**, 726–736 (2010)

[Xu12]   Z. Xu, *Intuitionistic Fuzzy Aggregation and Clustering* (Springer, Berlin, 2012)

[Xu14]   Z. Xu, *Intuitionistic Preference Modeling and Interactive Decision Making* (Springer, Berlin, 2014)

[XC11]   Z. Xu, Q. Chen, A multi-criteria decision making procedure based on interval-valued intuitionistic fuzzy Bonferroni means. J. Syst. Sci. Syst. Eng. **20**, 217–228 (2011)

[XY06]   Z. Xu, R.R. Yager, Some geometric aggregation operators based on intuitionistic fuzzy sets. Int. J. General Syst. **35**, 417–433 (2006)

[XY11]   Z. Xu, R.R. Yager, Intuitionistic fuzzy Bonferroni means. IEEE Trans. Syst., Man, Cybern., Part B: Cybern. **41**, 568–578 (2011)

[Xue97]   G. Xue, An efficient algorithm for minimizing a sum of Euclidean norms with applications. SIAM J. Optim. **7**, 1017–1036 (1997)

[Xue00]   G. Xue, An efficient algorithm for minimizing a sum of pnorms. SIAM J. Optim. **10**, 551–579 (2000)

[Yag88]   R.R. Yager, On ordered weighted averaging aggregation operators in multicriteria decision making. IEEE Trans. Syst., Man Cybern. **18**, 183–190 (1988)

[Yag09]   R.R. Yager, OWA aggregation of intuitionistic fuzzy sets. Int. J. General Syst. **38**, 617–641 (2009)

[Yag13]   R.R. Yager, Pythagorean fuzzy subsets, in *Proceedings of the Joint IFSA World Congress and NAFIPS Annual Meeting*. Edmonton, Canada (2013), pp. 57–61

[YA13]   R.R. Yager, A.M. Abbasov, Pythagorean membership grades, complex numbers, and decision making. Int. J. Intell. Syst. **28**, 436–452 (2013)

[Zad73]   L.A. Zadeh, Outline of a new approach to the analysis of complex systems and decision processes interval-valued fuzzy sets. IEEE Trans. Syst., Man Cybern. SMC-**3**, 28–44 (1973)

[Zad08]   L.A. Zadeh, Is there a need for fuzzy logic? Inf. Sci. **178**, 2751–2779 (2008)

[Zad09]   L.A. Zadeh, Toward extended fuzzy logic—a first step. Fuzzy Sets Syst. **160**, 3175–3181 (2009)

[ZXNL10]   H. Zhao, Z. Xu, M. Ni, S. Liu, Generalized aggregation operators for intuitionistic fuzzy sets. Int. J. Intell. Syst. **25**, 1–30 (2010)

Index

A

Absorbing element, 13, 24, 64, 93, 102, 150
Accuracy, 306
Additive aggregation function, 59
Additive fuzzy measure, 154, 159, 164
Additive generator, 159
Aggregation function, 3
  additive, 59, 150
  associative, 15, 24, 31
  averaging, 9, 10, 13, 24, 30, 31, 55, 183
  bipolar, 31, 32, 216
  bisymmetric, 16
  choosing, 32
  classes, 9
  composition, 29
  conjunctive, 5, 9, 13, 27, 31
  continuous, 21, 150
  decomposable, 15, 76
  disjunctive, 9, 13, 27, 31
  double, 217
  dual, 19
  extended, 4, 15
  fitting, 32
  homogeneous, 17, 59, 64, 67, 150
  idempotent, 10, 55, 150
  intuitionistic, 308
  kernel, 23, 59, 150, 174
  linear, 17
  Lipschitz, 24, 59, 214
  1-Lipschitz, 23
  locally internal, 17, 255
  mixed, 9, 28
  on lattice, 308
  penalty based, 183, 185
  p-stable, 23, 213
  recursive, 15
  representable, 308
  self-dual, 20, 65, 164
  self-identity, 17, 239, 247
  shift-invariant, 17, 150
  strictly increasing, 11, 58, 76
  symmetric, 10, 24, 31, 58, 216
  transformation, 29
  with flying parameter, 210
Aggregation operator, 4
α-quantile, 194, 243
A-median, 25, 138
Annihilator, 13
Anonymity, 11
Archimedean t-conorm, 159
Archimedean t-norm, 207
Arithmetic mean, 5, 10, 15, 58, 165, 209, 309, 310
Arithmetico-geometric mean, 90
Associativity, 15, 24
Attitudinal character, 56, 103
Automorphism, 18
Average orness value, 57
Averaging aggregation function, 9, 13, 24, 30, 55, 186

B

Bajraktarevic mean, 89, 211, 263
Balanced fuzzy measure, 154
Banzhaf index, 162
Banzhaf interaction index, 162
Basic probability assignment, 157

© Springer International Publishing Switzerland 2016
G. Beliakov et al., *A Practical Guide to Averaging Functions*,
Studies in Fuzziness and Soft Computing 329,
DOI 10.1007/978-3-319-24753-3

Basis-exponential mean, 72
Basis-radical mean, 72
Belief measure, 157, 159
Bi-capacity, 216
Bijection, 18
Binary tree extension, 93
Bisymmetry, 16, 76
Bivariate mean, 91
Bonferroni mean, 84, 227, 320
    generalized, 229
Boolean measure, 154, 165
Bounded sum, 5
Bregman loss function, 195
BUM function, 120
Buoyancy, 105, 170

C
Capacity, 145
Cartesian product of medians, 329
Cauchy mean, 88
Chain, 307
Choquet integral, 26, 60, 102, 111, 113, 124,
        145, 149, 176, 193, 212, 216
    calculation, 150
    dual, 164
    entropy, 164
    generalized, 173
    induced, 177
    orness value, 163
Coalition, 147, 160, 288, 291
Comparable, 21
Compound mean, 90
Cone monotonicity, 281
Conjunctive aggregation function, 5, 9, 13,
        27
Consistency, 239, 316
Continuity, 18, 21
Contraharmonic mean, 83
Convex function, 59, 154
Convex optimization, 41
Convexity, 67
Counter-harmonic mean, 83, 107, 211
Covering graph, 327

D
Decomposability, 15, 76
Decomposable fuzzy measure, 159
Density based mean, 271
Density based median, 272
Deviation mean, 194
Diagonal, 10, 207
Dirac measure, 147, 159

Directional monotonicity, 282
Discrete fuzzy measure, 146
Disjunctive aggregation function, 9, 13, 27
Dispersion, 76, 105, 117, 164
Dissimilarity function, 187, 188
Distance, 22, 192
Distorted probability, 109, 111, 158
Distributive lattice, 307, 328
Double aggregation function, 217
Dual OWA, 101
Dual power mean, 65
Dual product, 27
Duality, 18, 19, 65, 66, 132, 150, 153, 164,
        174, 270, 285

E
Einstein sum, 28, 207
Entropic mean, 195
Entropy, 76, 105, 117, 164
Equidistant weights, 125
Estimator of location, 256, 280
Exponential mean, 69
Exponential OWA, 109
Extended aggregation function, 4, 15

F
Faithful penalty, 186, 242, 340
Fermat point, 330
Flying parameter, 210
Fuzzy integral, 173
Fuzzy measure, 26, 145, 147
    λ-, 158, 170
    2-additive symmetric, 165
    3-additive symmetric, 165
    additive, 60, 154, 159, 164
    balanced, 154
    Banzhaf index, 162
    Banzhaf interaction index, 162
    belief, 157, 159
    boolean, 154, 165, 176
    convex, 157
    decomposable, 159
    Dirac, 147, 159
    distorted probability, 158
    dual, 153
    fitting, 166
    interaction index, 161
    k-additive, 152, 160, 163, 168
    m-alternating, 157
    m-monotone, 157
    necessity, 156
    plausibility, 157, 159

possibility, 156
probability, 154
self-dual, 153
Shapley value, 160, 168
strongest, 147
subadditive, 153
submodular, 114, 153, 157, 159, 161, 169
Sugeno, 158, 170
superadditive, 153
supermodular, 114, 153, 157, 159, 161, 169
symmetric, 154, 165
totally monotone, 157
weakest, 147
Fuzzy set, 7

G
Game theory, 160
Generalized Choquet integral, 173
Generalized logarithmic mean, 86, 91
Generalized OWA, 107, 114, 126
Generating function, 80, 127
of RIM quantifier, 121
Geometric-Arithmetic mean inequality, 61
Geometric mean, 5, 10, 13, 25, 30, 60, 63, 69
Gini mean, 82, 89, 195, 263
Global monotonicity, 247
Graduation curve, 208
Graph, 327
Grouping function, 225

H
Hamacher t–norm, 31, 211
Harmonic mean, 25, 30, 60, 69
Heronian mean, 85
H-index, 176
Hirsch index, 176
Hodges-Lehmann estimator, 138
Homogeneous, 17, 59, 64, 93, 132, 185, 195, 209, 221
Huber function, 35
Hurwizc aggregation, 102

I
Idempotency, 8, 10, 55, 76, 94, 207
Idempotization, 207
Identric mean, 86
Importance, 160, 168
Incomparable, 21, 67
Indeterminancy, 306

Induced Choquet integral, 177
Induced OWA, 130
Induced OWG, 133
Induced Sugeno integral, 178
Interaction index, 161, 168
Interpolation, 208
Interval-valued fuzzy set, 305
Intuitionistic fuzzy set, 305

K
k-additive fuzzy measure, 152, 160, 163, 168
Kernel aggregation function, 23, 150, 174
k-intolerant, 237
k-Lipschitz, 222
Kolmogorov-Nagumo, 76
k-tolerant, 237

L
Lagrangean mean, 88
λ-fuzzy measure, 158, 170
Lattice, 305, 307, 327
Least median of squares (LMS), 258
Least trimmed absolute deviations (LTA), 259
Least trimmed squares (LTS), 258
Least Winsorized squares, 260
Lehmer mean, 82, 83, 195, 261, 296
Linear aggregation function, 17
Linguistic OWA, 217
Linguistic OWG, 217
Linguistic quantifier, 120
Linguistic variable, 217
Lipschitz condition, 22, 23, 40, 59, 67, 213, 214, 222, 286
Lipschitz function, 59
Locally internal, 17, 255
Logarithmic mean, 86, 91
Lukasiewicz t-conorm, 316
Lukasiewicz t-norm, 28, 159

M
Majority, 288
Maximum, 5, 15, 24
Maximum entropy OWA (MEOWA), 117
Maximum likelihood, 201
Mean, 24
    arithmetic, 5, 10, 15, 25, 58, 165, 209
    arithmetico-geometric, 90
    Bajraktarevic, 89, 211
    basis-exponential, 72
    basis-radical, 72

bivariate, 91
Bonferroni, 84, 227, 320
   generalized, 229
Chauchy, 88
classical, 58
compound, 90
contraharmonic, 83
counter-harmonic, 83, 107, 211
density based, 271
deviation, 194
dual power, 65
entropic, 195
exponential, 69
fitting, 34
geometric, 5, 10, 13, 25, 30, 60, 63, 69
Gini, 82, 89, 263
harmonic, 25, 30, 60, 69
Heronian, 85
identric, 86
Lagrangean, 88
Lehmer, 82, 83, 261, 296
logarithmic, 86, 91
non-strict, 76
power, 62, 66, 69
quadratic, 63
quasi-arithmetic, 65, 88, 242
radical, 72
root-power, 62
Stolarsky, 86, 88
trigonometric, 69
trimmed, 257
Tsallis q-exponential, 73
weighted arithmetic, 25, 27, 34, 59, 164
weighted quasi-arithmetic, 65
weights, 73
Measure, 145
Median, 25, 102, 137, 174, 193, 242, 256,
   319, 327
   a-median, 138
   quasi-median, 138
   weighted, 139
Median graph, 328
Membership value, 7
Migrativity, 218, 222
Minimax disparity OWA, 119
Minimum, 5, 15, 24
Minimum variance OWA (MVOWA), 118
Minkowski gauge, 191, 199
Mixed aggregation function, 9, 28
Mixture function, 89, 211, 265
Möbius transformation, 147, 157, 160, 162,
   171
Mode, 251, 257, 275

Monotonicity, 2, 89, 251, 265
   directional, 292
   global, 246
   weak, 252
MYCIN's aggregation function, 28, 31

N
Natural distance, 327, 337
Natural extension, 308
Neat OWA, 106, 211
Necessity measure, 156
Negation, 18
   standard, 18
   strict, 18
   strong, 18
Neutral element, 12, 24, 31, 64, 102, 150
   strong, 12
Norm, 22, 34, 161, 170
Nullnorm, 28, 138

O
One divisor, 14
Optimal interpolation, 40, 214, 286
Order preservation, 35, 80
Order statistic, 102, 141, 150
Ordered weighted averaging (OWA), 26, 27,
   101, 138, 141, 146, 150, 165, 170,
   193, 200, 212, 225, 228, 260, 309,
   311
   central, 256
   dispersion, 105, 117
   dual, 101, 106
   entropy, 105, 117
   exponential, 109
   fitting, 114
   generalized, 107, 114, 126, 173
   geometric, 107
   harmonic, 107
   induced, 130
     generalized, 133
   linguistic, 217
   maximum entropy, 117
   minimax disparity, 119
   minimum variance, 118
   neat, 106, 211
   ordinal, 140
   orness measure, 103
   power-based, 107
   quadratic, 108
   radical, 109
   reverse, 101, 132
   self-dual, 104

trigonometric, 108
weighted, 109
Ordered weighted geometric function (OWG), 107
induced, 133
Ordered weighted harmonic function (OWH), 107
Ordered weighted maximum, 176
Ordered weighted minimum, 176
Ordered weighted power-based function, 107
Ordinal OWA, 140
Orness measure, 56, 60, 64, 68, 103, 163
Outlier, 35
Overlap function, 218

P
Partial conjunction, 58, 234
Partial disjunction, 58, 234
Pascal triangle, 74
Penalty, 184, 242, 253, 260, 276, 327
failthful, 186
Penalty based function, 184
Permutation, 11
3-Π function, 29
Plausibility measure, 157, 159
Poset, 307
Possibility measure, 156
Power mean, 62, 66, 69, 242
Pre-aggregation function, 296
Preservation of inputs ordering, 116, 127
Probabilistic sum, 27
Probability measure, 154, 158
Product, 5, 10, 15, 27
Product lattice, 306
Projection, 16
PROSPECTOR's aggregation function, 28, 31
p-stable, 23, 213

Q
Quadratic mean, 63
Quadratic OWA, 108
Quantifier, 74, 120, 165
Quasi-arithmetic mean, 65, 88, 173, 193, 207, 238, 242
calculation, 73
Quasi-median, 138
Quasi-penalty, 253

R
Radical mean, 72
Radical OWA, 109
Recursive, 15
Representable aggregation function, 308
Restricted dissimilarity function, 188
Reverse OWA, 101
RIM quantifier, 120
Robust statistics, 256
Root-power mean, 62

S
Score, 306, 309
Self-dual, 20, 65, 164
Self-dual fuzzy measure, 153
Self-identity, 17, 239, 240, 247
Shapley value, 160, 168
Shift-invariant, 17, 60, 93, 132, 185, 209, 212, 254
Shorth, 257, 286
Sierpinski family, 75
Spline, 111, 123, 128, 209
Stability, 79, 239
Standard negation, 18, 24
Stolarsky mean, 86, 88
Stress function, 121
Strict monotonicity, 11, 76
Strict negation, 18
Strong negation, 18, 24
fixed point, 19
Stronger, 21
Subadditive function, 154, 170
Subadditive fuzzy measure, 153
Submodular function, 67
Submodular fuzzy measure, 114, 153, 159, 161, 169
Substitutive criteria, 163, 168
Sugeno fuzzy measure, 158, 170
Sugeno integral, 26, 165, 174, 299, 300
calculation, 174
Sugeno-Weber t-conorm, 159
Superadditive function, 154
Superadditive fuzzy measure, 153
Supermodular function, 67
Supermodular fuzzy measure, 114, 153, 159, 161, 169
Symmetric fuzzy measure, 154, 165
Symmetry, 10, 24, 31, 216

T
Transform, 278
Triangular conorm, 27, 159

Lukasiewicz, 316
    probabilistic sum, 27
    Sugeno-Weber, 159
Triangular norm, 27, 226, 237
    Hamacher, 31, 211
    Lukasiewicz, 28, 159
    product, 27
Trigonometric mean, 69
Trigonometric OWA, 108
Trimmed mean, 257

U
Uninorm, 28

W
Weak monotoniciy, 252
Weaker, 21

Weight generating function, 120, 123, 126, 129
Weighted arithmetic mean, 27, 34, 59, 146, 150, 164
Weighted maximum, 174, 175
Weighted mean, 93, 210
Weighted median, 139
Weighted minimum, 174, 175
Weighted OWA (WOWA), 109
Weighted quasi-arithmetic mean, 65
Weighting triangle, 73
Weighting vector, 25, 59, 73, 176, 211
Weights dispersion, 76, 105, 117, 164

Z
Zero divisor, 14
Zeta transform, 148

Printed in the United States
By Bookmasters